Fabio Menna, Fabio Remondino and
Hans-Gerd Maas (Eds.)

Sensors and Techniques for 3D Object Modeling in Underwater Environments

MDPI

This book is a reprint of the Special Issue that appeared in the online, open access journal, *Sensors* (ISSN 1424-8220) from 2015–2016 (available at: http://www.mdpi.com/journal/sensors/special_issues/3DOM).

Guest Editors
Fabio Menna
3D Optical Metrology unit, Bruno Kessler Foundation
Italy

Fabio Remondino
3D Optical Metrology unit, Bruno Kessler Foundation
Italy

Hans-Gerd Maas
Institute of Photogrammetry and Remote Sensing, TU Dresden
Germany

Editorial Office	*Publisher*	*Managing Editors*
MDPI AG	Shu-Kun Lin	Lin Li and Limei Huang
Klybeckstrasse 64		
Basel, Switzerland		

1. Edition 2016

MDPI • Basel • Beijing • Wuhan • Barcelona

ISBN 978-3-03842-222-8 (Hbk)
ISBN 978-3-03842-223-5 (PDF)

Table of Contents

Chapter 1: Geometric Modeling and Photogrammetric Camera Calibration

Chapter 2: New 3D Imaging Sensors and Data Processing for Underwater Applications

Chapter 3: Sensor Integration and Data Processing

V

List of Contributors

Gianfranco Bianco DIMEG, University of Calabria, Via P. Bucci 46/C–Rende, Cosenza 87036, Italy.

Christian Bräuer-Burchardt Fraunhofer Institute Applied Optics and Precision Engineering, Albert-Einstein-Str. 7, D-07745 Jena, Germany.

Fabio Bruno DIMEG, University of Calabria, Via P. Bucci 46/C–Rende, Cosenza 87036, Italy.

Filipe Castro Ship Reconstruction Laboratory 4352 TAMU, Texas A & M University, College Station, TX 77843, USA.

Jim Chandler School of Civil and Building Engineering, Loughborough University, Loughborough, Leicestershire LE11 3TU, UK.

Pierre Charbonnier Cerema Dter Est, Image Processing and Optical Methods Research Team, 11 rue Jean Mentelin, B.P. 9, Strasbourg 67035, France.

Bertrand Chemisky COMEX, COmpanie Maritime d'EXpertise 36 boulevard des Océans, 13009 Marseille, France.

Colin Devey GEOMAR Helmholtz Centre for Ocean Research Kiel, RD4/RD2, Wischhofstr. 1-3, 24148 Kiel, Germany.

Pierre Drap Aix Marseille Université, CNRS, ENSAM, Université De Toulon, LSIS UMR 7296,13397 Marseille, France.

Francesco Fassi Politecnico di Milano, ABC Dep. 3DSurvey Group, via Ponzio 31, Milano 20133, Italy.

Edgar Ferreira School of Civil and Building Engineering, Loughborough University, Loughborough, Leicestershire LE11 3TU, UK.

Philippe Foucher Cerema Dter Est, Image Processing and Optical Methods Research Team, 11 rue Jean Mentelin, B.P. 9, Strasbourg 67035, France.

Timmy Gambin Archaeology Centre (Car Park 6), University of Malta, Msida MSD 2080, Malta.

Lamia Gaoua Aix Marseille Université, CNRS, ENSAM, Université De Toulon, LSIS UMR 7296,13397 Marseille, France.

Francesco Giordano Dipartimento di Scienze e Tecnologie, Università degli Studi di Napoli "Parthenope", Centro Direzionale, Isola C4, 80143 Napoli, Italy.

Jens Greinert GEOMAR Helmholtz Centre for Ocean Research Kiel, RD4/RD2, Wischhofstr. 1-3, 24148 Kiel, Germany.

Pierre Grussenmeyer ICube Laboratory UMR 7357, Photogrammetry and Geomatics Group, INSA Strasbourg, 24 Boulevard de la Victoire, Strasbourg 67084, France.

Samuel Guillemin ICube Laboratory UMR 7357, Photogrammetry and Geomatics Group, INSA Strasbourg, 24 Boulevard de la Victoire, Strasbourg 67084, France.

Matthias Heinze Fraunhofer Institute Applied Optics and Precision Engineering, Albert-Einstein-Str. 7, D-07745 Jena, Germany.

Bilal Hijazi Aix Marseille Université, CNRS, ENSAM, Université De Toulon, LSIS UMR 7296,13397 Marseille, France.

Josef Jansa Department of Geodesy and Geoinformation, Technische Universität Wien, Gusshausstrasse 27-29, Vienna 1040, Austria.

Anne Jordt GEOMAR Helmholtz Centre for Ocean Research Kiel, Kiel 24148, Germany.

Reinhard Koch Department of Computer Science, Kiel University, Kiel 24118, Germany.

Mathieu Koehl ICube Laboratory UMR 7357, Photogrammetry and Geomatics Group, INSA Strasbourg, 24 Boulevard de la Victoire, Strasbourg 67084, France.

Kevin Köser GEOMAR Helmholtz Centre for Ocean Research Kiel, Kiel 24148, Germany; GEOMAR Helmholtz Centre for Ocean Research Kiel, RD4/RD2, Wischhofstr. 1-3, 24148 Kiel, Germany.

Peter Kühmstedt Fraunhofer Institute Applied Optics and Precision Engineering, Albert-Einstein-Str. 7, D-07745 Jena, Germany.

Tom Kwasnitschka GEOMAR Helmholtz Centre for Ocean Research Kiel, RD4/RD2, Wischhofstr. 1-3, 24148 Kiel, Germany.

Antonio Lagudi DIMEG, University of Calabria, Via P. Bucci 46/C–Rende, Cosenza 87036, Italy.

Hans-Gerd Maas TU Dresden, Institute of Photogrammetry and Remote Sensing, Helmholtzstr. 10, D-01069 Dresden, Germany.

Miquel Massot-Campos Department of Mathematics and Computer Science, University of the Balearic Islands, Cra de Valldemossa km 7.5, Palma de Mallorca 07122, Spain.

Gaia Mattei Dipartimento di Scienze e Tecnologie, Università degli Studi di Napoli "Parthenope", Centro Direzionale, Isola C4, 80143 Napoli, Italy.

Fabio Menna 3D Optical Metrology unit, Bruno Kessler Foundation (FBK), via Sommarive 18, Trento 38123, Italy.

Djamal Merad Aix Marseille Université, CNRS, ENSAM, Université De Toulon, LSIS UMR 7296,13397 Marseille, France.

Emmanuel Moisan Cerema Dter Est, Image Processing and Optical Methods Research Team, 11 rue Jean Mentelin, B.P. 9, Strasbourg 67035, France; ICube Laboratory UMR 7357, Photogrammetry and Geomatics Group, INSA Strasbourg, 24 Boulevard de la Victoire, Strasbourg 67084, France.

Maurizio Muzzupappa DIMEG, University of Calabria, Via P. Bucci 46/C–Rende, Cosenza 87036, Italy.

Mohamad Motasem Nawaf Aix Marseille Université, CNRS, ENSAM, Université De Toulon, LSIS UMR 7296,13397 Marseille, France.

Erica Nocerino 3D Optical Metrology unit, Bruno Kessler Foundation (FBK), via Sommarive 18, Trento 38123, Italy.

Gunther Notni Fraunhofer Institute Applied Optics and Precision Engineering, Albert-Einstein-Str. 7, D-07745 Jena, Germany; Technical University Ilmenau, Ehrenbergstraße 29, D-98693 Ilmenau, Germany.

Gabriel Oliver-Codina Department of Mathematics and Computer Science, University of the Balearic Islands, Cra de Valldemossa km 7.5, Palma de Mallorca 07122, Spain.

Claudio Parente Dipartimento di Scienze e Tecnologie, Università degli Studi di Napoli "Parthenope", Centro Direzionale, Isola C4, 80143 Napoli, Italy.

Francesco Peluso Dipartimento di Scienze e Tecnologie, Università degli Studi di Napoli "Parthenope", Centro Direzionale, Isola C4, 80143 Napoli, Italy.

Norbert Pfeifer Department of Geodesy and Geoinformation, Technische Universität Wien, Gusshausstrasse 27-29, Vienna 1040, Austria.

Fabio Remondino 3D Optical Metrology unit, Bruno Kessler Foundation (FBK), via Sommarive 18, Trento 38123, Italy.

Marcel Rothenbeck GEOMAR Helmholtz Centre for Ocean Research Kiel, RD4/RD2, Wischhofstr. 1-3, 24148 Kiel, Germany.

Ewelina Rupnik Department of Geodesy and Geoinformation, Technische Universität Wien, Gusshausstrasse 27-29, Vienna 1040, Austria.

Mauro Saccone Aix Marseille Université, CNRS, ENSAM, Université De Toulon, LSIS UMR 7296,13397 Marseille, France.

Raffaele Santamaria Dipartimento di Scienze e Tecnologie, Università degli Studi di Napoli "Parthenope", Centro Direzionale, Isola C4, 80143 Napoli, Italy.

Ingo Schmidt Fraunhofer Institute Applied Optics and Precision Engineering, Albert-Einstein-Str. 7, D-07745 Jena, Germany.

Timm Schoening GEOMAR Helmholtz Centre for Ocean Research Kiel, RD4/RD2, Wischhofstr. 1-3, 24148 Kiel, Germany.

Julien Seinturier COMEX, COmpanie Maritime d'EXpertise 36 boulevard des Océans, 13009 Marseille, France.

Koji Shiono School of Civil and Building Engineering, Loughborough University, Loughborough, Leicestershire LE11 3TU, UK.

Mark Shortis School of Mathematical and Geospatial Sciences, RMIT University, GPO Box 2476, Melbourne 3001, Australia.

Jean-Christophe Sourisseau Aix Marseille Université, CNRS, Ministère de la Culture et de la Communication, CCJ UMR 7299, 13094 Aix En Provence, France.

Anja Steinführer GEOMAR Helmholtz Centre for Ocean Research Kiel, RD4/RD2, Wischhofstr. 1-3, 24148 Kiel, Germany.

Jan Sticklus GEOMAR Helmholtz Centre for Ocean Research Kiel, RD4/RD2, Wischhofstr. 1-3, 24148 Kiel, Germany.

Lars Triebe GEOMAR Helmholtz Centre for Ocean Research Kiel, RD4/RD2, Wischhofstr. 1-3, 24148 Kiel, Germany.

Jens Schneider von Deimling GEOMAR Helmholtz Centre for Ocean Research Kiel, Kiel 24148, Germany.

Rene Wackrow School of Civil and Building Engineering, Loughborough University, Loughborough, Leicestershire LE11 3TU, UK.

Tim Weiß GEOMAR Helmholtz Centre for Ocean Research Kiel, RD4/RD2, Wischhofstr. 1-3, 24148 Kiel, Germany.

Emanuel Wenzlaff GEOMAR Helmholtz Centre for Ocean Research Kiel, RD4/RD2, Wischhofstr. 1-3, 24148 Kiel, Germany.

Claudius Zelenka Department of Computer Science, Kiel University, Kiel 24118, Germany.

About the Guest Editors

Fabio Menna received a PhD in Photogrammetry in 2009 from Parthenope University of Naples, Italy and is currently a product researcher at the 3D Optical Metrology Unit (http://3dom.fbk.eu) of the Bruno Kessler Foundation (http://www.fbk.eu) in Trento, Italy. His research activities concern the design of new 3D imaging systems based on photogrammetry for applications in cultural heritage documentation, industrial metrology, and underwater photogrammetry. He has authored 50 publications and has received five awards for best papers at different conferences. He was the organizer of the first ISPRS/CIPA workshop "Underwater 3D Recording and Modeling", and has tutored more than 15 summer schools and presented many tutorials. He served as secretary of ISPRS Technical Commission V (2012-2016) and is currently acting as chair of WGII/8 Underwater data acquisition and processing of new technical Commission II of the ISPRS.

Fabio Remondino received his PhD in Photogrammetry in 2006 from ETH Zurich, Switzerland and now leads the 3D Optical Metrology Unit (http://3dom.fbk.eu) of the Bruno Kessler Foundation (http://www.fbk.eu), a public research center in Trento, Italy. His research interests include geospatial data collection and processing, heritage documentation, 3D modeling, sensor and data integration. He is the author of over 150 scientific publications in journals and presented at international conferences, has written five books, and eight Special Issues in journals. He has received 10 awards for best papers at conferences and organized 26 scientific events and 29 summer schools and tutorials. He is currently acting as President of ISPRS Technical Commission II "Photogrammetry", President of EuroSDR Commission I "Data Acquisition", and Vice-President of CIPA Heritage Documentation.

Hans-Gerd Maas received a PhD in Photogrammetry in 1992 from ETH Zurich and is presently full professor for Photogrammetry, at TU Dresden, Germany (http://www.tu-dresden.de/ipf/photo/). His research interests are in the fields of close range photogrammetry and laser scanning for environmental monitoring, 3D motion analysis, 3D object reconstruction and deformation measurement. He is the author of more than 200 publications and organizer of several scientific workshops, and has served as President of ISPRS Technical Commission V (2004–2008).

Preface to "Sensors and Techniques for 3D Object Modeling in Underwater Environments"

The Special Issue "Sensors and Techniques for 3D Object Modeling in Underwater Environments" originates from the ISPRS/CIPA Workshop "UNDERWATER 3D RECORDING & MODELING—Experiences in Data Acquisition, Calibration, Orientation, Modeling & Accuracy Assessment" (http://3dom.fbk.eu/files/underwater/index.html) which was held in April 2015, in Italy. The main workshop's scope was to bring together scientists, developers and advanced users in underwater 3D recording and to encourage cooperation and practice sharing. The workshop was focused on, but not limited to, topics such as optical-based 3D surveying techniques, underwater/multi-media photogrammetry, bathymetric LiDAR, SONAR-based techniques. The workshop emphasized the importance of strict sensor calibration and geometric modeling for delivering accurate results, but it also underlined how an appropriate data processing, sensor integration and multidisciplinary cooperation are fundamental for gathering reliable results in underwater scenarios.

Water is the most important element for human beings as it is essential for life. Besides its key role in the Earth's ecosystem, water plays an important role in human activities as well. Since remote times, humans have been connected to water bodies either in their natural form such as oceans, lakes, rivers, and wetlands or its manmade counterpart such as structures, flumes, channels, basins, dams, *etc*. History demonstrates that economic, expansive and social purposes have driven the realization of means for studying, exploiting, exploring and navigating the water bodies. Water covers approximately 71% of the planet's surface witnessing underwater many traces of past human activities that have been well preserved for centuries thanks to natural physio-chemical factors such as dim light, low temperatures as well as reduced oxygen. Oceans' seafloors, as well as some lakes, enclose reservoirs of energy in the form of natural deposits of fossil fuels whose size is still not completely known. Energy is required for human existence itself and the study of water biodiversity is of crucial importance for alimentary reasons: Seafood, for example, under responsible fishing, is considered one of the potential solutions to world hunger.

There is a plethora of reasons drawing attention to underwater environments for which a comprehensive list cannot easily be attained. Nevertheless, a common shared factor among underwater activities is the need for accurate spatial measurements that are required for accomplishing tasks such as recording, documenting, positioning, as well as sizing fauna and flora or for 3D modeling of natural and man-made environments. Nowadays, interest in underwater environments is very high as demonstrated by the ever increasing demand for applications. This results from technical achievements in diving apparatus, photographic techniques and underwater manned and unmanned vehicles (Remotely Operated Underwater Vehicle, Autonomous Underwater Vehicle, robots, *etc.*). Nevertheless, the complexity of photogrammetric underwater operations remains very high if compared to the corresponding counterpart on the mainland. Water is a medium inherently different from air and the first essential difference resides in the medium density. The access to underwater environments implies the use of special equipment that must be waterproof and resist high pressure. Furthermore, depending on water turbidity, sunlight is absorbed rapidly and selectively depending on wavelengths. Consequently, most optical-based sensors need suitable illumination to work properly. When the water is not sufficiently transparent, SONAR-based systems represent the most effective solution. Research, design and development of new techniques and procedures for system calibration and assessment of results quality remains a challenging and open issue. The variety of phenomena involved in 3D modeling underwater demands the joint participation of several and different technical knowledge to face new requirements. Shared problems can be better tackled through a more cooperative approach born from different experiences collected in different fields.

This book contains the outcomes of the aforementioned MDPI *Sensors'* Special Issue, composed of 13 peer-reviewed articles that collate viewpoints, each related to a different aspect of the underwater environment, starting from the actual border surface that separates it from the air, to the deeper ocean expanses. Applications including freshwater bodies on the mainland are also treated. Archaeology, civil engineering, biology, industrial and science lab metrology are the most reported application fields. The book is divided in three main chapters related to the most salient key-points of the different contributions published in the Special Issue.

Chapter 1: Geometric Modeling and Photogrammetric Camera Calibration

Strict geometric modeling and camera calibration remains a vivid topic in underwater photogrammetry, embracing different geometric and radiometric issues arising whenever an image of an object located underwater is taken.

In *Calibration Techniques for Accurate Measurements by Underwater Camera Systems* [pp. 3–25], a review of current approaches for the calibration of underwater camera systems is provided in theoretical and practical terms. The accuracy, reliability, validation and stability of underwater camera system calibration is also discussed. Samples of results from published reports are provided to demonstrate the range of possible accuracies for the measurements produced by underwater camera systems.

A flexible, yet strict geometric model for the handling of refraction effects on the optical path is shown in *On the Accuracy Potential in Underwater/Multimedia Photogrammetry* [pp. 26–40]. The model can be implemented as a module into photogrammetric standard tools, such as spatial resection, spatial intersection, bundle adjustment or epipolar line computation. The module is especially well suited for applications where an object in water is observed by cameras in the air through one or more planar glass interfaces. Several aspects, which are relevant for an assessment of the accuracy potential in underwater/multimedia photogrammetry, are discussed. These aspects include network geometry and interface planarity issues, as well as effects caused by refractive index variations and dispersion and diffusion under water. All these factors contribute to a rather significant degradation of the geometric accuracy potential in underwater/multimedia photogrammetry. In practical experiments, a degradation of the quality of results by a factor of two could be determined under relatively favorable conditions.

While several camera pressure housings made in different materials, shapes and sizes are available on the market and are being used for photogrammetric applications, a deep understanding of how their manufacture affects image formation and metric performances still needs to be further investigated. In *Geometric and Optic Characterization of a Hemispherical Dome Port for Underwater Photogrammetry* [pp. 41–67], a geometric investigation of a consumer grade underwater camera housing, manufactured by NiMAR and equipped with a 7″ dome port is presented. After a review of flat and dome ports, the work analyzes, using simulations and real experiments, the main optical phenomena involved when operating a camera underwater. Specific aspects which deal with photogrammetric acquisitions are considered in some laboratory tests and in a swimming pool. Results and considerations are presented and commented on.

Imaging systems have an indisputable role in revealing vegetation posture under diverse flow conditions, image sequences being generated from off-the-shelf digital cameras. Such sensors are cheap but introduce a range of distortion effects, a trait only marginally tackled in hydraulic studies focusing on water-vegetation dependencies. To bridge this gap, researchers present a simple calibration method to remove both camera lens distortion and refractive effects of water in *Camera Calibration for Water-Biota Research: The Projected Area of Vegetation* [pp. 68–79]. The effectiveness of the method is illustrated using the variable projected area, computed for both simple and complex shaped objects. Results demonstrate the significance of correcting images using a combined lens distortion and refraction model, prior to determining projected areas and further data analyses.

The selection of a 3D sensing system to be used in underwater applications is non-trivial. In *Optical Sensors and Methods for Underwater 3D Reconstruction* [pp. 83–129], state of the art optical sensors and methods for 3D reconstruction in underwater environments is presented. The techniques to obtain range-data have been listed and explained, together with the different sensor hardware that makes them possible. The literature has been reviewed, and a classification has been proposed for the existing solutions. New developments, commercial solutions and previous reviews on this topic have also been gathered and considered.

In *The Bubble Box: Towards an Automated Visual Sensor for 3D Analysis and Characterization of Marine Gas Release Sites* [pp. 130–156], a new underwater 3D scanning device, based on the fringe projection technique, is presented. It has a weight of about 10 kg and the maximal water depth for application of the scanner is 40 m. It covers an underwater measurement volume of 250 mm × 200 mm × 120 mm. The surface of the measurement objects is captured with a lateral resolution of 150 µm in a third of a second. An extended camera model which takes refraction effects into account, as well as a proposal of an effective, low-effort calibration procedure for underwater optical stereo scanners, is shown. Calibration evaluation results are presented and examples of first underwater measurements are given.

Several acoustic and optical techniques have been used for characterizing natural and anthropogenic gas leaks (carbon dioxide, methane) from the ocean floor. In *Underwater 3D Surface Measurement Using Fringe Projection Based Scanning Devices* [pp. 157–177], the authors introduce a wide baseline stereo-camera deep-sea sensor bubble box that observes bubbles from two orthogonal directions using calibrated cameras. Besides the setup and the hardware of the system, the authors provide a discussion about appropriate calibration and the different automated processing steps—deblurring, detection, tracking, and 3D fitting—that are crucial to arrive at a 3D ellipsoidal shape and rise speed of each bubble. The obtained values for single bubbles can be aggregated into statistical bubble size distributions or fluxes for extrapolation, based on diffusion and dissolution models and large scale acoustic surveys. An evaluation of the method is given through a controlled test setup with ground truth information.

The work in *Sinusoidal Wave Estimation Using Photogrammetry and Short Video Sequences* [pp. 178–212] presents a method to model the shape of the sinusoidal shape of regular water waves generated in a laboratory flume. The waves are traveling in time and render a smooth surface, with no white caps or foam. Two methods are proposed, treating the water as a diffuse and specular surface, respectively. The devised approaches are validated against the data received from a capacitive level sensor and on physical targets floating on the surface; the outcomes of which agree to a high degree.

In *Adjustment of Sonar and Laser Acquisition Data for Building the 3D Reference Model of a Canal Tunnel* [pp. 213–246], the authors present a method for the construction of a full 3D model of a canal tunnel by combining terrestrial laser (for its above-water part) and sonar (for its underwater part) scans collected from static acquisitions. Above- and under-water point clouds are co-registered to directly generate the full 3D model of the canal tunnel. Faced with the lack of overlap between both models, the authors introduce a robust algorithm that relies on geometrical entities and partially-immersed targets, which are visible in both the laser and sonar point clouds. A full 3D model, visually promising, of the entrance of a canal tunnel is obtained.

Chapter 3: Sensor Integration and Data Processing

3D modeling of the bottom of shallow waters is still a crucial open topic. The use of maritime vessels capable of carrying out bathymetric measurements is limited by the depth of the waters, so only small crafts are suitable. In *Integrating Sensors into a Marine Drone for Bathymetric 3D Surveys in Shallow* [pp. 249–269], an open prototype of an unmanned surface vessel (USV), named MicroVeGA, is described. The focus is on the main instruments installed on-board: a differential Global Position System (GPS) system and single beam echo sounder; inertial platform for attitude control; ultrasound obstacle-detection system with temperature control system; emerged and submerged video acquisition system. Experiments performed in two coastal sites showed the benefits of integrating existing low cost sensors and technologies.

The work in *Underwater Photogrammetry and Object Modeling: A Case Study of Xlendi Wreck in Malta* [pp. 270–315] presents a photogrammetry-based approach for deep-sea underwater surveys conducted from a submarine and guided by knowledge-representation combined with a logical approach (ontology). Two major issues are discussed in the paper: the first concerns deep-sea surveys using photogrammetry from a submarine; the second issue involves the extraction of known artefacts present on the site. This aspect of the research is based on an *a priori* representation of the knowledge involved using systematic reasoning.

The integration of underwater 3D data captured by acoustic and optical systems is a promising technique in various applications such as mapping or vehicle navigation. It allows for compensating the drawbacks of the low resolution of acoustic sensors and the limitations of optical sensors in bad visibility conditions. Aligning these data is a challenging problem. The authors of *An Alignment Method for the Integration of Underwater 3D Data Captured by a Stereovision System and an Acoustic Camera* [pp. 316–343] present a multi-sensor registration for the automatic integration of 3D data acquired from a stereovision system and a 3D acoustic camera in close-range acquisition. The effectiveness of the method has been demonstrated in this first experimentation of the proposed 3D opto-acoustic camera.

Underwater photogrammetry and, in particular, systematic visual surveys of the deep sea are by far less developed than similar techniques on land or in space. The main challenges are the rough conditions with extremely high pressure, the accessibility of target areas (container and ship deployment of robust sensors, then diving for hours to the ocean floor), and the limitations of localization technologies (no GPS). The absence of natural light complicates energy budget considerations for deep diving flash-equipped drones. Refraction effects influence geometric image formation considerations with respect to field of view and focus, while attenuation and scattering degrade the radiometric image quality and limit the effective visibility. To improve these issues, the authors of *DeepSurveyCam—A Deep Ocean Optical Mapping System* [pp. 344–366] present an AUV-based optical system intended for autonomous visual mapping of large areas of the seafloor (some square kilometers) in up to 6000 m water depth.

Acknowledgments

We would like to thank all authors who have submitted their manuscripts which compose the presented chapters and to all reviewers for their valuable work during the reviewing process.

Fabio Menna, Fabio Remondino and Hans-Gerd Maas
Guest Editors

Chapter 1:
Geometric Modeling and Photogrammetric Camera Calibration

Calibration Techniques for Accurate Measurements by Underwater Camera Systems

Mark Shortis

Abstract: Calibration of a camera system is essential to ensure that image measurements result in accurate estimates of locations and dimensions within the object space. In the underwater environment, the calibration must implicitly or explicitly model and compensate for the refractive effects of waterproof housings and the water medium. This paper reviews the different approaches to the calibration of underwater camera systems in theoretical and practical terms. The accuracy, reliability, validation and stability of underwater camera system calibration are also discussed. Samples of results from published reports are provided to demonstrate the range of possible accuracies for the measurements produced by underwater camera systems.

Reprinted from *Sensors*. Cite as: Shortis, M. Calibration Techniques for Accurate Measurements by Underwater Camera Systems. *Sensors* **2015**, *15*, 30810–30827.

1. Introduction

A recent report by the World Wildlife Fund [1] notes a sharp decline in marine biodiversity, caused by overfishing, coastal development and climate change. This decline is having a significant impact on the health of the marine ecosystems and threatens the survival of common seafood choices such as tuna, shrimp, whiting and salmon. The highest impact has been on these and many other highly utilised species caught in commercial or subsistence fisheries, with populations falling by 50% during 1970 to 2010.

The sustainability of wild fish stocks has been an ongoing concern that has been subject to many studies and reviews over the last few decades (for example, see [2]). Fishing has been shown to result in substantial changes in species composition and population distributions of target and non-target fish [3]. Over-fishing, especially of top level predators such as tuna and sharks, can result in unpredictable changes in marine ecosystems. In an era of increasing catch effort to maintain the dietary contribution of seafood, early detection of the impacts of over-fishing or detrimental changes in the environment is critical.

In response to declining wild fish stocks and increasing catch effort to land the same biomass, many countries have developed aquaculture industries to maintain levels of seafood dietary contribution [4]. Species such as tuna, tilapia and salmon are most commonly farmed due to their market acceptance, rapid growth and favourable food conversion rates [5]. For species subject to catch quotas, such as Southern Bluefin Tuna, the annual biomass of the catch must be estimated [6]. Once the fish are established in the aquaculture facility, monitoring of the biomass is essential for farm managers to optimise feed regimes and harvest strategies.

The age and biomass of fish can be reliably estimated based on length measurement and a length-weight or length-age regression [7,8]. When combined with spatial or temporal sampling in marine ecosystems, or counts of fish in an aquaculture cage or a trawl net, the distribution of lengths can be used to estimate distributions of or changes in biomass, and shifts in or impacts on population distributions. Underwater camera and video systems are now widely employed as a non-contact, non-invasive technique to capture accurate length information [9] and thereby estimate biomass or population distributions. Underwater camera and video systems have the further advantages that the measurements are repeatable and impartial [10], sample areas can be very accurately estimated [11] and the accuracy of the length measurements vastly improves the statistical power of the population estimates when sample counts are very low [12].

Underwater stereo-video systems have been used in the assessment of wild fish stocks with a variety of cameras and modes of operation [13–16], in pilot studies to monitor length frequencies of fish in aquaculture cages [6,17,18] and in fish nets during capture [19]. Commercial systems such as the AKVAsmart, formerly VICASS [20], and the AQ1 AM100 [18] are widely used in aquaculture and fisheries.

Marine conservation and fisheries stock assessment dominate the application of accurate measurement by underwater stereo systems, based on citations [9,14]. However there are many other applications of single camera and stereo systems reported in the literature. Stereo camera systems were used to conduct the first accurate sea bed mapping applications [21,22] and surveys of shipwrecks using either a frame [23] or towed body systems [24]. Single and stereo cameras have been used for monitoring of submarine structures, most notably to support energy exploration and extraction in the North Sea [25,26], underwater inspection of ship hulls [27] and structures [28], archaeological mapping of shipwrecks from submersibles [29], virtual modeling of archaeological sites [30], mapping of seabed topography [22,31], reconstruction of complex 3D structures [32] and inshore sea floor mapping [33,34].

A video camera has been used to measure the shape of fish pens [35] and a stereo camera has been used to map cave profiles [36]. Digital still cameras have been used underwater for mapping of artefacts in a ship wreck [37] and the estimation of sponge volumes [38]. Sea floor monitoring has also been carried out in deep water using continuously recorded stereo video cameras combined with a high resolution digital still camera [39]. A network of digital still camera images has been used to accurately characterise the shape of a semi-submerged ship hull [40].

The common factor for all of these applications of underwater imagery is a designed or specified level of accuracy. Video surveys for biomass or population distributions are directly dependent on the accuracy of the length measurements. Any inaccuracy will lead to significant errors in the estimated biomass [41] or a bias in the population distribution [12]. Other applications such as structural monitoring or seabed mapping must achieve a certain level of accuracy for the surface shape.

Calibration of any camera system is essential to achieve accurate and reliable measurements. Small errors in the perspective projection must be modelled and eliminated to prevent the introduction of systematic errors into the measurements. In the underwater environment, the calibration of the cameras is of even greater importance because the effects of refraction through the air, housing and water interfaces must be incorporated.

Compared to in-air calibration, camera calibration under water is subject to the additional uncertainty caused by attenuation of light through the housing port and water media, as well as the potential for small errors in the refracted light path due to modelling assumptions or non-uniformities in the media. Accordingly, the precision and accuracy of calibration underwater is always expected to be degraded relative to an equivalent calibration in air. Experience demonstrates that, because of these effects, underwater calibration is more likely to result in scale errors in the measurements.

2. Calibration Approaches

In a limited range of circumstances calibration may not be necessary. If a high level of accuracy is not required, and the object to be measured approximates a two dimensional planar surface, a very straightforward solution is possible.

Correction lenses or dome ports such as those described in [31,42] can be used to provide a near-perfect central projection under water by eliminating the refraction effects. Any remaining, small errors or imperfections can either be corrected using a grid or graticule placed in the field of view, or simply accepted as a small deterioration in accuracy. The correction lens or dome port has the further advantage that there is little, if any, degradation of image quality near the edges of the port. Plane camera ports exhibit loss of contrast and intensity at the extremes of the field of view due to acute angles of incidence and greater apparent thickness of the port material.

This simplified approach has been used, either with correction lenses or a pre-calibration of the camera system, to carry out two dimensional mapping. A portable control frame with a fixed grid or target reference is imaged before deployment or placed against the object to measured, to provide both calibration corrections as well as position and orient the camera system relative to the object. Typical applications of this approach are ship wreck mapping [23], sea

5

floor characterisation surveys [31], length measurements in aquaculture [17] and monitoring of sea floor habitats [43].

However if accuracy is a priority, and especially if the object to be measured is a three dimensional surface, then a comprehensive calibration is essential. The correction lens approach assumes that the camera is a perfect central projection and that the entrance pupil of the camera lens coincides exactly with the centre of curvature of the correction lens. Any simple correction approach, such as a graticule or control frame placed in the field of view, will be applicable only at the same distance. Any significant extrapolation outside of the plane of the control frame will inevitably introduce systematic errors.

The alternative approach of a comprehensive calibration translates a reliable technique from in air into the underwater environment. Close range calibration of cameras is a well-established technique that was pioneered by [44], extended to include self-calibration of the camera(s) by [45] and subsequently adapted to the underwater environment [46,47]. The mathematical basis of the technique is described in [48].

The essence of this approach is to capture multiple, convergent images of a fixed calibration range or portable calibration fixture (see Figure 1) to determine the physical parameters of the camera calibration. A typical calibration range or fixture is based on discrete targets to precisely identify measurement locations throughout the camera fields of view from the many photographs (see Figure 1). The targets may be circular dots or the corners of a checkerboard. Coded targets or checkerboard corners on the fixture can be automatically recognised using image analysis techniques [49,50] to substantially improve the efficiency of the measurements and network processing. The ideal geometry and a full set of images for a calibration fixture are shown in Figures 2 and 3 respectively.

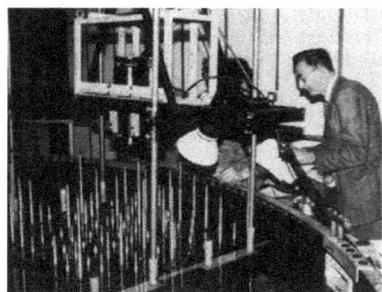

Figure 1. Typical portable calibration fixture ((**Left**), courtesy of NOAA) and test range ((**Right**), from [25]).

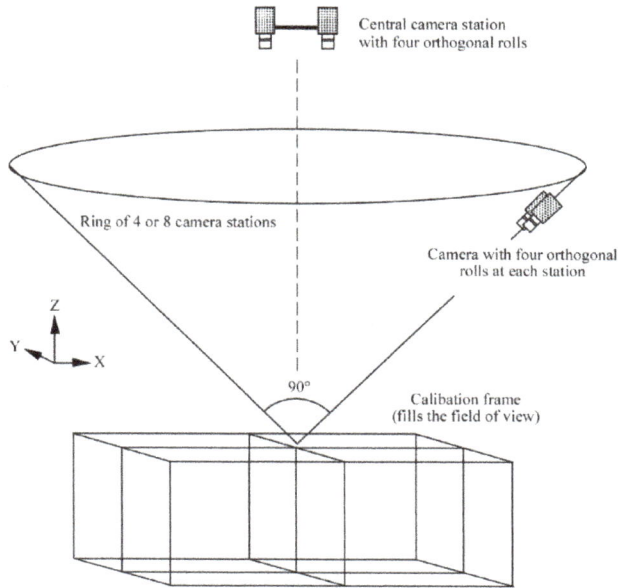

Figure 2. The ideal geometry for a self-calibration network.

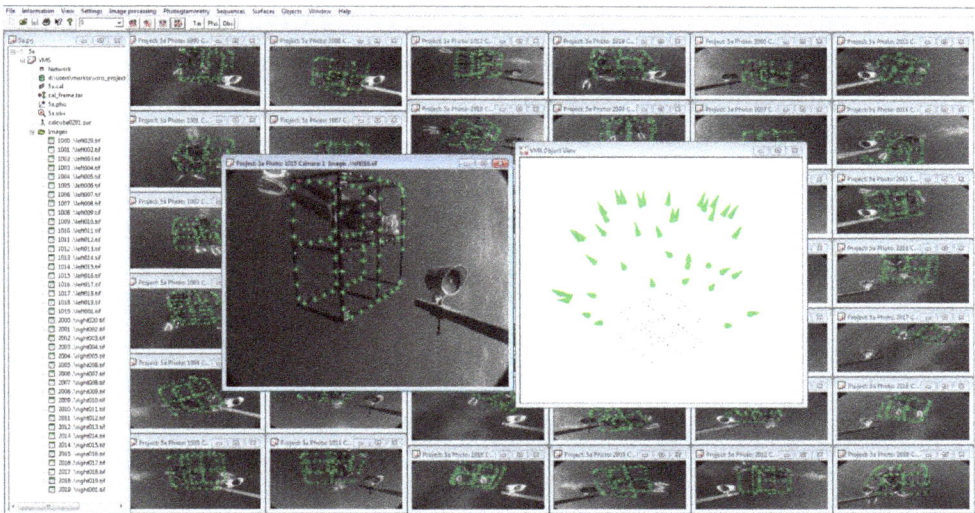

Figure 3. A full set of calibration images from an underwater stereo-video system, processed using Vision Measurement System (www.geomsoft.com/VMS). Both the cameras and the object have been rotated to acquire the convergent geometry of the network.

A fixed test range, such as the "Manhattan" object shown in Figure 1, has the advantage that accurately known target coordinates can be used in a pre-calibration approach, but the disadvantage that the camera system has to be transported to the range and then back to the deployment location. In comparison, accurate information for the positions of the targets on a portable calibration fixture is not required, as coordinates of the targets can be derived as part of a self-calibration approach. Hence it is immaterial if the portable fixture distorts or is dis-assembled between calibrations, although the fixture must retain its dimensional integrity during the image capture. Scale within the 3D measurement space is determined by introducing distances measured between pre-identified targets into the self-calibration network [51]. The known distances between the targets must be reliable and accurate, so known lengths are specified between targets on the rigid arms of the fixture or between the corners of the checkerboard.

In practice, cameras are most often pre-calibrated using a self-calibration network and a portable calibration fixture in a venue convenient to the deployment. The refractive index of water is insensitive to temperature, pressure or salinity [31], so the conditions prevailing for the pre-calibration can be assumed to be valid for the actual deployment of the system to capture measurements. The assumption is also made that the camera configurations, such as focus and zoom, and the relative orientation for a multi camera system, are locked down and undisturbed. A close proximity between the locations of the calibration and the deployment minimises the risk of a physical change to the camera system.

The process of self-calibration of underwater cameras is straightforward and rapid. The calibration can take place in a swimming pool, in an on-board tank on the vessel or, conditions permitting, adjacent to, or beneath, the vessel. The calibration fixture can be held in place and the cameras maneuvered around it, or the calibration fixture can be manipulated whilst the cameras are held in position, or a combination of both approaches can be used (see Figure 3). For example, a small 2D checkerboard may be manipulated in front of an ROV stereo-camera system held in a tank. A large, towed body system may be suspended in the water next to a wharf and a large 3D calibration fixture manipulated in front of the stereo video cameras. In the case of a diver-controlled stereo-camera system, a 3D calibration fixture may be tethered underneath the vessel and the cameras moved around it.

There are very few examples of *in-situ*, self-calibrations of camera systems, because this type of approach is not readily adapted to the dynamic and uncontrolled underwater environment. Nevertheless, there are some examples of a single camera or stereo-pair *in-situ* self-calibration [27,35,37,38]. In most cases a pre-calibration is conducted to determine an initial estimate of the calibration of the camera system.

3. Calibration Algorithms

Calibration of a camera system is necessary for two reasons. First, the internal geometric characteristics of the cameras must be determined [44]. In photogrammetric practice, camera calibration is most often defined by physical parameter set (see Figure 4) comprising principal distance, principal point location, radial [52] and decentring [53] lens distortions, plus affinity and orthogonality terms to compensate for minor optical effects [54,55]. The principal distance is formally defined as the separation, along the camera optical axis, between the lens perspective centre and the image plane. The principal point is the intersection of the camera optical axis with the image plane.

Second, the relative orientation of the cameras with respect to one another, or the exterior orientation with respect to an external reference, must be determined. Also known as pose estimation, both the location and orientation of the camera(s) must be determined. For the commonly used approach of stereo cameras, the relative orientation effectively defines the separation of the perspective centres of the two lenses, the pointing angles (omega and phi rotations) of the two optical axes of the cameras and the roll angles (kappa rotations) of the two focal plane sensors (see Figure 5).

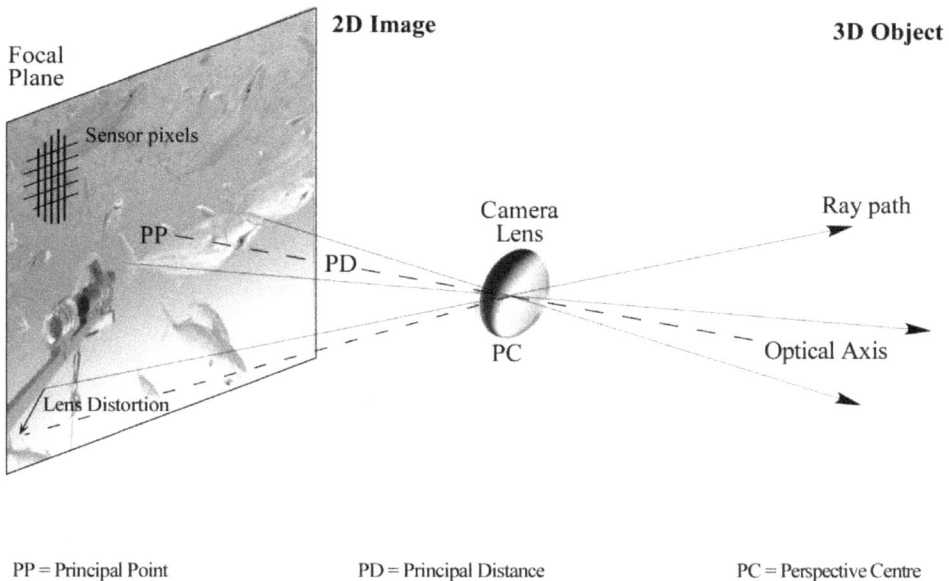

PP = Principal Point   PD = Principal Distance   PC = Perspective Centre

Figure 4. The geometry of perspective projection based on physical calibration parameters.

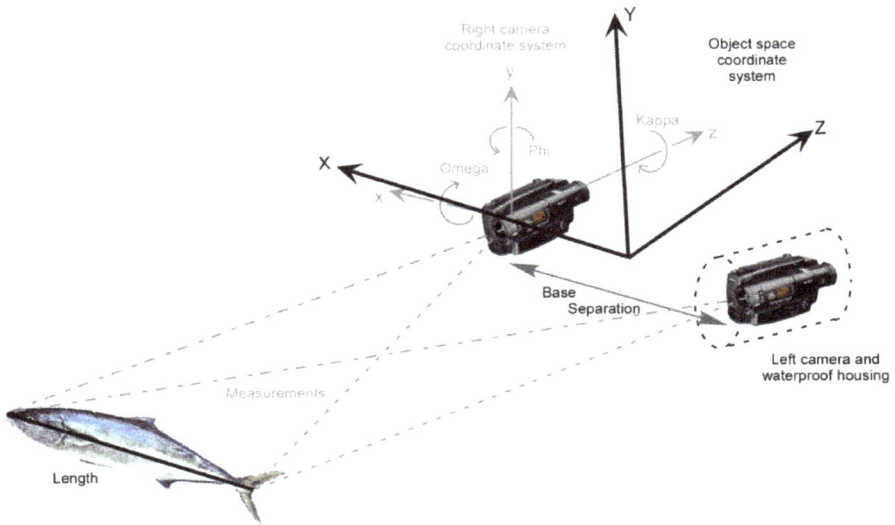

Figure 5. Schematic view of a stereo-image measurement of a length from 3D coordinates.

In the underwater environment the effects of refraction must be corrected or modelled to obtain an accurate calibration. The entire light path, including the camera lens, housing port and water medium, must be considered. By far the most common approach is to correct the refraction effects using absorption by the physical camera calibration parameters. Assuming that the camera optical axis is approximately perpendicular to a plane or dome camera port, the primary effect of refraction through the air-port and port-water interfaces will be radially symmetric around the principal point [56]. This primary effect can be absorbed by the radial lens distortion component of the calibration parameters. Figure 6 shows a comparison of radial lens distortion from calibrations in air and in water for the same camera. There will also be some small, asymmetric effects caused by, for example, alignment errors between the optical axis and the housing port, and perhaps non-uniformities in the thickness or material of the housing. These secondary effects can be absorbed by calibration parameters such as the decentring lens distortion and the affinity term. Figure 7 shows a comparison of decentring lens distortion from calibrations in air and in water of the same camera. Similar changes in the lens distortion profiles are demonstrated in [46,57].

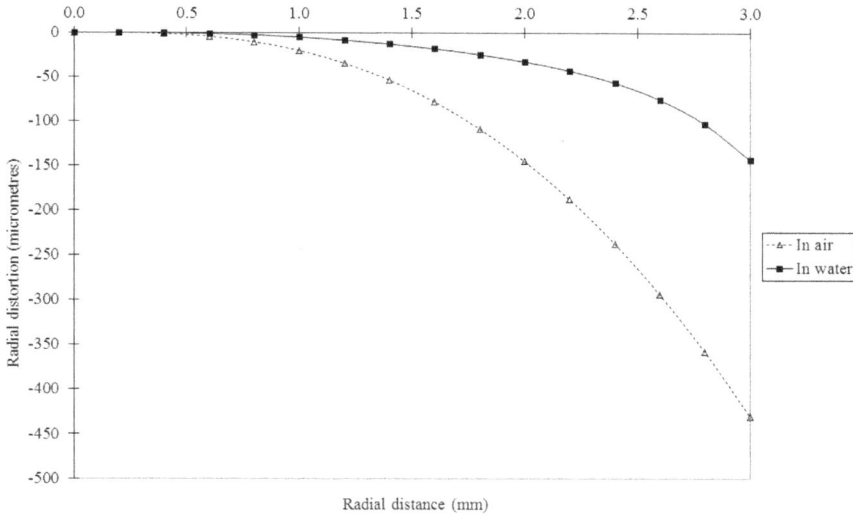

Figure 6. Comparison of radial lens distortion from in-air and in-water calibrations of a GoPro Hero4 camera operated in HD video mode.

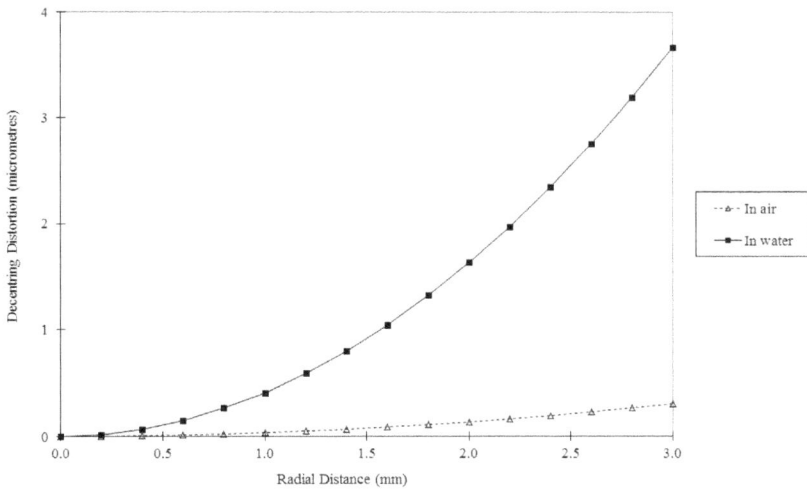

Figure 7. Comparison of decentring lens distortion from in-air and in-water calibrations of a GoPro Hero4 camera operated in HD video mode. Note the much smaller range of distortion values (vertical axis) compared to Figure 6.

Table 1 shows some of the calibration parameters for the in air and in water calibrations of two GoPro Hero4 camera. The ratios of the magnitudes of the parameters indicate whether there is a contribution to the refractive effects. As

11

could be expected, for a plane housing port the principal distance is affected directly, whilst changes in parameters such as the principal point location and the affinity term may include the combined influences of secondary effects, correlations with other parameters and statistical fluctuation. These results are consistent for the two cameras, consistent with other cameras tested, and [57,58] present similar outcomes from in air *versus* in water calibrations for flat ports. Very small percentage changes to all parameters, including the principal distance, are reported in [59] for housings with dome ports. Increases in principal distance of 1% to 25% for dome and flat ports are reported in [32]. All of these results are generally in accord with the expected physical model of the refraction.

Table 1. Comparison of parameters from in air and in water calibrations for two GoPro Hero4 camera used in HD video mode.

Camera	GoPro Hero4 #1			GoPro Hero4 #2		
Parameter	In Air	In Water	Ratio	In Air	In Water	Ratio
PPx (mm)	0.080	0.071	0.88	−0.032	−0.059	1.82
PPy (mm)	−0.066	−0.085	1.27	−0.143	−0.171	1.20
PD (mm)	3.676	4.922	1.34	3.658	4.898	1.34
Affinity	−6.74E−03	−6.71E−03	1.00	−6.74E−03	−6.84E−03	1.01

The disadvantage of the absorption approach for the refractive effects is that there will always be some systematic errors which are not incorporated into the model. The effect of refraction invalidates the assumption of a single projection centre for the camera [60], which is the basis for the physical parameter model. The errors are most often manifest as scale changes when measurements are taken outside of the range used for the calibration process. Experience over many years of operation demonstrates that, if the ranges for the calibration and the measurements are commensurate, then the level of systematic error is generally less than the precision with which measurements can be extracted. This masking effect is partly due to the elevated level of noise in the measurements, caused by the attenuation and loss of contrast in the water medium.

The alternative to the simple approach of absorption is the more complex process of geometric correction, effectively an application of ray tracing of the light paths through the refractive interfaces. A two phase approach is developed in [61] for a stereo camera housing with concave lens covers. An in air calibration is carried out first, followed by an in water calibration that introduces 11 lens cover parameters such as the centre of curvature of the concave lens and, if not known from external measurements, refractive indices for the lens covers and water. A more general geometric correction solution is developed for plane port housings in [62]. Additional unknowns in the solution are the distance between the camera perspective centre and the housing, and the normal of the plane housing port, whilst the port thickness and refractive indices must be known. Using ray tracing, [63] develops a general solution to refractive surfaces that, in theory, can accommodate any shape of camera housing port. The shape of the refractive surface and the refractive indices must be known.

A variation on the geometric correction is the perspective centre shift or virtual projection centre approach. A specific solution for a planar housing port is developed in [64]. The parameters include the standard physical parameters, the refractive indices of glass and water, the distance between the perspective centre and the port, the tilt and direction of the optical axis with respect to the normal to the port, and the housing interface thickness. A modified approach neglects the direction of the optical axis and the thickness of thin ports, as these factors can be readily absorbed by the standard physical parameters. Again a two phase process is required, first a "dry" calibration in air and then a "wet" calibration in water [64]. A similar principle is used in [65], also with a two phase calibration approach.

The advantage of these techniques is that, without the approximations in the models, the correction of the refractive effects is exact. The disadvantages are the requirements for two phase calibrations and known data such as refractive indices. Further, in some cases the theoretical solution is specific to a housing type, whereas the absorption approach has the distinct advantage that it can be used with any type of underwater housing.

As well as the common approaches described above, some other investigations are worthy of note. The Direct Linear Transformation (DLT) algorithm [66] is used with three different techniques in [67]. The first is essentially an absorption approach, but used in conjunction with a sectioning of the object space to minimise the remaining errors in the solution. A double plane correction grid was applied in the second approach. In the last technique a formal refraction correction model is included with the requirements that the camera-to-interface distance and the refractive index must be known. The solutions presented in [67] suggest that both the absorption and refraction correction approaches can be used successfully in association with different calibration algorithms, either linear models such as DLT [66], multi-stage linear solutions [68,69] or non-linear models based on the standard physical parameters [44].

A review of refraction correction methods for underwater imaging is given in [60]. The perspective camera model, ray-based models and physical models are analysed, including an error analysis based on synthetic data. The analysis demonstrates that perspective camera models incur increasing errors with increasing distance and tilt of the refractive surfaces, and only the physical model of refraction correction permits a complete theoretical compensation.

Once the camera calibration is established, single camera systems can be used to acquire measurements when used in conjunction with reference frames [29] or sea floor reference marks [37]. For multi-camera systems the relative orientation is required as well as the camera calibration. The relative orientation can be included in the self-calibration solution as a constraint [70] or can be computed as a post-process based on the camera positions and orientations for each set of synchronised exposures [47]. In either case, it is important to detect and eliminate outliers, usually caused by lack of synchronisation, that would otherwise unduly influence the calibration solution or the relative orientation computation. Outliers caused by synchronisation effects are more common for systems based on camcorders or video cameras in separate housings, which typically use an external device such as a flashing LED light to synchronise the images to within one frame [47].

In the case of post-processing, the exterior orientations for the sets of synchronised exposures are initially in the frame of reference of the calibration fixture, so each set must be transformed into a local frame of reference with respect to a specific baseline between the cameras. In the case of stereo cameras, the local frame of reference is adopted as the centre of the baseline between the camera perspective centres, with the axes aligned with the baseline direction and the mean optical axis pointing direction (see Figure 5). The final parameters for the precise relative orientation are adopted as the mean values for all sets in the calibration network, after any outliers have been detected and eliminated.

4. Calibration Reliability and Stability

The reliability and accuracy of the calibration of underwater camera systems is dependent on a number of factors. Chief amongst the factors are the geometry and redundancy for the calibration network. A high level of redundant information, provided by many target image observations on many exposures, produces high reliability so that outliers in the image observations can be detected and eliminated. An optimum three dimensional geometry is essential to minimise correlations between the parameters and ensure that the camera calibration is an accurate representation of the physical model [45]. However it should be noted that it is not possible to eliminate all correlations between the calibration parameters. Correlations are always present between the three radial distortion terms and between the principal point and two decentring terms.

The accuracy of the calibration parameters is enhanced if the network of camera and target locations meets the following criteria:

(1) The camera and target arrays are three dimensional in nature. Two dimensional arrays are a source of weak network geometry. Three dimensional arrays minimise correlations between the internal camera calibration parameters and the external camera location and orientation parameters.

(2) The many, convergent camera views approach a $90°$ intersection at the centre of the target array. A narrowly grouped array of camera views will produce shallow intersections, weakening the network and thereby decreasing the confidence with which the calibration parameters are determined.

(3) The calibration fixture or range fills the field of view of the camera(s) to ensure that image measurements are captured across the entire format. If the fixture or range is small and centred in the field of view then the radial and decentring lens distortion profiles will be defined very poorly because measurements are captured only where the signal is small in magnitude.

(4) The camera(s) are rolled around the optical axis for different exposures so that $0°, 90°, 180°$ and $270°$ orthogonal rotations are spread throughout the calibration network. A variety of camera rolls in the network also minimises correlations between the internal camera calibration parameters and the external camera location and orientation parameters.

If these four conditions are met, the self-calibration approach can be used to simultaneously and confidently determine the camera calibration parameters, camera exposure locations and orientations, and updated target coordinates [45].

In recent years there has been an increasing adoption of a calibration technique using a small 2D checkerboard and a freely available Matlab solution [71]. The main advantages of this approach are the simplicity of the calibration fixture and the rapid measurement and processing of the captured images, made possible by the automatic recognition of the checkerboard pattern [50]. A practical guide to the use of this technique is provided in [72].

However the small size and 2D nature of the checkerboard limits the reliability and accuracy of measurements made using this technique [41]. The technique is equivalent to a test range calibration rather than a self-calibration, because the coordinates of the checkerboard corners are not updated. Any inaccuracy in the coordinates, especially if the checkerboard has variations from a true 2D plane, will introduce systematic errors into the calibration. Nevertheless, the 2D fixture can produce a calibration suitable for measurements at short ranges and with modest accuracy requirements. AUV and diver operated stereo camera systems pre-calibrated with this technique have been used to capture fish length measurements [16,72] and tested for the 3D re-construction of artefacts [59].

The stability of the calibration for underwater camera systems has been well documented in published reports [73,74]. As noted previously, the basic camera settings such as focus and zoom must be consistent between the calibration and deployments, usually ensured through the use of tape or a locking screw to prevent the settings from being inadvertently altered. For cameras used in air, other factors are handling of the camera, especially when the camera is rolled about the optical axis or a zoom lens is being employed, and the quality of the lens mount. Any distortion of the camera body or movement of the lens or optical elements will result in variation of the relationship between the perspective centre and the imager at the focal plane, which will disturb the calibration [75]. Fixed focal length lenses are preferred over zoom lenses to minimise the instabilities.

However the most significant sensitivity for the calibration stability of underwater camera systems is the relationship between the camera lens and the housing port. Rigid mountings of the cameras in the housings is critical to ensure that the total optical path from the image sensor to the water medium is consistent [73]. Testing and validation has shown that the camera calibration is only reliable if the cameras in the housings are mounted on a rigid connection to the camera port [74]. This applies to both within a single deployment and between multiple, separate deployments of the camera system. Unlike correction lenses and dome ports, a specific position and alignment within the housing is not necessary, but the distance and orientation of the camera lens relative to the housing port must be consistent. The most reliable option is a direct, mechanical linkage between the camera lens and the housing port that can consistently re-create the physical relationship. The consistency of distance and orientation is especially important for portable camcorders because they must be regularly removed from the housings to retrieve storage media and replenish batteries.

Finally, for multi-camera systems, in air or in water, the camera housings must have a rigid mechanical connection to a base bar to ensure that the separation and relative orientation of the cameras is also consistent. Perturbation of the separation or relative orientation often results in apparent scale errors which can be readily confused with refractive effects. Figure 8 shows some results of repeated calibrations of a GoPro Hero 2 stereo-video system. The variation in the parameters between consecutive calibrations demonstrates a comparatively stable relative orientation but a more unstable camera calibration caused by a non-rigid mounting of the camera in the housing. Note that these tests were based on video frames captured with a motionless camera and calibration object in order to avoid any motion effects from the rolling shutter used by GoPro cameras [76]. Rapid motion should be avoided for GoPro cameras when capturing video for calibration or measurement.

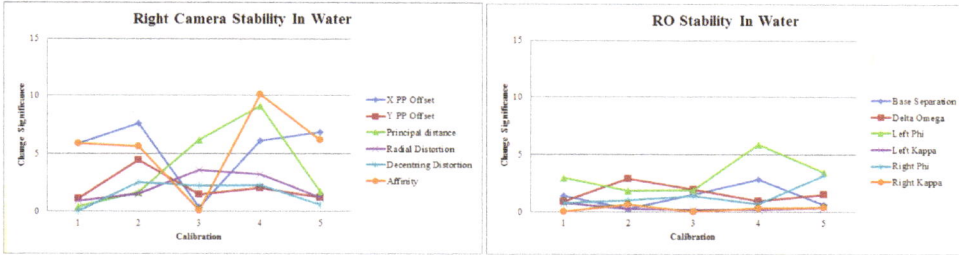

Figure 8. Stability of the right camera calibration parameters (**Left**) and the relative orientation parameters (**Right**) for a GoPro Hero 2 stereo-video system. The vertical axis is the change significance of individual parameters between consecutive calibrations [73].

5. Calibration and Validation Results

The first evaluation of a calibration is generally the internal consistency of the network solution that is used to compute the calibration parameters, camera locations and orientations, and if applicable, updated target coordinates. The "internal" indicator is the Root Mean Square (RMS) error of image measurement, a metric for the internal "fit" of the least squares estimation solution [48]. Note that in general the measurements are based on an intensity weighted centroid to locate the centre of each circular target in the image [77].

To allow comparison of different cameras with different spacing of the light sensitive elements in the CMOS or CCD imager, the RMS error is expressed in fractions of a pixel. In ideal conditions in air, the RMS image error is typically in the range of 0.03–0.1 pixels [77]. In the underwater environment, the attenuation of light and loss of contrast, along with small non-uniformities in the media, degrades the RMS error into the range of 0.1–0.3 pixels (see Table 2). This degradation is a combination of a larger statistical signature for the image measurements and the influence of small, uncompensated systematic errors. In conditions of poor lighting or poor visibility the RMS error deteriorates rapidly [72].

The second metric that is commonly used to compare the calibration, especially for in air operations, is the proportional error, expressed as the ratio of the magnitude of the average precision of the 3D coordinates of the targets to the largest 3D Euclidian distance contained within the volume of the object. This "external" indicator provides a standardised, relative measure of precision in the object space. In the circumstance of a camera calibration, the largest 3D distance is the diagonal span of the test range volume, or the diagonal span of the volume envelope of all imaged locations of the calibration fixture. Whilst the RMS image error may be favourable, the proportional error may be relatively poor if the object is contained within a small volume or the geometry of the calibration network is poor. Table 2 presents a sample of some results for the precision of calibrations. It is evident that the proportional error can vary substantially, however an average figure is approximately 1:5000.

As a consequence of the potential misrepresentation by proportional error, independent testing of the accuracy of underwater camera systems is essential to ensure the validity of 3D locations, length, area or volume measurements. For stereo and multi camera systems, the primary interest is length measurements that are subsequently used to estimate biomass or age. One validation technique is to use known distances on the rigid components of the calibration fixture [6], however this has some limitations. As already noted, the circular, discrete targets are dissimilar to the natural feature points of a fish snout or tail, and are measured by different techniques. The variation in size and angle of the distance on the calibration fixture may not correlate well with the size and orientation of fish when measured. In particular, measurements of fish are often taken at greater ranges than that of the calibration fixture, partly due to expediency in surveys and partly because the calibration fixture must be close enough to the cameras to fill a reasonable portion of the field of view. Given the approximations in the refraction models, it is important that accuracy validations are carried out at ranges greater than the average range to the calibration fixture. Further, it has been demonstrated that the accuracy of length measurements is dependent on the separation of the cameras in a multi camera system [41] and significantly affected by the orientation of the fish relative to the cameras [47,78]. Accordingly, validation of underwater video measurement systems is typically carried out by introducing a known length, either a rod or a fish silhouette, which is measured manually at a variety of ranges and orientations within the field of view (see Figure 9).

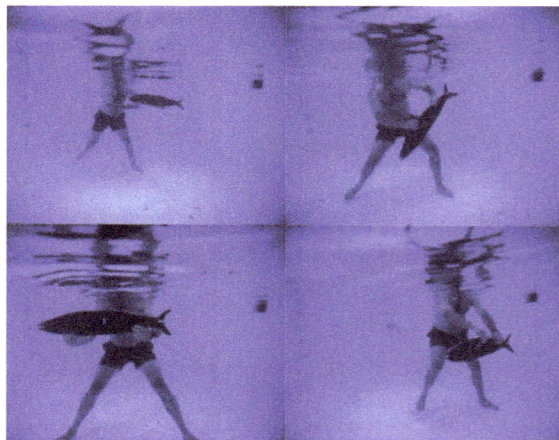

Figure 9. Example of a fish silhouette validation in a swimming pool (courtesy of Prof. E. S. Harvey).

Table 2. A sample of some published results for the precision of underwater camera calibrations. Note that [35] used observations of a mobile fish pen and the measurements used by [61] were made to the nearest whole pixel.

Technique	RMS Image Error (pixels)	RMS XYZ Error (mm)	Proportional Error
Absorption [47,73]	0.1–0.3	0.1–0.5	1:3000–1:15,000
Absorption [35]	0.3	40–200	1:500
Geometric correction [61]	1.0	10	1:210
Perspective shift [64]	0.3	2.0	1:1000
Absorption [40]	0.2–0.25	1.9	1:32,000

In the best case scenario of clear visibility and high contrast targets, the RMS error of validation measurements is typically less than 1 mm over a length of 1 m, equivalent to a length accuracy of 0.1%. In realistic, operational conditions using fish silhouettes or validated measurements of live fish, length measurements have an accuracy of 0.2% to 0.7% [6,11,41,64,78]. The accuracy is somewhat degraded if a simple correction grid is used [17] or a simplified calibration approach is adopted [72]. A sample of published validation results is given in Table 3.

Table 3. A sample of some published results for the validation of underwater camera calibrations.

Technique	Validation	Percentage Error
Absorption [47]	Length measurement of silhouettes or rods throughout the volume	0.2%–0.7%
Lens distortion grid [17]	Caliper measurements of Chinook Salmon	1.5%
Absorption [6]	Caliper measurements of Southern Bluefin Tuna	0.2%
Perspective shift [64]	Flat reference plate and straight line re-construction	0.4%
Absorption [40]	Similarity transformation between above and below water networks	0.3%
Radial lens distortion correction [72]	Distances on checkerboard	0.9%–1.5%
Absorption [41]	Length measurements of a rod throughout the volume	0.5%
Perspective shift [65]	Flat reference plate and distance between spheres	0.4%–0.7%

Validations of biomass estimates of Southern Bluefin Tuna measured in aquaculture pens and sponges measured in the field have shown that volume or biomass can be estimated with an accuracy of the order of a few percent. The Southern Bluefin Tuna validation was based on distances such as body length and span, made by a stereo-video system and compared to a length board and caliper system of manual measurement. Each Southern Bluefin Tuna in a sample of 40 fish was also individually weighed. The stereo-video system produced an estimate of better than 1% for the total biomass [6]. Triangulation meshes on the surface of simulated and live specimens were used to estimate the volume of sponges. The resulting errors were 3%–5%, and no worse than 10%, for individual sponges [38]. Greater variability is to be expected for the estimates of the sponge volumes, because of the uncertainty associated with the assumed shape of the unseen substrate surface beneath each sponge.

By the very nature of conversion from length to weight, errors can be amplified significantly. Typical regression functions are power series with a near cubic term [7,8,41]. Accordingly, inaccuracies in the calibration and the precision of the measurement may combine to produce unacceptable results. A simulation is employed by [41] to demonstrate clearly that the predicted error in the biomass of a fish, based on the error in the length, deteriorates rapidly with range from the cameras, especially with a small 2D calibration fixture and a narrow separation between the stereo cameras. Errors in the weight in excess of 10% are possible, reinforcing the need for validation testing throughout the expected range of measurements. Validation at the most distant ranges, where errors in biomass can approach 40%, is critical to ensure that an acceptable level of accuracy is maintained.

6. Conclusions

This paper has presented a review of different calibration techniques that incorporate the effects of refraction from the camera housing and the water medium. Calibration of underwater camera systems is essential to ensure the accuracy and reliability of measurements of marine fauna, flora or artefacts. Calibration is a key process to ensure that the analysis of biomass, population distribution or dimensions is free of systematic errors.

Irrespective of whether an implicit absorption or an explicit refractive model is used in the calibration of underwater camera systems, it is clear from the sample of validation results that an accuracy of the order of 0.5% of the measured dimensions can be achieved. Less favourable results are likely when approximate methods, such as 2D planar correction grids, are used. The configuration of the underwater camera system is a significant factor that has a primary influence on the accuracy achieved. However the advantage of photogrammetric systems is that the configuration can be readily adapted to suit the required measurement accuracy.

Further investigation of different calibration algorithms is warranted to assess the merits of the various approaches. Otherwise confounding factors, such as the size of the calibration fixture, the range of locations and the image measurement technique, should be common to all calibration techniques to gain a valid comparison. The evaluation of such testing should be based on a consistent and rigorous validation process to ensure that all techniques are compared on the same basis.

Acknowledgments: The author gratefully acknowledges sustained research collaborations with Euan S. Harvey, Curtin University, Australia, and Stuart Robson, University College London, England, for the contributions they have made to the development of underwater camera calibration and validation techniques.

Conflicts of Interest: The author declares no conflicts of interest.

References

1. World Wildlife Fund, 2015. Living Blue Planet Report. Available online: http://awsassets.wwf.org.au/downloads/mo038_living_blue_planet_report_16sep15.pdf (accessed on 29 October 2015).
2. Pauly, D.; Christensen, V.; Guenette, S.; Pitcher, T.J.; Sumaila, U.R.; Walters, C.J.; Watson, R.; Zeller, D. Towards sustainability in world fisheries. *Nature* **2002**, *418*, 689–695.
3. Watson, D.L.; Anderson, M.J.; Kendrick, G.A.; Nardi, K.; Harvey, E.S. Effects of protection from fishing on the lengths of targeted and non targeted fish species at the Houtman Abrolhos Islands, Western Australia. *Mar. Ecol. Prog. Ser.* **2009**, *384*, 241–249.
4. Duarte, C.M.; Holmer, M.; Olsen, Y.; Soto, D.; Marbà, N.; Guiu, J.; Black, K.; Karakassis, I. Will the Oceans Help Feed Humanity? *BioScience* **2009**, *59*, 967–976.
5. Naylor, R.L.; Goldberg, R.J.; Primavera, J.H.; Kautsky, N.; Beveridge, M.C.; Clay, J.; Folk, C.; Lubchenco, J.; Mooney, H.; Troell, M. Effect of aquaculture on world fish supplies. *Nature* **2000**, *405*, 1017–1024.
6. Harvey, E.S.; Cappo, M.; Shortis, M.R.; Robson, S.; Buchanan, J.; Speare, P. The accuracy and precision of underwater measurements of length and maximum body depth of Southern Bluefin Tuna (Thunnus maccoyii) with a stereo-video camera system. *Fish. Res.* **2003**, *63*, 315–326.
7. Pienaar, L.V.; Thomson, J.A. Allometric weight-length regression model. *J. Fish. Res. Board Can.* **1969**, *26*, 123–131.
8. Santos, M.N.; Gaspar, M.B.; Vasconcelos, P.; Monteiro, C.C. Weight–length relationships for 50 selected fish species of the Algarve coast (southern Portugal). *Fish. Res.* **2002**, *59*, 289–295.
9. Shortis, M.R.; Harvey, E.S.; Abdo, D.A. A review of underwater stereo-image measurement for marine biology and ecology applications. In *Oceanography and Marine Biology: An Annual Review*; Gibson, R.N., Atkinson, R.J.A., Gordon, J.D.M., Eds.; CRC Press: Boca Raton, FL, USA, 2009; Volume 47.
10. Murphy, H.M.; Jenkins, G.P. Observational methods used in marine spatial monitoring of fishes and associated habitats: A review. *Mar. Freshw. Res.* **2010**, *61*, 236–252.
11. Harvey, E.S.; Fletcher, D.; Shortis, M.R.; Kendrick, G. A comparison of underwater visual distance estimates made by SCUBA divers and a stereo-video system: Implications for underwater visual census of reef fish abundance. *Mar. Freshw. Res.* **2004**, *55*, 573–580.
12. Harvey, E.S.; Fletcher, D.; Shortis, M.R. Improving the statistical power of visual length estimates of reef fish: Comparison of divers and stereo-video. *Fish. Bull.* **2001**, *99*, 63–71.
13. Santana-Garcon, J.; Newman, S.J.; Harvey, E.S. Development and validation of a mid-water baited stereo-video technique for investigating pelagic fish assemblages. *J. Exp. Mar. Biol. Ecol.* **2014**, *452*, 82–90.
14. Mallet, D.; Pelletier, D. Underwater video techniques for observing coastal marine biodiversity: A review of sixty years of publications (1952–2012). *Fish. Res.* **2014**, *154*, 44–62.

15. McLaren, B.W.; Langlois, T.J.; Harvey, E.S.; Shortland-Jones, H.; Stevens, R. A small no-take marine sanctuary provides consistent protection for small-bodied by-catch species, but not for large-bodied, high-risk species. *J. Exp. Mar. Biol. Ecol.* **2015**, *471*, 153–163.

16. Seiler, J.; Williams, A.; Barrett, N. Assessing size, abundance and habitat preferences of the Ocean Perch Helicolenus percoides using a AUV-borne stereo camera system. *Fish. Res.* **2012**, *129*, 64–72.

17. Petrell, R.J.; Shi, X.; Ward, R.K.; Naiberg, A.; Savage, C.R. Determining fish size and swimming speed in cages and tanks using simple video techniques. *Aquac. Eng.* **1997**, *16*, 63–84.

18. Phillips, K.; Rodriguez, V.B.; Harvey, E.; Ellis, D.; Seager, J.; Begg, G.; Hender, J. *Assessing the Operational Feasibility of Stereo-Video and Evaluating Monitoring Options for the Southern Bluefin Tuna Fishery Ranch Sector*; Fisheries Research and Development Corporation Report: Canberra, Australia, 2009.

19. Rosen, S.; Jörgensen, T.; Hammersland-White, D.; Holst, J.C. DeepVision: A stereo camera system provides highly accurate counts and lengths of fish passing inside a trawl. *Can. J. Fish. Aquat. Sci.* **2013**, *70*, 1456–1467.

20. Shieh, A.C.R.; Petrell, R.J. Measurement of fish size in Atlantic salmon (salmo salar l.) cages using stereographic video techniques. *Aquac. Eng.* **1998**, *17*, 29–43.

21. Hale, W.B.; Cook, C.E. Underwater microcontouring. *Photogramm. Eng.* **1962**, *28*, 96–98.

22. Pollio, J. Underwater mapping with photography and sonar. *Photogramm. Eng.* **1971**, *37*, 955–968.

23. Hohle, J. Reconstruction of an underwater object. *Photogramm. Eng.* **1971**, *37*, 948–954.

24. Pollio, J. Remote underwater systems on towed vehicles. *Photogramm. Eng.* **1972**, *38*, 1002–1008.

25. Leatherdale, J.D.; Turner, D.J. Underwater photogrammetry in the North Sea. *Photogramm. Rec.* **1983**, *11*, 151–167.

26. Baldwin, R.A. An underwater photogrammetric measurement system for structural inspection. *Int. Arch. Photogramm.* **1984**, *25*, 9–18.

27. O'Byrne, M.; Pakrashi, V.; Schoefs, F.; Ghosh, B. A comparison of image based 3D recovery methods for underwater inspections. In Proceedings of the 7th European Workshop on Structural Health Monitoring, Nantes, France, 8–11 July 2014; pp. 671–678.

28. Negahdaripour, S.; Firoozfam, P. An ROV stereovision system for ship-hull inspection. *IEEE J. Ocean. Eng.* **2006**, *31*, 551–564.

29. Bass, G.F.; Rosencrantz, D.M. The ASHREAH—A pioneer in search of the past. In *Submersibles and Their Use in Oceanography and Ocean Engineering*; Geyer, R.A., Ed.; Elsevier: Amsterdam, The Netherlands, 1977; pp. 335–350.

30. Drap, P.; Seinturier, J.; Scaradozzi, D.; Gambogi, P.; Long, L.; Gauch, F. Photogrammetry for virtual exploration of underwater archeological sites. In Proceedings of the 21st International Symposium, CIPA 2007: AntiCIPAting the Future of the Cultural Past, Athens, Greece, 1–6 October 2007.

31. Moore, E.J. Underwater photogrammetry. *Photogramm. Rec.* **1976**, *8*, 748–763.

32. Bianco, G.; Gallo, A.; Bruno, F.; Muzzupappa, M. A comparison between active and passive techniques for underwater 3D applications. *Int. Arch. Photogramm. Remote Sens. Spat. Inf. Sci.* **2011**, *34*, 357–363.

33. Newton, I. Underwater Photogrammetry. In *Non-Topographic Photogrammetry*; Karara, H.M., Ed.; American Society for Photogrammetry and Remote Sensing: Bethesda, MD, USA, 1989; pp. 147–176.

34. Doucette, J.S.; Harvey, E.S.; Shortis, M.R. Stereo-video observation of nearshore bedforms on a low energy beach. *Mar. Geol.* **2002**, *189*, 289–305.

35. Schewe, H.; Moncreiff, E.; Gruendig, L. Improvement of fish farm pen design using computational structural modelling and large-scale underwater photogrammetry. *Int. Arch. Photogramm. Remote Sens.* **1996**, *31*, 524–529.

36. Capra, A. Non-conventional system in underwater photogrammetry. *Int. Arch. Photogramm. Remote Sens.* **1992**, *29*, 234–240.

37. Green, J.; Matthews, S.; Turanli, T. Underwater archaeological surveying using Photomodeler, VirtualMapper: Different applications for different problems. *Int. J. Naut. Archaeol.* **2002**, *31*, 283–292.

38. Abdo, D.A.; Seager, J.W.; Harvey, E.S.; McDonald, J.I.; Kendrick, G.A.; Shortis, M.R. Efficiently measuring complex sessile epibenthic organisms using a novel photogrammetric technique. *J. Exp. Mar. Biol. Ecol.* **2006**, *339*, 120–133.

39. Shortis, M.R.; Seager, J.W.; Williams, A.; Barker, B.A.; Sherlock, M. Using stereo-video for deep water benthic habitat surveys. *Mar. Technol. Soc. J.* **2009**, *42*, 28–37.

40. Menna, F.; Nocerino, E.; Troisi, S.; Remondino, F. A photogrammetric approach to survey floating and semi-submerged objects. In Proceedings of the SPIE 8791, Videometrics, Range Imaging, and Applications XII, and Automated Visual Inspection (87910H), Munich, Germany, 14–16 May 2013.

41. Boutros, N.; Harvey, E.S.; Shortis, M.R. Calibration and configuration of underwater stereo-video systems for applications in marine ecology. *Limnol. Oceanogr. Methods* **2015**, *13*, 224–236.

42. Ivanoff, A.; Cherney, P. Correcting lenses for underwater use. *J. Soc. Motion Pict. Telev. Eng.* **1960**, *69*, 264–266.

43. Chong, A.K.; Stratford, P. Underwater digital stereo-observation technique for red hydrocoral study. *Photogramm. Eng. Remote Sens.* **2002**, *68*, 745–751.

44. Brown, D.C. Close range camera calibration. *Photogramm. Eng.* **1971**, *37*, 855–866.

45. Kenefick, J.F.; Gyer, M.S.; Harp, B.F. Analytical self calibration. *Photogramm. Eng. Remote Sens.* **1972**, *38*, 1117–1126.

46. Fryer, J.G.; Fraser, C.S. On the calibration of underwater cameras. *Photogramm. Rec.* **1986**, *12*, 73–85.

47. Harvey, E.S.; Shortis, M.R. A system for stereo-video measurement of sub-tidal organisms. *Mar. Technol. Soc. J.* **1996**, *29*, 10–22.

48. Granshaw, S.I. Bundle adjustment methods in engineering photogrammetry. *Photogramm. Rec.* **1980**, *10*, 181–207.

49. Shortis, M.R.; Seager, J.W. A practical target recognition system for close range photogrammetry. *Photogramm. Rec.* **2014**, *29*, 337–355.

50. Zhang, Z. A flexible new technique for camera calibration. *IEEE Trans. PAMI* **2000**, *22*, 1330–1334.

51. El-Hakim, S.F.; Faig, W. A combined adjustment of geodetic and photogrammetric observations. *Photogramm. Eng. Remote Sens.* **1981**, *47*, 93–99.

52. Ziemann, H.; El-Hakim, S.F. On the definition of lens distortion reference data with odd-powered polynomials. *Can. Surv.* **1983**, *37*, 135–143.

53. Brown, D.C. Decentring distortion of lenses. *Photogramm. Eng.* **1966**, *22*, 444–462.

54. Fraser, C.S.; Shortis, M.R.; Ganci, G. Multi-sensor system self-calibration. In Proceedings of the SPIE 2598, Videometrics IV, Philadelphia, PA, USA, 25–26 October 1995; pp. 2–18.

55. Shortis, M.R. Multi-lens, multi-camera calibration of Sony Alpha NEX 5 digital cameras. In Proceedings of the CD-ROM, GSR_2 Geospatial Science Research Symposium, Melbourne, Australia, 10–12 December 2012.

56. Li, R.; Tao, C.; Zou, W.; Smith, R.G.; Curran, T.A. An underwater digital photogrammetric system for fishery geomatics. *Int. Arch. Photogramm. Remote Sens.* **1996**, *31*, 319–323.

57. Lavest, J.M.; Rives, G.; Lapresté, J.T. Underwater camera calibration. In *Computer vision—ECCV 2000*; Vernon, D., Ed.; Springer: Berlin/Heidelberg, Germany, 2000; pp. 654–668.

58. Rahman, T.; Anderson, J.; Winger, P.; Krouglicof, N. Calibration of an underwater stereoscopic vision system. In *OCEANS 2013 MTS/IEEE-San Diego: An Ocean in Common*; IEEE Computer Society: San Diego, CA, USA, 2013.

59. Bruno, F.; Bianco, G.; Muzzupappa, M.; Barone, S.; Razionale, A.V. Experimentation of structured light and stereo vision for underwater 3D reconstruction. *ISPRS J. Photogramm. Remote Sens.* **2011**, *66*, 508–518.

60. Sedlazeck, A.; Koch, R. Perspective and non-perspective camera models in underwater imaging-Overview and error analysis. In *Outdoor and Large-Scale Real-World Scene Analysis*; Dellaert, F., Frahm, J.-M., Pollefeys, M., Leal-Taixé, L., Rosenhahn, B., Eds.; Springer: Berlin/Heidelberg, Germany, 2012; pp. 212–242.

61. Li, R.; Li, H.; Zou, W.; Smith, R.G.; Curran, T.A. Quantitative photogrammetric analysis of digital underwater video imagery. *IEEE J. Ocean. Eng.* **1997**, *22*, 364–375.

62. Jordt-Sedlazeck, A.; Koch, R. Refractive calibration of underwater cameras. In *Computer Vision-CCV 2012*; 12th European Conference on Computer Vision; Springer: Berlin/Heidelberg, Germany, 2012; pp. 846–859.

63. Kotowski, R. Phototriangulation in multi-media photogrammetry. *Int. Arch. Photogramm. Remote Sens.* **1988**, *27*, 324–334.

64. Telem, G.; Filin, S. Photogrammetric modeling of underwater environments. *ISPRS J. Photogramm. Remote Sens.* **2010**, *65*, 433–444.

65. Bräuer-Burchardt, C.; Kühmstedt, P.; Notni, G. Combination of air- and water-calibration for a fringe projection based underwater 3D-scanner. In *Computer Analysis of Images and Patterns*; Azzopardi, G., Petkov, N., Eds.; Springer International: Charn, Switzerland, 2015; pp. 49–60.

66. Abdel-Aziz, Y.I.; Karara, H.M. Direct linear transformation into object space coordinates in close-range photogrammetry. In Proceedings of the ASPRS Symposium on Close-Range Photogrammetry, Urbana, IL, USA, 28–29 January 1971; pp. 1–18.

67. Kwon, Y.H.; Casebolt, J.B. Effects of light refraction on the accuracy of camera calibration and reconstruction in underwater motion analysis. *Sports Biomech.* **2006**, *5*, 95–120.

68. Heikkila, J.; Silvén, O. A four-step camera calibration procedure with implicit image correction. In Proceedings of the IEEE Computer Society Conference on Computer Vision and Pattern Recognition, San Juan, Puerto Rico, 17–19 June 1997; pp. 1106–1112.

69. Tsai, R.Y. A versatile camera calibration technique for high-accuracy 3D machine vision metrology using off-the-shelf TV cameras and lenses. *IEEE J. Robot. Autom.* **1987**, *3*, 323–344.

70. King, B.R. Bundle adjustment of constrained stereo pairs-Mathematical models. *Geomat. Res. Australas.* **1995**, *63*, 67–92.

71. Bouguet, J. Camera Calibration Toolbox for MATLAB. California Institute of Technology. Available online: http://www.vision.caltech.edu/bouguetj/calib_doc/index.html (accessed on 28 October 2015).

72. Wehkamp, M.; Fischer, P. A practical guide to the use of consumer-level still cameras for precise stereogrammetric *in situ* assessments in aquatic environments. *Underw. Technol.* **2014**, *32*, 111–128.

73. Harvey, E.S.; Shortis, M.R. Calibration stability of an underwater stereo-video system: Implications for measurement accuracy and precision. *Mar. Technol. Soc. J.* **1998**, *32*, 3–17.

74. Shortis, M.R.; Miller, S.; Harvey, E.S.; Robson, S. An analysis of the calibration stability and measurement accuracy of an underwater stereo-video system used for shellfish surveys. *Geomat. Res. Australas.* **2000**, *73*, 1–24.

75. Shortis, M.R.; Beyer, H.A. Calibration stability of the Kodak DCS420 and 460 cameras. In Proceedings of the SPIE 3174, Videometrics. V, San Fiego, CA, USA, 30–31 July 1997.

76. Liang, C.-K.; Peng, Y.-C.; Chen, H.; Li, S.; Pereira, F.; Shum, H.-Y.; Tescher, A.G. Rolling shutter distortion correction. In Proceedings of the SPIE 5960, Visual Communications and Image Processing 2005, Beijing, China, 12–15 July 2005.

77. Shortis, M.R.; Clarke, T.A.; Robson, S. Practical testing of the precision and accuracy of target image centring algorithms. In Proceedings of the SPIE 2598, Videometrics IV, Philadelphia, PA, USA, 25–26 October 1995; pp. 65–76.

78. Harvey, E.S.; Shortis, M.R.; Stadler, M.; Cappo, M. A comparison of the accuracy of measurements from single and stereo-video systems. *Mar. Technol. Soc. J.* **2002**, *36*, 38–49.

On the Accuracy Potential in Underwater/Multimedia Photogrammetry

Hans-Gerd Maas

Abstract: Underwater applications of photogrammetric measurement techniques usually need to deal with multimedia photogrammetry aspects, which are characterized by the necessity of handling optical rays that are refracted at interfaces between optical media with different refractive indices according to Snell's Law. This so-called multimedia geometry has to be incorporated into geometric models in order to achieve correct measurement results. The paper shows a flexible yet strict geometric model for the handling of refraction effects on the optical path, which can be implemented as a module into photogrammetric standard tools such as spatial resection, spatial intersection, bundle adjustment or epipolar line computation. The module is especially well suited for applications, where an object in water is observed by cameras in air through one or more planar glass interfaces, as it allows for some simplifications here. In the second part of the paper, several aspects, which are relevant for an assessment of the accuracy potential in underwater/multimedia photogrammetry, are discussed. These aspects include network geometry and interface planarity issues as well as effects caused by refractive index variations and dispersion and diffusion under water. All these factors contribute to a rather significant degradation of the geometric accuracy potential in underwater/multimedia photogrammetry. In practical experiments, a degradation of the quality of results by a factor two could be determined under relatively favorable conditions.

Reprinted from *Sensors*. Cite as: Maas, H.-G. On the Accuracy Potential in Underwater/Multimedia Photogrammetry. *Sensors* **2015**, *15*, 18140–18152.

1. Introduction

Photogrammetric tasks requiring the tracing of optical rays through multiple optical media with different refractive indices are called "multimedia photogrammetry" in the photogrammetric literature. This denomination has already been coined at a time, when the term "multimedia" was completely unknown in its contemporary meaning of the combined use of digital media such as text, images, film, animation and audio.

Multimedia photogrammetry is characterized by the refraction of optical rays at the transition between optical media with different refractive indices, which can be modeled by Snell's Law. An early treatment of this issue can be found in [1], who worked on the relative orientation of stereo aerial images of underwater scenes on analogue plotters and coined the term "two-media photogrammetry". References [2,3] showed an analytical solution and took the step from two- to multimedia photogrammetry, replacing the straight imaging rays by polygons. Kotowski [4] developed a ray-tracing method for tracing rays through an arbitrary number of parameterized interfaces, which was implemented in a bundle adjustment. It allows for handling both image invariant and object invariant interfaces.

In photogrammetry, we can distinguish three major categories of applications of multimedia techniques:

- In aerial photogrammetry, photo bathymetry is a technique to derive models of the sea floor from stereo imagery, provided limited depth and sufficient water transparency [3,5]. The air-water transition can be modeled on the basis of Snell's Law. Most implementations herein assume the water surface to be horizontal and planar, with waves on the water surface leading to significant errors [6].
- In underwater photogrammetry, cameras (with suitable housing) are used underwater. Some of these cameras are equipped with lenses specially designed for underwater imaging. As an alternative, cameras may be equipped with a planar front window, which can geometrically be treated as an image invariant interface. Typical application examples are in archaeology [7], the recording of ship wrecks, marine biology [8], measurements in nuclear power stations [9] or in the measurement of the shape of fishing nets [10].
- Many applications in industrial/technical close range photogrammetry deal with objects or processes in liquids, which are observed by cameras situated outside the observation vessel, imaging the scene through a planar window (e.g., 3D flow velocity measurement techniques [11,12]). The ray path herein is a twice-broken beam, which is refracted when passing through the three optical media interfaces air-glass-liquid (or vice versa).

Also in lidar bathymetry, which is used to determine underwater topography by airborne laser scanning [13,14], geometric models are used which can be derived from the above categories.

In the following, a multimedia model first introduced by Maas [11] will be shown, which can flexibly be integrated as a module in standard tools of photogrammetry. Subsequently, several extensions of the model will be discussed. The second part of the paper addresses several factors degrading the accuracy potential of underwater/multimedia photogrammetry.

2. A standard Model for Multimedia Close Range Photogrammetry

Many applications of multimedia close range photogrammetry in industrial-technical applications require the observation of objects or processes in a liquid through a glass window, which can be considered a plane parallel plate. This configuration allows for some algorithmic and computational simplifications, which form the basis for a flexible—yet strict—multimedia photogrammetry model (see also [11,15]). The model can be integrated as a module into the collinearity equations and can thus be used in photogrammetric standard procedures such as spatial resection, forward intersection, bundle adjustment or epipolar line computation. Like almost all approaches shown in the literature, the model assumes homogeneity and isotropy of the optical media.

The collinearity condition per definition connects image coordinates, camera projection center and object point coordinates. Its basic assumption, that image point, camera projection center and object point form a straight line, is not fulfilled any more in multimedia photogrammetry due to the refraction of the rays at the multimedia interfaces. The approach proposes a radial shift of an underwater object point with respect to the camera nadir point in a way that the collinearity condition if re-established. This radial shift is implemented as a correction term into observation equations derived from the collinearity equation. Simplifications can be achieved when defining the coordinate system in a way that the X/Y-plane is identical with one of the interface planes glass/water or air/glass (see Figure 1).

The procedure can be explained as shown in Figure 1: An object point P is imaged onto image point $p\prime$ through the water-glass and glass-air interfaces. Obeying to Snell's Law, the imaging ray is refracted twice on its path and thus not suited for the collinearity condition. If P were radially shifted to $\overline{P}$ in a plane parallel to the X/Y-plane, the collinearity condition could be applied with $\overline{P}$ like in the standard one-media case. Therefore the goal is to compute the radial shift R relative to the nadir point N ($R > 0$ if $n_2 > n_1$ and $n_3 > n_1$). This will, for instance, allow using the radially shifted point in a spatial resection for camera orientation and calibration. Typical values might be $n_1(air) = 1.0$, $n_2(glass) = 1.5$, $n_3(water) = 1.34$ (cmp. Equation (4)).

The calculation of the radial shift R can be derived from Figure 1:

$$R = Z_0\tan\beta_1 + t\tan\beta_2 + Z_P\tan\beta_3 \qquad (1)$$

and

$$\overline{R} = (Z_0 + t + Z_P)\tan\beta_1 \qquad (2)$$

Snell's Law connects the incidence angles:

$$n_1\sin\beta_1 = n_2\sin\beta_2 = n_3\sin\beta_3 \qquad (3)$$

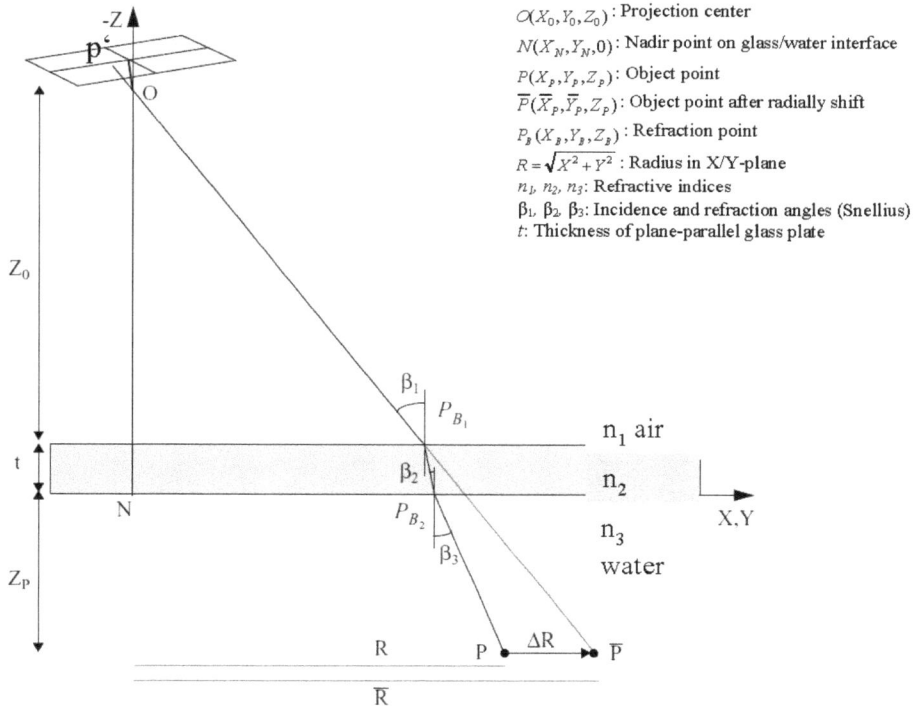

Figure 1. Radial shift for multimedia effect compensation [11].

The thickness of the glass plate t and its refractive index n_2 are usually assumed to be known and fixed. The refractive index of water depends on the optical wavelength as well as water temperature, salinity and depth and can be obtained from an empirical formula as used in [2]:

$$n_w = 1.338 + 4 \times 10^{-5}(486 - \lambda + 0.003d + 50S - T) \qquad (4)$$

(with n_w = refractive index of water, d = water depth (m), λ = wave length (nm), T = water temperature (°C), S = water salinity (%)).

A closed solution of the above equation system is not possible due to the trigonometric functions. Therefore an iterative procedure is being used, wherein P itself is chosen as a first approximation of $\overline{P}$:

$$\overline{R}_{(0)} = \sqrt{(X_P - X_0)^2 + (Y_P - Y_0)^2} \qquad (5)$$

29

For the 1. iteration we get the incidence angle in medium 1 from (Equation (2))

$$\beta_1 = \arctan(\frac{\overline{R}_{(0)}}{Z_0 + t + Z_P}) \tag{6}$$

and subsequently the incidence and refractive angles in the other media from (Equation (3))

$$\beta_2 = \arcsin(\frac{n_1}{n_2}\sin\beta_1) \quad \beta_3 = \arcsin(\frac{n_1}{n_3}\sin\beta_1) \tag{7}$$

This yields a correction term

$$\Delta R = R - (Z_0\tan\beta_1 + t\tan\beta_2 + Z_P\tan\beta_3) \tag{8}$$

and the radial shift for the 1. Iteration

$$\overline{R}_{(1)} = \overline{R}_{(0)} + \Delta R \tag{9}$$

with $\overline{R}_{(1)}$ we get new incidence and refractive angles β1, β2, β3, which can be used to compute a new R etc., until $R < \epsilon$ (e.g., with $\epsilon = 0.0001$ mm).

Switching back from polar to Cartesian coordinates after the last iteration, we get the coordinates of the radially shifted point $\overline{P}$:

$$\overline{X}_P = X_0 + (X_P - X_0)\frac{\overline{R}}{R} \tag{10}$$

$$\overline{Y}_P = Y_0 + (Y_P - Y_0)\frac{\overline{R}}{R}$$

$$\overline{Z}_P = Z_P$$

$\overline{P}$ can then be used in the collinearity equation instead of P, so that the equation can be used as an observation equation in spatial resection, forward intersection (with two or more images) or bundle adjustment. This offers the great advantage that existing photogrammetric software solutions can be extended by a multimedia module handling the radial shift procedure, without any modification in the core software tools. That means that the whole multimedia problem is simply out-sourced into the radial shift computation module.

The procedure can easily be extended to an arbitrary number of parallel interfaces. It should be noted that the approach is generic with respect to the camera viewing direction and not limited to viewing directions perpendicular to a planar glass interface (as is commonly the case in underwater photogrammetry models). The model can also be deduced from the generalized model shown in [9] with the simplifications shown here. A related approach is has been shown in [16], who very vividly connects it with the "apparent places" as known from astronomical geodesy.

3. Computational Acceleration

As stated above, there is no closed solution to obtain the radial shift from the above equation system. There is only a ray-tracing based straight-forward solution limited to forward intersection, which avoids the procedure via the radial shift [11]. However, this solution is restricted to two cameras and does not include an adjustment, thus not making proper use of the redundant information.

The computation time in the iterative procedure for determining the radial shift parameter in the strict solution as shown in Section 2 may be reduced by about 50% by introducing an over-compensation factor [11]. A much more efficient reduction of the computational effort can be achieved by outsourcing the multimedia calculations into a lookup-table. This may for instance be relevant in photogrammetric 3D-PTV (particle tracking velocimetry) systems [17], where the coordinates of several thousand neutrally buoyant tracer particles in a liquid flow have to be determined from image sequences of three or four cameras over several seconds or minutes at 25 Hz imaging rate. In the processing of these image sequences, millions of forward intersections have to be computed, each of them requiring the iterative multimedia shift procedure. In a lookup-table based solution, the problem can be reduced to the initialization of a two-dimensional lookup-table with the depth Z_P and the radial nadir point distance R of a point P as entry parameters and the radial shift ratio $\overline{R}/R$ as a result. Due to the reference to the nadir point, one lookup-table per camera has to be established. The lookup-table entries can be generated in a two-dimensional (Z_P, R) raster using the iterative model shown above. If the lookup-tables are initialized at a sufficient density, the relative radial shift of each point can easily be obtained by bilinear interpolation in the lookup-tables. The loss of accuracy caused by this interpolation-based procedure depends on the density of the initialization of the lookup-tables. [11] shows that less than 2000 lookup-table entries provide a good basis for handling the multimedia geometry without significant loss of accuracy in a typical 3D-PTV constellation.

4. Epipolar Lines in Multimedia Photogrammetry

An additional effect in multimedia photogrammetry is the fact that epipolar lines are not straight lines anymore, as can be seen from Figure 2. As a consequence, epipolar line based image matching techniques have to deal with curved epipolar lines, and the computation of rectified normal images as a pre-processing step for applying dense image matching techniques gets more complicated. The actual amount of bending depends on the proportion of the optical path lengths in air and water. While the effect may be negligible in the case of underwater photogrammetry

31

(where the camera is oriented perpendicular to a glass window and the path length in air is short), it will usually be relevant in industrial/technical applications of close range photogrammetry, where objects or processes in liquids are observed by cameras situated outside the observation vessel. In [11], a procedure to approximate the bended epipolar line by a polygon on the basis of intersections of a twice-broken beam with several depth planes is shown.

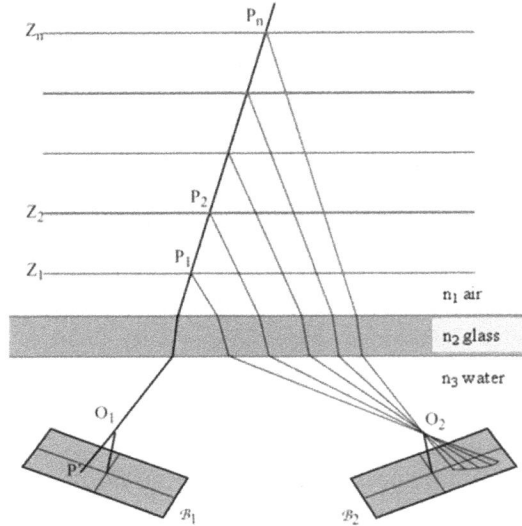

Figure 2. Epipolar geometry in multimedia environment.

5. Model Extensions and Variations

The model as shown above allows for some extensions. For instance, reference [18] shows the simultaneous determination of the refractive index of the liquid by introducing it as an additional unknown into a multimedia bundle adjustment procedure. This may be relevant, if the refractive index is unknown, for instance due to unknown salinity or temperature of the liquid. Both parameters may cause significant changes in the refractive index of water ($n_w = 0.002$ per percent salinity, $n_w = 0.00004$ per °C in temperature according to (Equation (4)). In an experiment on the validation of refractive index determinability, a standard deviation $\sigma n_w = 0.00015$ could be achieved. This value is better than the sensitivity of many optical refractometers. It allows for instance for the determination of the salinity with a standard deviation of less than 0.1%, thus giving the option of examining the properties of the liquid itself in multimedia photogrammetry applications.

Reference [19] showed an approach to observing phenomena in a plexi-glass combustion engine. The multimedia photogrammetry interfaces herein can be modeled as a plane and a cylinder (Figure 3). The parameters of both interfaces are introduced as unknowns into bundle adjustment, with the adjustment designed as a two-step procedure in order to de-correlate parameters.

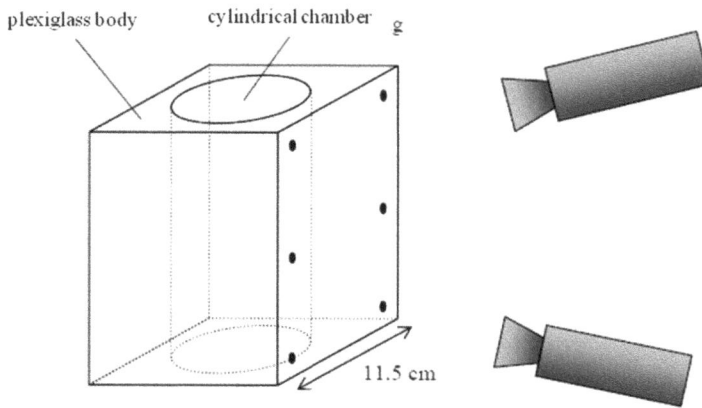

Figure 3. Photogrammetric measurement inside a glass engine [19].

Reference [20] showed a sophisticated multimedia photogrammetry model, which is also integrated into a bundle adjustment program and allows for the simultaneous determination of the geometric parameters of an arbitrary number of (not necessarily planar) interfaces in addition to the determination of the refractive index. In an experiment imaging a 3D target field under water with cameras in air through a planar interface, he was able to determine 3D object point coordinates, camera orientation and calibration parameters, planar interface geometry parameters as well as the refractive index simultaneously.

Wolff [21] introduced a new representation and taxonomy of optical systems, wherein the projection center may be a point, a line or a plane, and shows the applicability to multimedia photogrammetry in processing data of an experiment on the photogrammetric reconstruction of fluvial sediment surfaces.

Several authors discuss simplified models of underwater photogrammetry with camera and object under water and the camera viewing perpendicularly to a planar interface. Telem *et al.* [22] avoid the strict modeling of multimedia geometry by absorbing the multimedia photogrammetry effects (which show a radial symmetric behavior if the camera viewing direction is perpendicular to the planar glass interface) by the camera constant and radial lens distortion parameters. Lavest *et al.* [23] state that the effective focal length in underwater photogrammetry is approximately equal to the focal length in air, multiplied by the refractive index of water. Agrafiotis *et al.* [24] extend this model by also considering the dependency on the percentages of air and water within the total camera-to-object distance. Obviously, these models only hold for underwater photogrammetry cases with a camera viewing perpendicularly onto a planar interface.

6. Accuracy Aspects

The nature of underwater imaging and the necessity of applying geometric multimedia photogrammetry models imply several aspects degrading the accuracy potential of underwater photogrammetry. Therefore, despite strict geometric modeling, the accuracy potential in underwater/multimedia photogrammetry will usually be significantly worse than in conventional photogrammetry. Some important degrading factors are briefly discussed in the following sections:

Network geometry: The refraction according to Snell's Law reduces the opening angle of a camera when viewing from air into water due to the higher refractive index of water. As one can see from Figure 4, the refraction may also lead to a smaller ray intersection angle in 3D coordinate determination from stereo imagery and thus degrade the depth coordinate precision when imaging through the optical media air-(glass)-water [18].

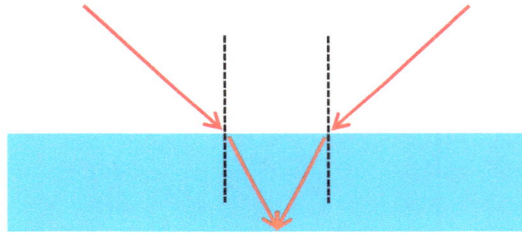

Figure 4. Forward intersection angle in two-media photogrammetry.

Interface planarity: Deviations from planarity in the glass interface between the optical media air and water will lead to variations in the surface normal vectors. This directly translates into errors in the local incidence angles, consequently leading to 3D object coordinate errors. The size of this effect depends on the quality of the glass of the "planar" interface. The effect may be rather large if low quality glass is being used. Simultaneous modeling of the glass interface geometry is rather complex and will often lead to an over-parameterization of the system. Obviously, the effect is much worse when omitting the glass interface and observing objects under water with a camera in air through the spatio-temporally changing wave pattern of an open water surface (which is the standard case in photo bathymetry [6]).

Refractive index: Local inhomogeneities of the refractive index of the liquid (for instance due to temperature or salinity gradients within the liquid) will lead to multiply curved optical paths to be handled in photogrammetric tools, which can hardly be modeled. Practical experiments in [11] showed that, while the simultaneous determination of a homogeneous refractive index in multimedia photogrammetry turned out to be possible (cmp. Section 5), the determination of a spatially resolved refractive index field failed due to extremely high correlations in the equation system.

Dispersion: The variation of the refractive index over the visible part of the electro-magnetic spectrum is 1.4% in water, while it is only 0.008% in air ([2], cmp. (Equation (4)). Shorter wavelength (blue) light experiences a stronger refraction than longer wavelength (red) light, leading to color seams (red towards the nadir point, blue outward) in RGB images or blur in black-and-white images (Figure 5). These blur effects will reduce the image quality as well as the image measurement precision potential. For standard solid state sensors having a larger sensitivity in the red than in the blue, the effects will even be asymmetric, thus leading to a systematic shift of the centroid of imaged targets. Using Bayer pattern based RGB cameras, interferences between the dispersion effect and the Bayer pattern have to be expected.

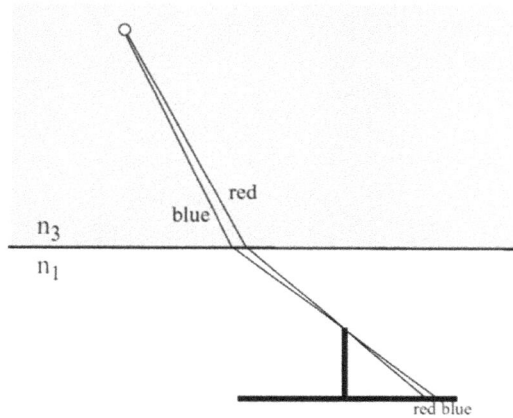

Figure 5. Effect or dispersion [18].

Yau *et al.* [25] suggest a model to cope with dispersion effects by handling wavelength-dependent pay paths in the calibration procedure through one or more planar layers perpendicular to the camera viewing direction.

Diffraction: Reference [26] has shown, that effects of diffraction cannot be assumed to be symmetric anymore in convergent camera configurations in multimedia close range photogrammetry, leading to a further decrease of image quality and image measurement precision.

Image focus: The best focus plane known from conventional photography is not planar anymore when imaging objects underwater. In limited depth-of-focus conditions, this may increase defocusing effects.

Lens design: As long as standard camera lenses are used, they will (especially in a convergent configuration) not be optimized for the optical system air-glass-water, again leading to a degradation of image quality and image measurement precision.

Water quality: Turbidity, small particles and gas bubbles in the water will cause absorption and diffusion effects, thus reducing image brightness and contrast. Especially in larger water depth and unfavorable turbidity conditions, this will also contribute to an impaired measurability of image coordinates [2].

7. Loss-of-Accuracy Validation

To show the accuracy degradation effects discussed in the previous section, some small experiments were conducted imaging a calibration target reference field used for the calibration of a photogrammetric system designed for thermo-capillar convection flow velocity field determination [27]. The 200×150 mm^2 calibration field was placed into the experimental cell made of graded glass. It was imaged by a black-and-white four-camera arrangement (Figure 6) first without water in the glass vessel (*i.e.*, optical path air-glass-air) and then with water in the vessel (*i.e.*, optical path air-glass-water). Between the two experiments, the camera settings remained unchanged; only the orientation angles ω and φ had to be re-adjusted in order to warrant an identical field of view in both experiments. Data processing was performed introducing some of the reference targets as control points and 39 targets as (unknown) check points for an external precision check.

Figure 6. 4-Camera system in thermo-capillar convection experiment [27].

The following three bundle adjustment computations were performed:

I   Processing of the air-glass-air case, camera orientation and calibration parameters introduced as unknowns.

36

II   Processing of the air-glass-water case, camera orientation parameters introduced as unknowns, camera calibration parameters taken from I (as camera settings were unchanged).

III   Processing of the air-glass-water case, camera orientation and calibration parameters introduced as unknowns.

Table 1. Results from multimedia photogrammetry validation experiment.

	$\hat{\sigma}_0$ (bundle)	Internal Object Point Precision $\hat{\sigma}_{XYZ}$	External Object Point Precision $\bar{\sigma}_{XYZ}$
I	0.49 μm	0.010/0.011/0.023 mm	0.013/0.011/0.024 mm
II	1.96 μm		0.031/0.072/0.153 mm
III	1.10 μm		0.021/0.034/0.044 mm

The following conclusions can be drawn from the results of the experiment as shown in Table 1:

- The results in (I) are according to the expectations: The standard deviation of unit weight obtained from the bundle adjustment is in the order of $^1/_{25}$ pixel, and the rather good congruence between the internal 3D object point precision parameters (obtained from the self-calibrating bundle adjustment) and the external precision parameters (obtained from 39 independent check points) proves the absence of errors in the geometric and stochastic model.

- In (II) with water filled into the vessel and the cameras re-oriented to capture the same field of view, the external precision figures are much worse (approximately by a factor 5) than in (I). This has to be contributed to the aspects discussed in the former section (except the planarity of the glass interface, as this was present in both experiments).

- In (III) with another self-calibration performed, results get significantly better than in (II) and are only by approximately a factor 2 worse than in (I). This can be explained by the fact that some of the effects discussed in Section 6 show a systematic nature, and that these effects are at least partly compensated by the camera self-calibration parameters. In fact, a significance test between the two parameter sets yielded a highly significant difference between the camera calibration parameters obtained in air and water, despite un-changed camera settings. The largest difference was found in the image shear parameter, which is thus taking a large amount of the systematic part of the errors introduced by the effects discussed in Section 6.

As a conclusion, one can state that the experiment proves the degradation of the accuracy in a multimedia environment. One can also see, that a self-calibration of the cameras in the actual environment leads to better results than pre-calibrated cameras, because errors coming from the multimedia environment show a partly systematic behavior and are partly compensated by camera calibration parameters. Although the conditions in this experiment were rather favorable (low depth, clear water, uniform temperature, zero salinity), the degradation of the geometric precision still amounts to approximately a factor two, with much stronger degradations to be expected under less favorable conditions.

These results correspond to results published in the literature. For instance, Menna *et al.* [28] also report a loss of precision (based on internal bundle adjustment standard deviations) by a factor two, with the largest loss in depth direction. Similar degradations will also have to be faced in other optical 3D underwater measurement techniques. For instance, Ekkel *et al.* [29] report a degradation (also under rather favorable conditions) of the accuracy of profile measurements with a laser triangulation system from 22 μm in air to 35 μm in sweet water. They report a further significant degradation of accuracy when applying the system in salt water.

8. Conclusions and Outlook

The paper has shown that strict geometric modeling in underwater and multimedia photogrammetry is possible by a flexible geometric model based on virtual underwater 3D points obtained by a nadir point and depth dependent radial shift of target points. The model depicts an elegant solution, which can be introduced as a module to strictly model ray paths in standard photogrammetric tasks such as spatial intersection, multiple-image forward intersection and self-calibrating bundle adjustment. It can be extended, for instance towards the simultaneous determination of the refractive index. Users of widely-used off-the-shelf photogrammetry software packages such as structure-from-motion tools, which are also becoming widely used in underwater photogrammetry for archaeology and ecology surveys, have to keep in mind that neglecting the effects caused by refraction in the imaging process (or trying to absorb it by standard lens distortion compensation parameters) contributes to a degradation of the quality of results.

Even with strict geometric modeling, underwater/multimedia photogrammetry is accompanied by some effects, which lead to a degradation of the accuracy potential of photogrammetric underwater 3D measurements. Although these effects are partly compensated by camera self-calibration parameters, a degradation by approximately a factor two was obtained in an experiment with rather controlled and favorable conditions. More research has to be performed to further reduce these degrading effects by further improved geometric modeling and adapted self-calibration schemes.

Acknowledgments: Some of the experiments reported in the paper were conducted during the authors' time at ETH Zurich. The support obtained in their hydro-mechanic lab is greatly acknowledged.

Conflicts of Interest: The author declare no conflict of interest.

References

1. Rinner, K. *Abbildungsgesetz und Orientierungsaufgaben in der Zweimedienphotogrammetrie*; Östereichische Zeitschrift für Vermessungswesen, Sonderheft 5; Eigenverlag: Wien, Österreich, 1948.
2. Höhle, J. Zur Theorie und Praxis der Unterwasser-Photogrammetrie. In *Schriften der DGK*; Verlag der Bayerischen Akademie der Wissenschaften: München, Germany, 1971; Volume Heft Nr. 163.
3. Okamoto, A.; Höhle, J. Allgemeines analytisches Orientierungsverfahren in der Zwei- und Mehrmedien-Photogrammetrie und seine Erprobung. *Bildmess. Luftbildwes.* **1972**, *40*, 1–12.
4. Kotowski, R. Phototriangulation in Multi-Media-Photogrammetry. *Int. Arch. Photogramm. Remote Sens.* **1988**, *XXVII*, 324–334.
5. Vanderhaven, G. Data Reduction and Mapping for Photobathymetry. In Proceedings of the ASP Coastal Mapping Symposium, Rockville, MD, USA, 14–16 August 1978; pp. 69–88.
6. Okamoto, A. Wave Influences in Two-Media Photogrammetry. *Photogramm. Eng. Remote Sens.* **1982**, *48*, 1487–1499.
7. Drap, P.; Seinturier, J.; Scaradozzi, D.; Gambogi, P.; Long, L.; Gauch, F. Photogrammetry for virtual exploration of underwater archaeological sites. In Proceedings of the XXI International CIPA Symposium, Athens, Greece, 1–6 October 2002.
8. Shortis, M.; Harvey, E.; Abdo, D. A review of underwater stereo-image measurement for marine biology and ecology applications. In *Oceanography and Marine Biology: An Annual Review*; Gibson, R.N., Atkinson, R.J.A., Gordon, J.D., Eds.; CRC Press: Boca Raton, FL, USA, 2009; Volume 47, pp. 257–292.
9. Przybilla, H.-J.; Kotowski, R.; Meid, A.; Weber, B. Geometric Quality Control in Nuclear Power Stations—A Procedure for High Precision Underwater Photogrammetry. *Int. Arch. Photogramm. Remote Sens.* **1988**, *27*, 513–526.
10. Zwart, P.R. Measuring the Shape of Fishing Nets—An Application of Three Media Close Range Photogrammetry. In Proceedings of the Symposium on the Applications of Close Range Photogrammetry, Melbourne, Australia, 17–18 November 1987.
11. Maas, H.-G. Digitale Photogrammetrie in der Dreidimensionalen Strömungsmesstechnik. Ph.D. Thesis, ETH Zürich, Zürich, Switzerland, 1992.
12. Maas, H.-G.; Grün, A. Digital photogrammetric techniques for high-resolution 3-D flow velocity measurements. *Opt. Eng.* **1995**, *34*, 1970–1976.
13. Irish, J.; Lillycrop, J. Scanning laser mapping of the coastal zone: The SHOALS system. *ISPRS J. Photogramm. Remote Sens.* **1999**, *54*, 123–129.
14. Mandlburger, G.; Pfennigbauer, M.; Steinbacher, F.; Pfeifer, N. Airborne Hydrographic LiDAR Mapping—Potential of a new technique for capturing shallow water bodies. In Proceedings of the 19th International Congress on Modelling and Simulation, Perth, Australia, 12–16 December 2011.

15. Maas, H.-G. Geometrische Modelle der Mehrmedienphotogrammetrie. *AVN* **2014**, *121*, 112–116.

16. Philips, J. *Ein Photogrammetrisches Aufnahmesystem zur Untersuchung Dynamischer Vorgänge im Nahbereich*; Veröffentlichungen des Geodätischen Instituts der RWTH Aachen, Heft 30: Aachen, Germany, 1981.

17. Maas, H.-G.; Grün, A.; Papantoniou, D. Particle tracking in threedimensional turbulent flow—Part I: Photogrammetric determination of particle coordinates. *Exp. Fluids* **1993**, *15*, 133–146.

18. Maas, H.-G. New developments in multimedia photogrammetry. In *Optical 3-D Measurement Techniques III*; Grün, A., Kahmen, H., Eds.; Wichmann Verlag: Karlsruhe, Germany, 1995.

19. Putze, T. Erweiterte Verfahren zur Mehrmedienphotogrammetrie komplexer Körper. In *Photogrammetrie—Laserscanning—Optische 3D-Messtechnik: Beiträge der Oldenburger 3D-Tage 2008*; Luhmann, Th., Ed.; Verlag Herbert Wichmann: Berlin, Germany, 2008.

20. Mulsow, C. A Flexible Multi-media Bundle Approach. *Int. Arch. Photogramm. Remote Sens. Spat. Inf. Sci.* **2010**, *38*, 472–477.

21. Wolff, K. Zur Approximation allgemeiner optischer Abbildungsmodelle und deren Anwendung auf Eine Geometrisch Basierte Mehrbildzuordnung am Beispiel einer Mehrmedienabbildung. Ph.D. Thesis, Universität Bonn, Bonn, Germany, 2006.

22. Telem, G.; Filin, S. Photogrammetric modeling of underwater environments. *ISPRS J. Photogramm. Remote Sens.* **2010**, *65*, 433–444.

23. Lavest, J.; Rives, G.; Lapresté, J. Underwater camera calibration. In *Computer Vision—ECCV 2000*; Vernon, D., Ed.; Springer: Berlin, Germany, 2000; pp. 654–668.

24. Agrafiotis, P.; Georgopoulos, A. Camera Constant in the Case of Two Media Photogrammetry. *Int. Arch. Photogramm. Remote Sens. Spat. Inf. Sci.* **2015**, *XL-5/W5*, 1–6.

25. Yau, T.; Gong, M.; Yang, Y.-H. Underwater camera calibration using wavelength triangulation. In Proceedings of the 2013 IEEE Conference Computer Vision and Pattern Recognition (CVPR), Portland, OR, USA, 23–28 June 2013; pp. 2499–2506.

26. Meid, A. Wissensgestützte Digitale Bildkoordinatenmessung in Aberrationsbehafteten Messbildern. Ph.D. Thesis, Universität Bonn, Bonn, Germany, 1991.

27. Stüer, H.; Maas, H.-G.; Virant, M.; Becker, J. A volumetric 3D measurement tool for velocity field diagnostics in microgravity experiments. *Meas. Sci. Technol.* **1999**, *10*, 904–913.

28. Menna, F.; Nocerino, E.; del Pizzo, S.; Ackermann, S.; Scamardella, A. Underwater photogrammetry for 3D modelling of floating objects: The case study of a 19-foot motor boat. In Proceedings of the 14th Congress of International Maritime Association of Mediterranean, Genoa, Italy, 13–16 September 2011.

29. Ekkel, T.; Schmik, J.; Luhmann, Th.; Hastedt, H. Precise laser-based optical 3D measurement of welding seams under water. *Int. Arch. Photogramm. Remote Sens. Spat. Inf. Sci.* **2015**, *XL-5/W5*, 117–122.

Geometric and Optic Characterization of a Hemispherical Dome Port for Underwater Photogrammetry

Fabio Menna, Erica Nocerino, Francesco Fassi and Fabio Remondino

Abstract: The popularity of automatic photogrammetric techniques has promoted many experiments in underwater scenarios leading to quite impressive visual results, even by non-experts. Despite these achievements, a deep understanding of camera and lens behaviors as well as optical phenomena involved in underwater operations is fundamental to better plan field campaigns and anticipate the achievable results. The paper presents a geometric investigation of a consumer grade underwater camera housing, manufactured by NiMAR and equipped with a 7″ dome port. After a review of flat and dome ports, the work analyzes, using simulations and real experiments, the main optical phenomena involved when operating a camera underwater. Specific aspects which deal with photogrammetric acquisitions are considered with some tests in laboratory and in a swimming pool. Results and considerations are shown and commented.

Reprinted from *Sensors*. Cite as: Menna, F.; Nocerino, E.; Fassi, F.; Remondino, F. Geometric and Optic Characterization of a Hemispherical Dome Port for Underwater Photogrammetry. *Sensors* **2016**, *16*, 48.

1. Introduction

Despite being a hostile environment both for humans and optical equipment, underwater measurement using photogrammetry can be feasible in several cases. Photogrammetry still represents the most useful recording technique currently available underwater due to its flexibility and ease of image data interpretation.

Nowadays the increasing number of demanding applications is growing constantly, mainly thanks to improvements of data processing software as well as technical achievements in diving apparatus, photographic equipment and underwater manned and unmanned vehicles. Furthermore, the widespread availability of scuba diving centers has expanded the knowledge and education about underwater environment even if only at a recreational level.

The general knowledge about underwater exploration and documentation has rapidly grown in the last few years and different users are demanding low-cost, quick, easy and fast 3D measurement solutions. Divers have developed the ability of taking notes and sketches underwater or acquiring photographs and videos. The popularity of automatic photogrammetric and structure from motion techniques in air among non-experts has stirred and also promoted experiments underwater, achieving impressive visual results [1–3] as much as serious concerns about safety. Indeed, appropriate training for in-water activities is crucial in order to assure safety. For this reason, diving courses for scientists who want to learn and safely practice digital recording techniques underwater have been proposed [4].

Since the beginning of underwater photography in 1850s by the pioneer William Bauer, it was noticeably obvious that the acquisition of underwater photographs would need severe modifications to photographic equipment. Nowadays, the causes of underwater photography problems are still often obscure to most users, and this is even more the case for non-experts in surveying and photogrammetry. In many projects, when conditions which guarantee safety for divers involved in the project are met, consumer cameras in their own underwater camera housings equipped with external strobe lights can be operated by divers.

Testing and investigating the geometrical characteristics of underwater consumer grade photographic equipment when used for photogrammetric applications would be advisable if accuracy and reliability matter. Professional results always rely on the control of all the technical parameters involved. The knowledge about photographic equipment and its behavior in different conditions is the first step to be investigated.

Whether they are in shallow or deep water, most underwater photogrammetric applications have to deal with a challenging optical environment due to water ripple reflections, light absorption and turbidity. Water is a medium inherently different from air and the first essential difference resides in the medium density. Seawater is nearly 800 times denser than air, and this influences the image formation underwater as the path of optical rays is altered. Density of seawater is not constant through depth, being a function of temperature, salinity and pressure. Although pressure is an extremely critical factor for very deep underwater inspections, its variation with depth affects any underwater optical system at whatever depth. Internal arrangement may be altered and subject to changes as the working depth varies.

By considering all the aforementioned constraining factors, underwater photogrammetry deals most of the time with close- and very close-range distances, usually maximum in the order of few meters. As for all close-range photogrammetric surveys carried out in air, the key parameters for network planning and acquisition—such as ground sample distance (GSD), baseline, image overlap, expected accuracy, nominal focal length, sensor resolution, aperture value, depth of field, *etc.*—must be known to plan the survey.

Paper Aims, Methods and Tests

In this contribution a geometric investigation of a consumer grade underwater camera housing equipped with a 7″ dome port is presented. The housing is specifically manufactured by NiMAR (Modena, Italy [5]) for a Nikon D300 camera. The system, composed by digital camera, lens, camera housing and dome port was already used by the authors for the underwater and in-air joint survey of the Costa Concordia ship gash [6].

The NiMAR NI303D is a pressure housing for the D300 camera (Nikon, Tokyo, Japan). It is made of polycarbonate and gives access to the most used buttons and functions of the camera. Through a bayonet at the front of the housing different ports, spherical and flat, can be mounted. As for many other housings made by other manufacturers, not all the camera functions are available due to the complexity in reaching levers or buttons over the camera body. Some missing functions are not a big problem most of the time, while some others may be a limitation for photogrammetric applications and make the operations more complex or longer to find proper workaround solutions.

The aim of the work is to analyze, using photogrammetric methods, the main optical phenomena which involve a camera operating underwater. Specific aspects which deal with photogrammetric acquisitions are considered and practical suggestions provided. Within the presented investigation, carried out with the Italian manufacturer of waterproof camera housings NiMAR, experimental setups are being designed and investigated to calibrate and test underwater camera housings for photogrammetric applications. Both simulations and tests underwater were carried out to design and implement systematic tests for underwater camera housings. Theoretical graphs and optical calculations for the underwater ports, both flat and dome are derived using the freely available WinLens 3D Basic and Predesigner and software application by Qioptiq (Goettingen, Germany [7]). Optical distortions for the camera-lens and underwater case-port system are based on formulas well known in optics and photogrammetry and coded by the authors in Matlab. Underwater acquisitions were performed in a 2 meter deep swimming pool.

2. Consumer Grade Flat and Dome Ports

Two types of lens port are employed in waterproof housing for underwater photography: flat and dome. Figure 1 features a variety of functional, fancy, professional and consumer-grade waterproof housings available on the market for any type of digital cameras and needs. Very low-cost consumer grade housings, such as the models shown in Figure 1b,c, can be made of non-rigid materials and consequently may be not well suited for photogrammetric acquisitions. Indeed, relative movements between the port and camera are likely to occur as pressure, temperature and even diver skills can continuously vary. The uncontrolled relative movements may cause optical distortions that are difficult to model and quantify a priori.

Figure 1. Waterproof housings for digital cameras and flashes: (**a**) Canon with flat port for compact cameras; (**b**) Watershot iPhone housing; (**c**) TteooBL waterproof bag; (**d**) Outex; (**e**) GoPro Hero 3 in its protective waterproof case (Christography); (**f**) Sealife with twin-flash (Hunteroc); (**g**) Equinox Housings for DSLR Nikon; (**h**) Ikelite for Canon; (**i**) Seacam with dome port.

When it comes to optic properties, whose knowledge is fundamental for both recreational and professional photography, as well as for photogrammetry, flat and dome ports have their inherent intrinsic pros and cons.

In the following sections, an overview on how the two types of port are addressed in the photogrammetric literature is provided (Section 2.1). Then, a brief description of flat port characteristics is reported (Section 2.2.) with the aim of pointing out the main differences with dome ports. Finally, an extensive and critical analysis on dome ports is provided (Section 2.3).

2.1. Underwater Camera Calibration—Literature Review

Flat ports have been intensely studied for photogrammetric underwater applications in relation to camera calibration, an issue that has been faced for almost 50 years [8]. Among the technical and scientific community, underwater photogrammetry is often called multimedia photogrammetry, where the term indicates that the light ray travels across different media: water where the object is immersed, glass or the material the port is made, and air where the camera-lens system works. The transition among these different elements causes a ray's deviation from the path that it would travel if it were travelling just in air (see Section 2.2). This deviation must be taken into account if a source of error in the measurement process wants to be eliminated. Two main approaches for handling this issue have been proposed in the literature: (i) the collinearity model is modified to take into account the rigorous geometric interpretation of light propagation in multimedia (camera housing-water), also known as ray tracing approach [9]; (ii) the refractive effect of the different interfaces is absorbed by camera calibration parameters using a standard pinhole camera model and a terrestrial-like self-calibration approach [10].

A variety of different methods for a rigorous modelling of underwater image formation has been proposed. Mulsow [11] proposed a multi-media bundle models, where refractive indices surface and even mathematical parameters of the interfaces are introduced as unknowns: the method is particularly suited for Particle Image Velocimetry (PIV) applications and needs to be tested in underwater environment [11]. Jordt-Sedlazeck and Koch [12] propose a light propagation model to be used for color correction along with a strict underwater camera calibration [12]. Treibitz *et al.* [13] include also the distance of the lens from the medium interface to take into account the position variation of the entrance pupil [13]. Agrafiotis and Georgopoulos included the percentages of air and water within the total camera-to-object distance as parameter in their calibration model [14].

The second camera calibration approach is justified by the evidence that the principal component of refractive effects is radial. Therefore, it can be implicitly compensated by the standard, odd-ordered polynomial model for radial distortion, whilst any residual effects from asymmetric components of the housing are partly or wholly absorbed into other parameters of the camera calibration, such as decentering lens distortion or the affinity term [15]. Moreover, the environmental conditions can be hardly modelled *a priori*: refractive index of water changes with depth, temperature and salinity, the shape of the camera housings and port may change with depth due to changing pressure levels [16]. Hence, calibrating the underwater system (camera and lens inside the waterproof case) at the predominant working conditions (depth, temperature, *etc.*) would provide more accurate and reliable results [15].

To avoid severe refraction effects, dome ports can be adopted [17]. In this case, the pinhole camera model would be completely fulfilled if the center of perspective of the camera-lens system were exactly placed in the dome surface center of curvature. In the authors' knowledge, few studies have tried to quantify the limit of the above statement.

Kunz and Singh [18] proposed a model for hemispherical port calibration to be performed in air and used underwater. The authors simulate the effect of a centering error of entrance pupil in a pressure housing with dome port but they do not provide any theoretical or practical evidence of the employed values [18].

Besides the type of lens port used, the mechanical stability of the whole system camera + lens + underwater housings is an important factor to be considered. Shortis *et al.* [19] investigated the camera-case stability, showing that significant variations in camera calibration parameters are found when removing the camera from the waterproof housing for example to download the images and then reassembling to use the system again straight after [19]. As a concluding remark, a degradation of the geometric accuracy by a factor two in underwater/multimedia photogrammetry should be expected [20].

2.2. Flat Lens Port

A flat port is essentially a flat plane of optically transparent glass or plastic in front of the lens, as depicted in Figure 1a–h. Flat ports are the most common waterproof housing of compact digital cameras, being their manufacturing less expensive than dome ports.

Neglecting the thickness of the port, the flat surface acts as boundary or interface between two different media: water outside the waterproof case and air inside. The two media are characterized by different refractive indices, bigger for the denser water and smaller for air. Consequently, the rays of light coming from object in water deviate from their original path when pass through port and reach the camera sensor. Specifically, each ray is bent toward the normal to the boundary surface according to the Snell's law. This phenomenon causes that the camera-lens field of view (FOV) is reduced, as shown in in Figure 2. The white box represents a waterproof housing with flat port with inside a Nikon D300 (DX format) mounting a 24 mm lens (Nikon); the cyan part represents the water. For the sake of simplicity, the thickness of the port in the figure is ignored. In red the nominal focal lens and FOV in air are shown (Figure 2b). Because of the presence of flat port, the FOV is narrowed and, conversely, the focal length is increased.

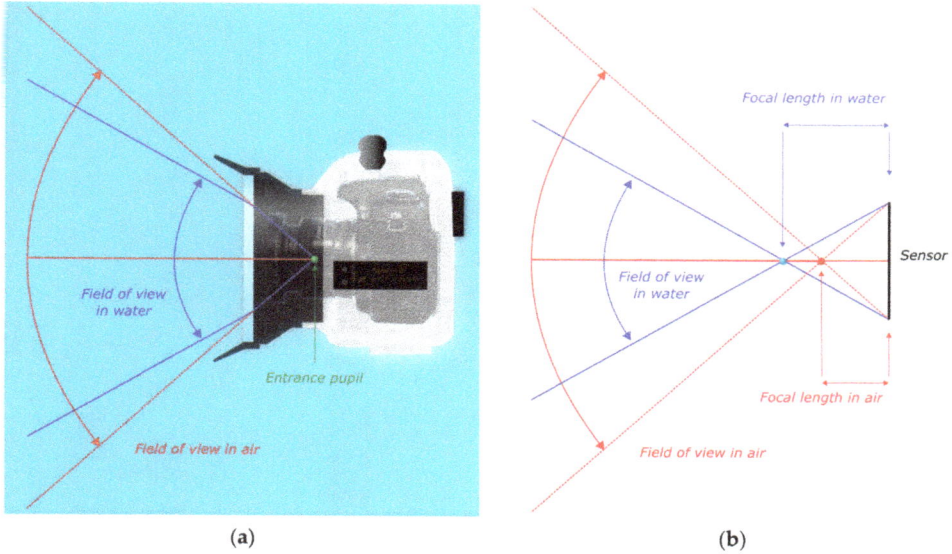

Figure 2. Effect of flat port—Reduction of FOV. (**a**) Optical rays (in red) from the camera when pass through the planar boundary are bent towards the normal to the flat surface, narrowing the field of view; (**b**) This phenomenon can be regarded as an increase of the nominal focal length.

Flat ports have also a limitation in terms of maximum field of view. Indeed a ray of light entering the glass of the port from water and then living the interface glass/air is subject to total internal reflection as the ray passes from a means with higher refraction index to another with lower one. By applying Snell's law, the critical angle can be calculated, leading to a maximum field of view for every flat port of about 96° as visible from Figure 3 where, as the entrance angle θ_w rises, the refracted angle is bent away from the normal to the flat port surface. When it reaches about 48° the ray does not enter the housing anymore as it is subject to total internal reflection (green ray). As expected the angle of refraction θ_a is 90° in this case.

As one would expect, by lowering the FOV and keeping the subject at the same distance from the camera, the magnification of the object becomes larger than in air by a factor approximately equal to the ratio between the refraction indices of water and air. Moreover the object appears closer to the camera. When a flat port is used, not only the FOV and focal length vary, but also the lens distortion is affected.

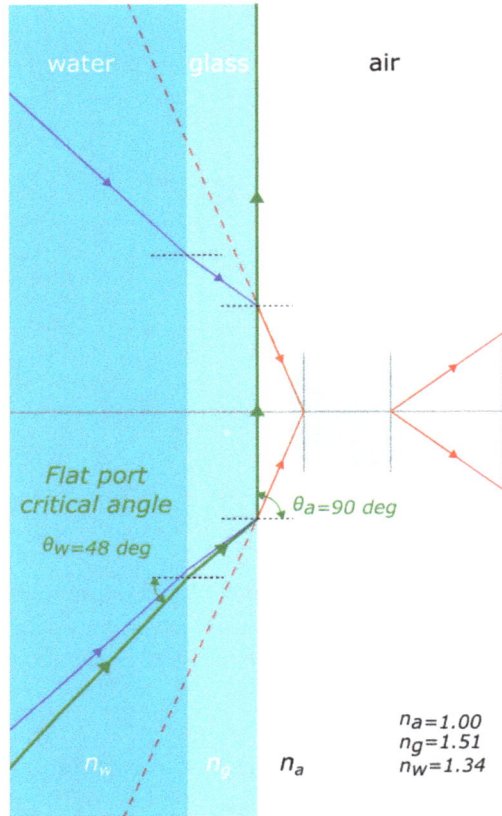

Figure 3. In a flat port the maximum field of view is limited by the total internal reflection.

Typically, a ray of light passing from water through a flat port introduces a pincushion radial symmetric distortion as depicted in Figure 4. The objective lens inside the housing outlined in the figure is supposed to be free from any type of geometric distortions and would reproduce an undistorted image of an object when placed in air. Conversely, placing the camera underwater, in a pressure housing with flat port, would produce a prominent pincushion distortion depicted in blue in Figure 4b. Under the assumption that the distance between the lens and flat port can be neglected, the image distortion factor D, expressed ad the ratio between r'/r, can be computed using the following formula [21]:

$$D = \left(\frac{n_w^2 - sin^2\, \theta_a}{1 - sin^2\, \theta_a} \right)^{1/2}$$

(1)

where:

- θ_a is the entrance angle in air;
- n_w is the refraction coefficient of water.

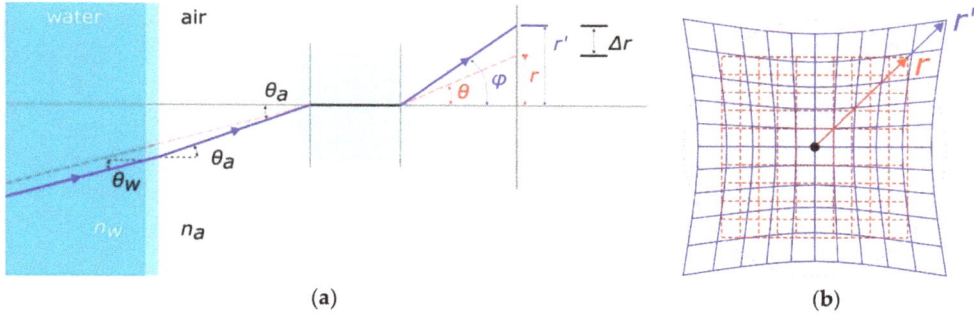

(a)   (b)

Figure 4. Effect of a flat port on image distortion: (**a**) section view; (**b**) image plane view.

In addition to radial distortion, flat lens port also introduces chromatic aberration. The rays of light, when refracted, are separated into different wavelengths of the visible color spectrum (component colors) that do not travel at the same speed and when passing through the flat port are differently bent. The separate components can overlap, causing a loss of sharpness and color saturation.

2.3. Spherical or Hemispherical Dome Lens Port

Spherical or hemispherical dome lens port, like the one shown in Figure 1i, solves the problems introduced by flat port, *i.e.*, FOV and focal length of the camera-lens system are preserved, peculiarity very crucial when dealing with wide angle lenses. Nevertheless, other issues arise when using dome port. A spherical dome port is a concentric lens that acts as an additional optical element to the camera lens. Indeed, it is a real lens, more precisely, a negative or diverging lens: both the focal length and image distance are negative so that the image is formed to the left, *i.e.*, in front of the dome. Such an image is called virtual image and is upright and smaller than the object (Figure 5—produced using the free optical design software WinLens 3D Basic by Qioptiq). The camera-lens system behind the dome port will actually focus not on the real object but on the virtual image produced by the dome port itself at a smaller distance than the object.

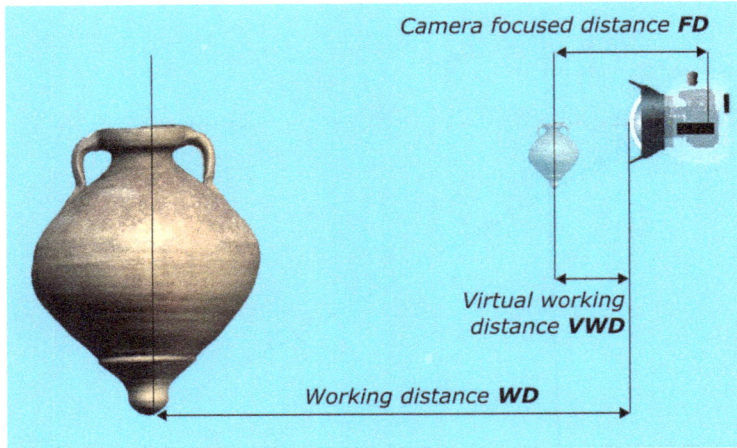

Figure 5. Effect of a dome port: a subject located at a working distance (WD) appears smaller and much closer to the camera—virtual working distance (VWD).

Being the dome port a spherical lens, it suffers from spherical aberrations *i.e.,* optical rays passing through the peripheral parts of the dome do not converge in the same focal point of the rays passing through the center. The result is that there is not a single image plane and a blurred image can be produced.

Another undesirable optical effect of a spherical dome port is the field curvature that causes flat object to be projected on a paraboloidal surface, known as Petzval surface (Figure 6). Being instead the image sensor flat, the consequence is that the object can appear not completely in focus or not uniformly sharp across the image.

In summary, differently from flat lens ports, dome ports:

1. preserve the FOV and focal length of the camera-lens system;
2. do not change significantly the shape of lens distortion;
3. cause the camera-lens system to focus much closer than in air;
4. can introduce spherical aberrations and field curvature, producing unsharp images.
5. increase the DOF.

Points 1 and 2 are completely verified if the entrance pupil of the camera lens coincides with the center of curvature of the dome port (Figure 5). The larger the distance between entrance pupil and center of curvature, the greater the geometric distortions and chromatic aberrations introduced.

The effect of the non-concentricity of the dome port and entrance pupil can be easily simulated using Winlens software. The behavior of the misalignment can be summarized as follows:

50

(1) a misalignment of the dome port on a plane orthogonal to the optical axis of the objective lens produces decentering distortions;

(2) a misalignment along the optical axis of the objective lens produces a pincushion-type radial symmetric distortion if the center of the pupil entrance is in front of the center of the spherical dome whereas a barrel-type radial symmetric distortion if behind.

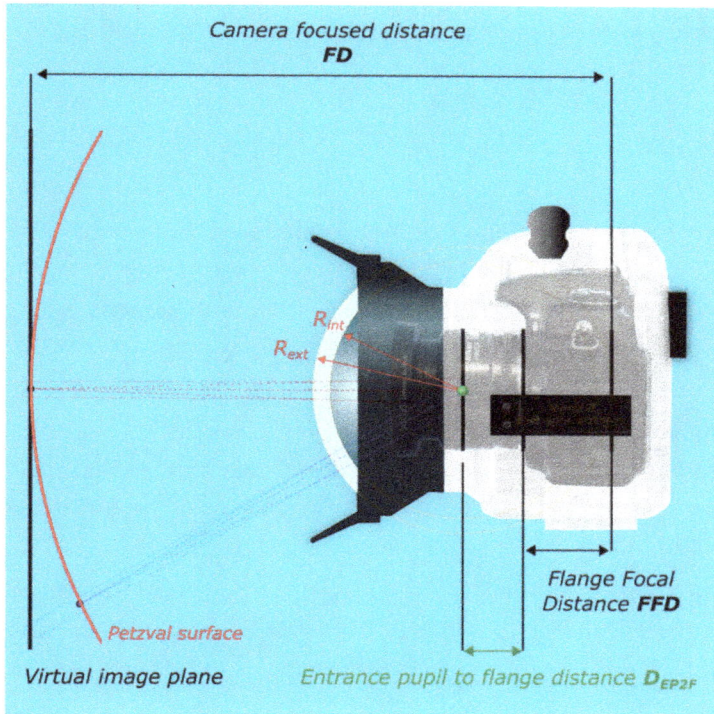

Figure 6. Main photographic elements in the image formation through a dome port.

Despite being a paramount point for photographers and photogrammetrists, especially for spherical photogrammetry applications, the position of the center of the entrance pupil is not provided by lens manufacturer. Its position changes when focusing and it is not easy to predict its motion as it depends on the optical design of the lens, most of the times only partially available. An objective lens is a system composed of many lenses through which the bundle of rays, entering the objective, converge and diverge in their passage. The amount of rays which enter the lens is controlled by a diaphragm known as aperture stop. The entrance pupil represents the center of perspective of the lens and can be seen as the virtual image of the hole materialized by the aperture stop seen from the front of the lens [21]. When placed inside the pressure housing, behind a spherical dome port whose radius is R, the virtual image of an object at infinity will be projected in front of the dome at about 4R from the center of the spherical dome or 3R from the dome glass. For a dome port of 8 cm radius the real world from infinity up to 1 m from the dome surface is compressed in a virtual space in front of the dome which extends approximately from 17 cm to 23 cm from the glass (see Section 4). It is clear that the camera must be able to focus at these distances to obtain a sharp image. The popular Gopro Hero sport camera (up to the version 2) suffered from out-of-focus images when placed underwater because of the small radius of its dome port and impossibility to focus on the virtual image due to fixed focus. This was a big issue for such a compact camera camera so that in the newer version a flat port was adopted thus reducing the field of view.

3. Laboratory Investigation and Geometric Characterization of NiMAR NI320 Dome Port and NI303D Waterproof Case

The aim of this investigation is to study the manufacturing of a commercial underwater housing with dome port (Figure 7), *i.e.*, to measure the following fundamental characteristics:

- position of the camera lens entrance pupil (Section 3.1);
- deviation of external and internal dome surfaces from an ideal spherical shell;
- misalignment between the center of curvature of the dome surfaces and entrance pupilof the camera lens.

The second analysis implies the evaluation of the radius and center of curvature of the two spheres fitting the external and internal dome surfaces respectively (Section 3.2).

Figure 7. NiMAR NI320 dome port and NI303D waterproof case.

The third requires a more difficult measurement process, described in Sections 3.3 and 3.4. All the measurements reported in the following parts are referred to the center of the camera flange (Figure 8). Table 1 summarizes the characteristics of the dome port as provided by the manufacturer.

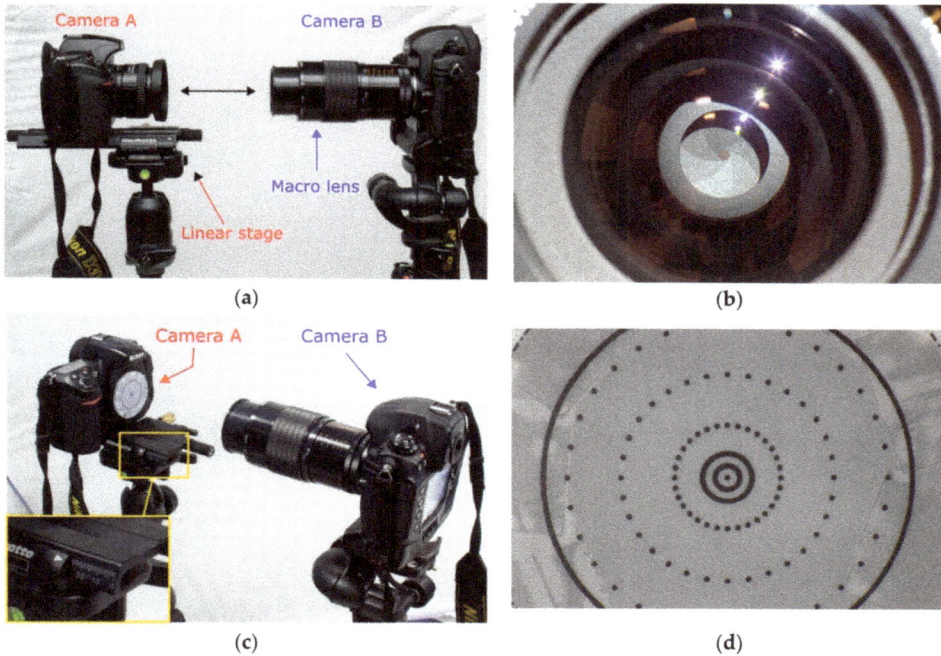

Figure 8. Measurement of the distance D_{EP2F} between the entrance pupil and the flange.

Table 1. Manufacturing parameters of the tested NiMAR hemispherical dome.

Product Code	Material	Nikkor Lenses That Can be Fitted in (mm)	Internal Diameter (mm)	External Diameter (mm)	Length (mm)	Weight (gr)
NI320	Crystal glass	20–24–28–35	94	176	91	671

It is worth to note that the same dome port is used with different lenses whose focal length varies from 20 mm to 35 mm. Consequently, also the position of the entrance pupil may vary inside the case. Moreover, the datasheet reports the dimensions of the black frame of the dome whereas no information about the radius of curvature, the center and thickness of the glass surface is usually available.

To recover the information needed to fully characterize the optics of the camera-lens and underwater case-dome systems, a reverse engineering process is carried out through the steps described in following subsections.

3.1. Measurement of Lens Entrance Pupil

The distance of the entrance pupil from the flange on the camera body was measured according to the following procedure:

(1) the objective lens is mounted on the camera and placed on a linear stage on a stable tripod which is then leveled accurately, this camera will be called camera A (Figure 8a);

(2) another camera B with a macro lens is placed on another tripod in front of camera A, aligned and leveled accurately to make the optical axes of cameras A and B as more collinear as possible (Figure 8a);

(3) the aperture stop of camera A is stopped down to close the iris (through for example bulb function or long exposure);

(4) the macro lens on camera B is used as collimator and focused on the iris of the objective of camera A (Figure 8b);

(5) the lens from camera A is removed and substituted with a flat and thin sheet with markers printed on it that will cover the flange of the camera body. The sheet is aligned and attached to the flange of the camera body A using thin double sided tape;

(6) the camera body A without lens is translated ahead toward the macro lens through the linear stage (Figure 8c) until the markers appears in focus (Figure 8d). Due to the limited depth of field of the camera B (macro), the repeatability of linear translation of camera body A is in the order of some tenths of millimeter;

(7) the linear translation is taken as offset (distance along the optical axis) of the entrance pupil from the camera flange.

In order to guarantee a better stability of the setup, the tripods were fixed with hot glue on the floor while to improve the readings of linear translations a caliper was used. The measurements were carried out for two lenses, the Nikkor 24 mm f/2.8 AF-D and the Nikkor 35 mm f/2.0 AF-D.

The entrance pupil distance from the flange, here indicated as D_{EP2F}, was measured with the lens focused at different distances (focus distance, FD). The subject to entrance pupil distance (D_{S2EP}) was also computed according to the following equation:

$$D_{S2EP} = FD - FFD - D_{EP2F} \qquad (2)$$

where:

- FD is the focus distance read on the lens barrel and it indicates the distance between the camera sensor and subject;
- FFD is the flange focal distance, equal to 46.5 mm for Nikon cameras;
- D_{EP2F} is the entrance pupil offset measured according to procedure described above.

Table 2 summarizes the computed distances.

Table 2. Measured distances between flange and entrance pupil distances.

Lens	FD @ Infinity	FD @ 300 mm		FD @ 250 mm	
	D_{EP2F}	D_{EP2F}	D_{S2EP}	D_{EP2F}	D_{S2EP}
Nikkor 24 mm f/2.8 AF-D	24.6 mm	26.9 mm	226.6 mm	not possible	not possible
Nikkor 35 mm f/2.0 AF-D	18.6 mm	25.4 mm	228.1 mm	26.9 mm	176.6 mm

The method here described allows the entrance pupil to be precisely located along the optical axis (Z component), but does not provide any information about potential in-plane offsets (XY components, *i.e.*, its position projected onto the flange plane).

The position of entrance pupil was measured also for the lens focused at infinity for the sake of completeness. However, it is worth to note that with a dome of radius comparable to the ones under investigation the virtual image of a subject very far away from the camera ($\approx @$ infinity) is formed at a distance less than 290 mm from the entrance pupil. As a consequence, in such a condition, the camera focus should not be set to infinity to avoid blurred or out of focus images.

3.2. 3D Model of NI320 Dome Port

An opaque coating was applied over the transparent glass dome NI320 (Figure 9a), which was surveyed using a triangulation based laser scanner (ShapeGrabber, Ottawa, ON, Canada [22]). The produced polygonal model, shown in Figure 9b, has a surface sampling step of about 0.25 mm.

Two spheres are fitted to the external (Figure 9c) and internal dome surfaces, respectively (Table 3), showing that the two centers of curvature do not coincide but have a distance less than 1.5 mm. The actual thickness of the spherical dome is computed as difference between the two radii and it is equal to 7.7 mm.

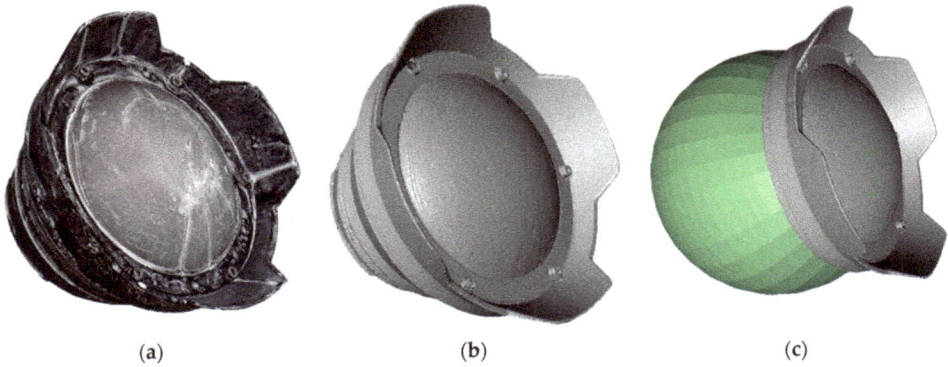

(a)   (b)   (c)

Figure 9. Reverse engineering of NI320 dome port: (**a**) opaque coating applied to avoid reflection and wrong measurements; (**b**) obtained polygonal model; (**c**) fitting of a sphere on the external dome surface.

Table 3. Fitting of external and internal dome surfaces.

	Fit Statistics		Sphere			
	Num. of Points	Stdv (mm)	Centre (mm)			Radius (mm)
			X	Y	Z	
External	≈170,000	>0.1	−0.4	−1.0	21.5	83.4
Internal	≈150,000	>0.1	−0.4	−1.5	20.3	75.7

3.3. Measurement of the Nikon D300 Camera and NI303D Pressure Housing Flange Centers

The flange is a metal ring on digital cameras and a plastic ring on underwater cases where the rear of the lens and the rear of the flat/dome port are respectively mounted. The position of the camera mounting flange with respect to the waterproof housing is fundamental in order to locate the lens entrance pupil (see Section 3.1). Analogously, the identification of the plane containing the housing flange is needed to find the position of the centers of dome surfaces relatively to the camera-lens + housing-port system.

To find the relative position between the two mounting flanges, *i.e.*, for the camera and underwater case, a photogrammetric survey was conducted (Figure 10). The two planes containing the elements of interest were materialized and identified through black and white dots measured in the images (Figure 10c). The target positions on the circular sheet attached to the camera flange were designed to be concentric with the Nikon F mount flange. The estimated centering error is less than 0.5 mm. The 3D coordinates of the triangulated dots were fitted through a least squares procedure to find the reference planes and centers of the flange for both camera and housing.

The coordinate measurement system was fixed in the center of the camera flange, with the XY plane coincident with the camera flange plane, and the Z axis along the optical axis, positive toward the housing flange (Figure 10d).

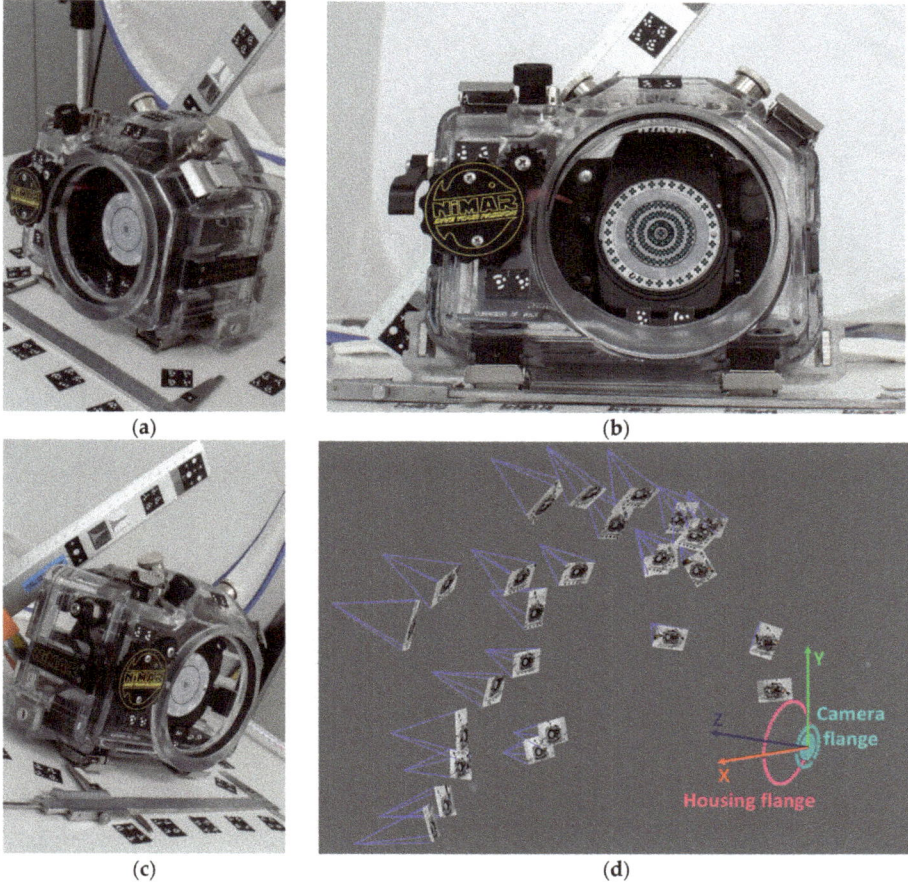

Figure 10. Photogrammetric survey of camera and pressure case flanges: (**a**) and (**b**) examples of acquired images; (**c**) dots identifying the camera flange markers in an image; (**d**) camera network and measured 3D points for the identification of camera and housing flanges with the established reference system.

3.4. Assembly of the Different Measurements

A plane was fitted to the rear of the dome port and aligned to the flange plane of the housing (magenta plane in Figure 11). From the measurements described in the previous steps, the relative position of key geometric and optic elements is known and the complete geometry of the camera-lens + housing-dome system is reconstructed.

The main outcome of the survey described above is that, in the worst condition, the position of the entrance pupil both for the 24 mm and 35 mm lens results maximum 5.4 mm head of the center of curvature of the dome surfaces (*i.e.*, closer to the dome), while the misalignment in the XY plane, due to the misalignment of the dome center of curvature, is about 1 mm (Table 4).

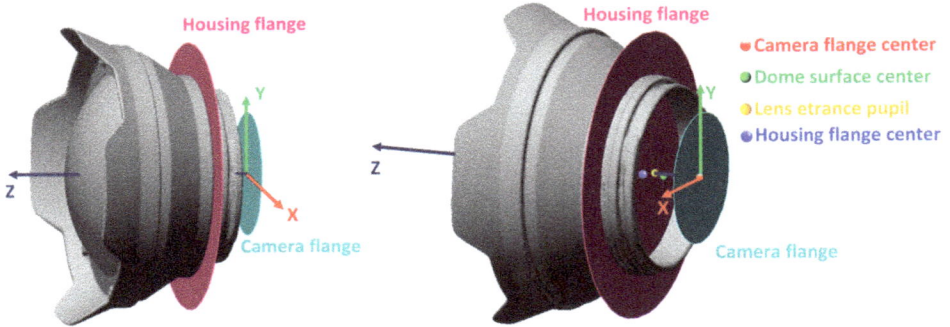

Figure 11. Two views of NI303D housing with NI320 dome port and reference planes and points.

Table 4. Coordinates of key geometric and optic elements.

Point Name	Position (mm)		
	X	Y	Z
Camera flange center	0.0	0.0	0.0
Dome surface center	−0.4	−1.0	21.5
Lens entrance pupil	0	0	26.9
Housing flange center	−0.4	−1.0	34.3

4. Optic Characterization

Using the data of the NI320 dome port obtained through the reverse engineering process described in Section 3, an optic characterization of the same view port is carried out using the optical ray tracing software Winlens. Using the values of the internal and external radius (Table 3) and the thickness computed as difference between the external and internal radii, under paraxial assumptions, positions of the virtual image *versus* the real object can be computed. The dome port glass is a N-BK7, very commonly used for underwater view port and characterized by a refraction coefficient of 1.5168.

Table 5 lists the values of working distances (*WD*) from a real object point underwater *versus* its virtual image distance (or subject to entrance pupil distance D_{S2EP}) from the entrance pupil (supposed to be placed in the dome center). Virtual working distances (VWD), as described in Figure 5, are also reported. For the sake of simplicity the distances are here considered positive, even if for convention they should be negative quantities.

Table 5. Real object distance *versus* its virtual image underwater for the NiMAR NI320 dome port.

WD (mm)	200	300	400	500	750	1000	3000	5000	10000	Infinity
D_{S2EP} (mm)	164.1	184.9	199.7	210.8	229.4	240.8	269.7	276.8	282.5	288.5
VWD (mm)	80.8	101.6	116.4	127.5	146.1	157.5	186.4	193.5	199.2	205.2

In Table 5 it is evident how the virtual image of a real object underwater is compressed in a very narrow virtual space just 20 cm deep in front of the dome glass. By using the values of D_{S2EP} from Table 2 corresponding to minimum focusing distances for the two Nikkor AF 24 and 35 mm lenses and closest corresponding values from Table 5, the minimum working distances WD for the two lenses result to be respectively −750 mm and −300 mm (Figure 12). For closer objects, additional close up lenses must be mounted to the front of the camera in order to produce sharp images.

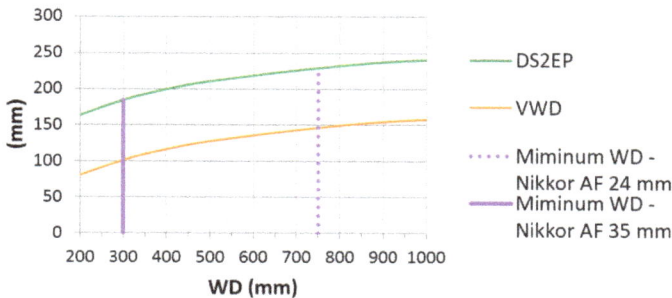

Figure 12. Variation of subject to entrance pupil distance (DS2EP) and virtual working distance (VWD) in function of the working distance (WD). The minimum WD for Nikkor 24 mm and 35 mm focal length are also drawn.

Underwater Variation of the DOF

This section presents a simulation which demonstrates and quantifies the increase of DOF underwater, one of the main characteristics of dome ports introduced in Section 2.3. The following analysis is performed, under paraxial assumptions, for

59

an aperture value f/8 and a circle of confusion of 15 microns (<3× pixel size). Let us consider a Nikon D300 with 24 mm lens in the NiMAR NI303D housing with the NI320 dome port and the subject to be at 1 m distance from the entrance pupil of the 24 mm lens. In this computation, the entrance pupil is supposed to be concentric to the spherical dome. Under these hypotheses, when underwater, a virtual image from the entrance pupil is formed at 238 mm, on which the camera has to be focused. Near and far sharp limits (NSL and FSL, respectively) values of the DOF result equal to −227 mm and −249 mm. These values in the virtual image space correspond to planes at −792 mm and −1355 mm distances from the camera in the real object space, leading to a total DOF of 563 mm. In air, for the same set up and the object at a 1 m distance from the entrance pupil, the total DOF would be 424 mm, with the NSL and FSL respectively at −831 mm and −1260 mm., The ratio between the DOF in water and in air approaches the ratio between the two refraction indexes, *i.e.*, 1.33–1.34 which corresponds to a relative increase of about 33%–34% of the DOF underwater.

5. Swimming Pool Tests

In the previous sections, the optic characteristics of dome ports have been described from a theoretical point of view and then proved with a real case example. In the followings, the influence of the view port on the photogrammetric system (camera + lens) is quantified in a real underwater scenario through some tests performed in a swimming pool. The results afterwards presented are part of some tests aimed at investigating the performance of consumer-grade underwater camera housings when used for photogrammetric purposes.

Camera Calibration

A first prototype of an underwater test-field made of a planar aluminum board was temporarily fixed on a wall of the pool (Figure 13a). The test-field was prepared with photogrammetric coded targets and some resolution targets (Figure 13b).

Two photogrammetric acquisitions for self-calibration were realized, one underwater and one in air. The camera with the 24 mm at f/8.0 was focused underwater at 1m using the autofocus system then the autofocus was disabled to keep the interior orientation parameters of the camera stable as much as possible. Between the two image acquisitions, the camera was not removed from the pressure housing to keep the assembly as much stable as possible. Table 6 reports the camera calibration parameters obtained from the two calibrations.

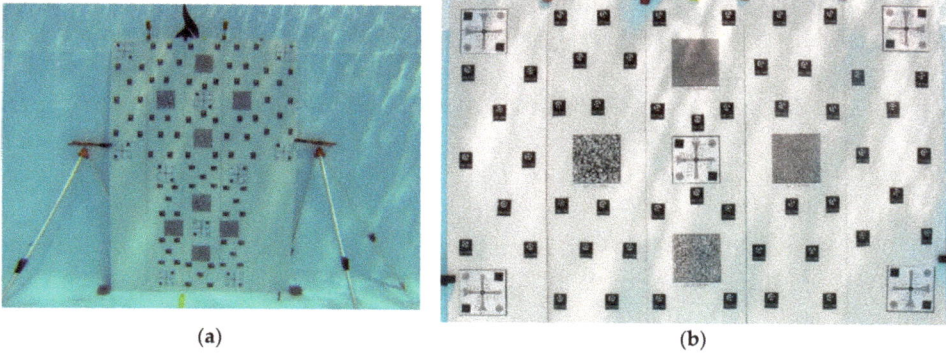

(a)   (b)

Figure 13. Test-field used in the swimming pool: **(a)** overall view of the complete board with stands; **(b)** part of the board used for the camera calibration.

Table 6. Comparison between camera calibration in water (UW) and in air. Some non-significant additional parameters were not computed during the self-calibration procedure.

Camera Calibration Parameters	AIR		UW	
	value	std	value	std
Principal distance (mm)	25.801	0.006	26.208	0.002
Principal Point x_0 (mm)	−0.026	0.002	−0.058	0.003
Principal Point y_0 (mm)	−0.144	0.003	−0.207	0.002
K1	1.842e-004	1.2e-006	1.663e-004	6.1e-007
K2	−3.030e-007	7.4e-009	−2.582e-007	3.4e-009
K3	-	-	-	-
P1	-	-	6.582e-006	1.2e-006
P2	-	-	1.620e-005	8.7e-007

As shown in Figure 14, the lens displays quite a pronounced barrel radial distortion both in air (red) and in water (blue). As previously anticipated by the reverse engineering of the dome, the advanced position of the entrance pupil of the lens respect to the dome center introduces a small pincushion compensation effect resulting in a less negative overall distortion (less barrel). A significant variation in the principal distance between in air and underwater calibrations is also observed. This change is expected as the closer is the lens to the dome surface, the less spherical is the portion of the surface of the dome the camera looks trough. The extreme limit is when the lens front is very close to the dome inner surface and the entrance pupil is much more ahead than in the case study of this paper: in this case the dome portion in the field of view of the camera approaches the one of a flat port with a consequent increase of the principal distance by a factor of about 1.33 as explained in Section 2.2.

Decentering distortion is introduced, due to the in-plane offset between lens entrance pupil and dome surface center. In air the decentering distortion parameters were not statistically significant thus were not adjusted for. As it can be observed in the graph, its magnitude in water is anyway very small compared to the radial component, as expected due to the smaller in-plane than along the axis misalignment.

The in-plane offset can also explain the difference in the coordinates of the principal points.

In Figure 15, the system distortions are visualized according to a color map (distortion map): the color represents the difference between the ideal pixel position (no distortion) and the actual pixel position due to the influence of radial and decentering distortions determined through camera calibration. The difference between the distortion map in air (Figure 15a) and in water (Figure 15b) is reported in Figure 15c. As expected, the maximum difference is reached at the borders, whose magnitude is comparable with the differences highlighted in the distortion curves. An asymmetric behavior can be also observed, likely due to the small in-plane misalignment between the lens entrance pupil and dome surface center of curvature, slightly bigger along the Y axis. The optic behavior of the dome port results to be well modelled by the pinhole camera and Brown's distortion model. It is worth to note that the experiments were carried out at a small depth; consequently, it may be expected that the influence of the pressure on the watertight case is considerably less critical and probably easier to be absorbed by standard calibration parameters than in deep water. To demonstrate the DOF increasing in underwater situations, a slant view of the test-field is analysed (Figure 16). An image used during the underwater calibration was compared with a very similar one acquired in the laboratory. Both cameras are in the pressure housing and focused so that an object is sharp at 1m from the entrance pupil of the lens.

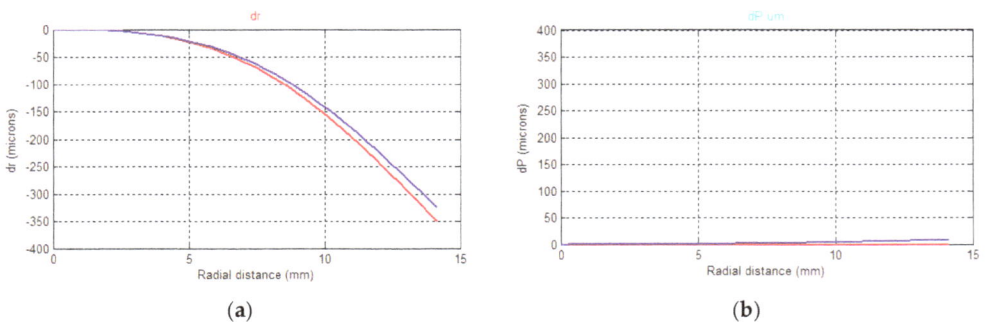

(a)   (b)

Figure 14. (a) Radial and (b) decentering distortion curves: the curves in red are related to the camera calibration in air, the curves in blue to the camera calibration underwater.

62

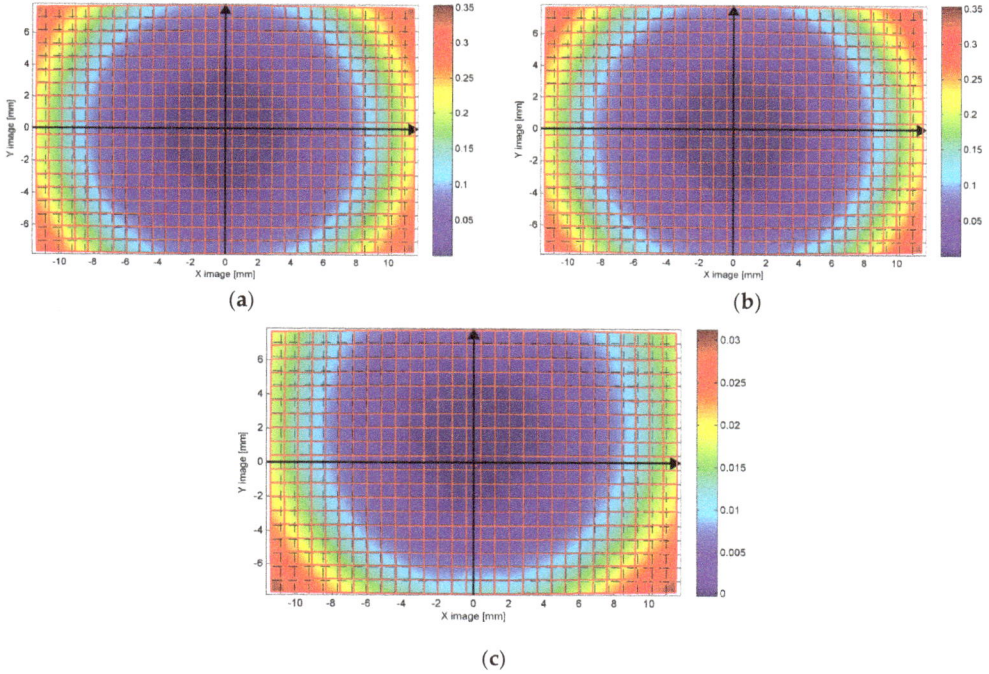

Figure 15. Distortion maps (difference in mm between ideal and actual distorted pixel position): (**a**) in air, (**b**) in water; and (**c**) difference air-water.

As visible in Figure 16 the increase of DOF underwater is quite evident and in accordance with what anticipated in Section 4. Indeed, the resolution target highlighted in green was at about 1m while the orange one was at 1.75 m from the camera. Whilst the target at 1 m is sharp in both the images taken underwater and in air, the one at 1.75 m from the camera is completely out of focus in the image taken in air.

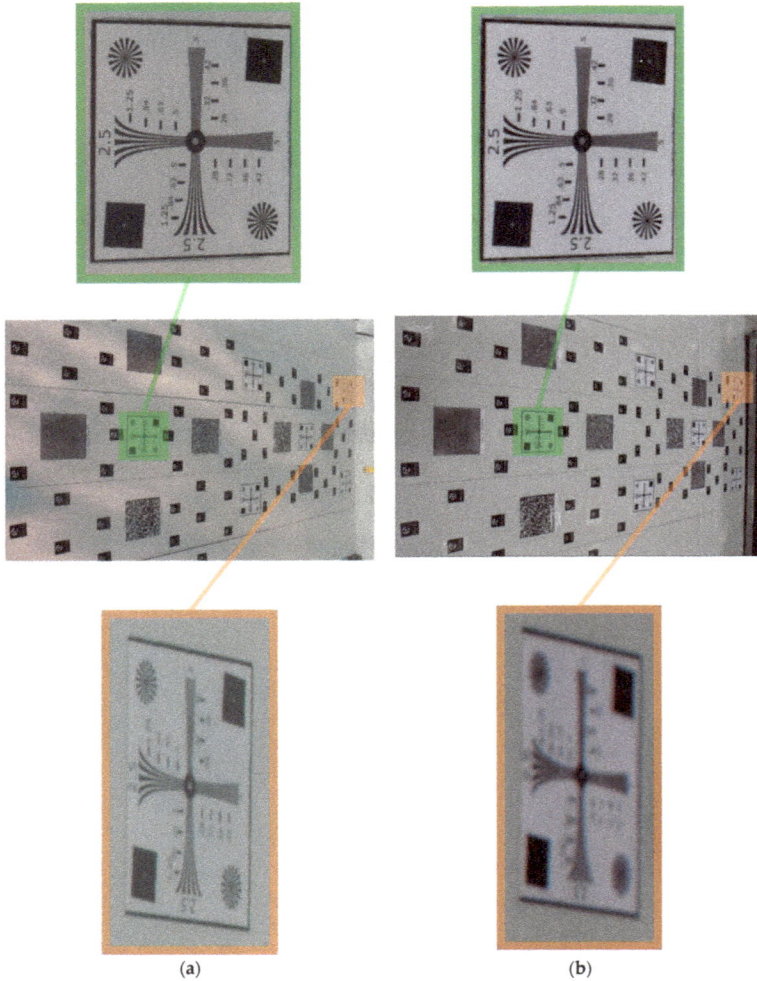

Figure 16. DOF variation between two pictures taken from very similar positions (**a**) underwater and (**b**) in air.

6. Conclusions and Future Developments

The paper presented the optic and geometric characterization of a consumer grade pressure camera housing (manufactured by NiMAR) that was successfully used in the underwater survey of the Costa Concordia gash [6]. The main advantages and drawbacks of flat and dome glasses were presented, in particular with respect to parameters used in the photogrammetric planning. Indeed, the concepts of virtual images generated by dome ports are often difficult to understand, especially for non-experts in surveying and photogrammetry and may lead to wrong or not optimized practices in the field (*i.e.*, focus fixed at infinity or pre-focused in air or too high aperture values for increasing the DOF). The 3D reverse engineering of a NiMAR dome port and its position relatively to the entrance pupil of the lens served as input for some computer simulations carried out through a freely available optical ray tracing software. The computer analyses anticipated some behaviors on the camera calibration parameters concerning the radial and decentering distortions when the entrance pupil is ahead or behind, or laterally displaced to the center of curvature of the dome. A DOF increasing in underwater scenarios was also anticipated trough computer analyses.

The final swimming pool tests sustained and demonstrated the validity of computer simulations. These tests will be soon followed by other analyses and tests underwater in swimming pool and in open water to deliver photogrammetric guidelines for underwater camera housing and applications.

The developed tests served to design a new modular calibration test-field whose base plane is shown in Figure 17. The test-field displays circular coded target on a slant square background for MTF measurements, DOF evaluation and geometric camera calibration. Some of the targets will be stuck on out-of-plane elements to make the test-field more suitable for camera calibration (depth variation).

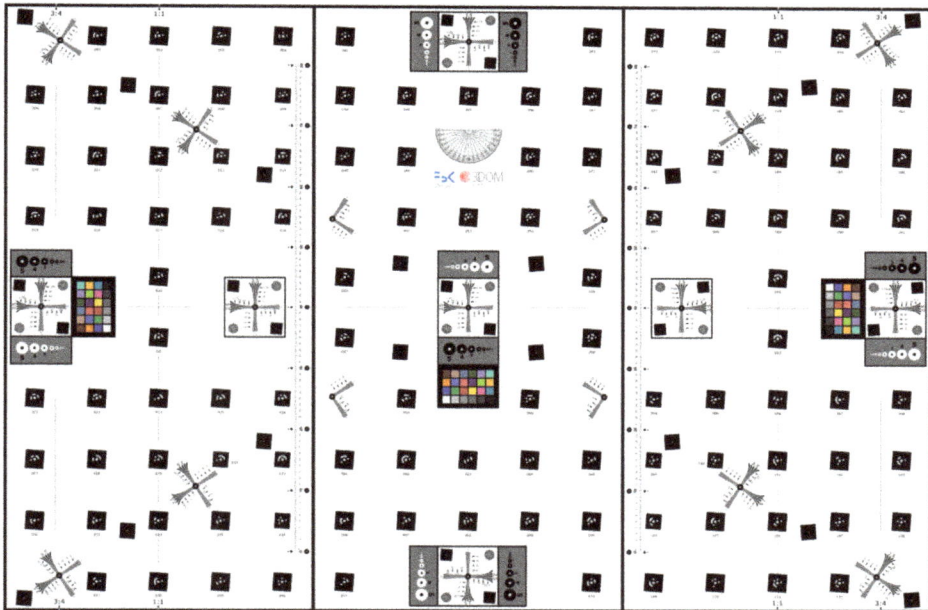

Figure 17. The newly designed modular test-field for resolution and DOF measurements as well as camera calibration. The test-field, being modular, can be mounted to accommodate different heights of the targets.

Acknowledgments: The authors would like to thank NiMAR which supported this research by providing some photographic underwater equipment and for the useful insights about pressure housings manufacturing techniques; Manfrotto company [23], a VITEC Group brand, for supporting with tripods, stands and laboratory photographic equipment; Salvatore Troisi (Parthenope University of Naples, Italy) for the useful discussions and for providing the NiMAR NI320 dome port and NI303D waterproof case. A special thank also to Pierre Toscani [24] for the precious suggestions on the measurement of the entrance pupil position.

Author Contributions: Fabio Menna and Erica Nocerino conceived the main ideas presented in this research work, performed the tests and analyses. Francesco Fassi contributed extensively in designing and performing the underwater acquisitions. Fabio Remondino significantly contributed to critically revise and finalize the research. All authors equally contributed in the paper writing.

Conflicts of Interest: The authors declare no conflict of interest.

References

1. Bruno, F.; Gallo, A.; de Filippo, F.; Muzzupappa, M.; Petriaggi, B.D.; Caputo, P. 3D documentation and monitoring of the experimental cleaning operations in the underwater archaeological site of baia (Italy). In Proceedings of the Digital Heritage International Congress (DigitalHeritage), Marseille, France, 28 October–1 November 2013; pp. 105–112.

2. Burns, J.H.R.; Delparte, D.; Gates, R.D.; Takabayashi, M. Utilizing underwater three-dimensional modeling to enhance ecological and biological studies of coral reefs. *Int. Arch. Photogramm. Remote Sens. Spatial Inf. Sci.* **2015**, *1*, 61–66.

3. Drap, P.; Merad, D.; Seinturier, J.; Mahiddine, A.; Peloso, D.; Boi, J.M.; Long, L.; Chemisky, B.; Garrabou, J. Underwater programmetry for archaeology and marine biology: 40 years of experience in Marseille, France. In Proceedings of the Digital Heritage International Congress (DigitalHeritage), Marseille, France, 28 October–1 November 2013; pp. 97–104.

4. Papadimitriou, K. Course outline for a scuba diving speciality "underwater survey diver". *Int. Arch. Photogramm. Remote Sens. Spatial Inf. Sci.* **2015**, *XL-5/W5*, 161–166.

5. NiMAR S.r.l. Housings for Underwater Photography and Underwater Videography. Available online: http://www.nimar.it/ (accessed on 30 October 2015).

6. Menna, F.; Nocerino, E.; Troisi, S.; Remondino, F. A photogrammetric approach to survey floating and semi-submerged objects. *Proc. SPIE* **2013**, *8791*.

7. Qioptiq Photonics GmbH & Co. KG. Photonics for Innovation. Available online: http://www.qioptiq.com/ (accessed on 30 October 2015).

8. Drap, P. Underwater photogrammetry for archaeology, special applications of photogrammetry. In *Special Applications of Photogrammetry*; InTech: Rijeka, Croatia, 2012; pp. 111–136.

9. Rongxing, L.; Haihao, L.; Weihong, Z.; Smith, R.G.; Curran, T.A. Quantitative photogrammetric analysis of digital underwater video imagery. *IEEE J. Oceanic Eng.* **1997**, *22*, 364–375.

10. Harvey, E.; Shortis, M. Calibration stability of and underwater stereo–video system: Implications for measurement accuracy and precision. *J. Mar. Technol. Soc.* **1998**, *2*, 3–17.

11. Mulsow, C. A flexible multi-media bundle approach. *Int. Arch. Photogramm. Remote Sens. Spatial Inf. Sci.* **2010**, *XXXVIII-5*, 472–477.

12. Jordt-Sedlazeck, A.; Koch, R. Refractive calibration of underwater cameras. In *Computer Vision—Eccv 2012*; Fitzgibbon, A., Lazebnik, S., Perona, P., Sato, Y., Schmid, C., Eds.; Springer: Berlin/Heidelberg, Germany, 2012; Volume 7576, pp. 846–859.

13. Treibitz, T.; Schechner, Y.Y.; Kunz, C.; Singh, H. Flat refractive geometry. *IEEE Trans. Pattern Anal. Mach. Intell.* **2012**, *34*, 51–65.

14. Agrafiotis, P.; Georgopoulos, A. Camera constant in the case of two media photogrammetry. *Int. Arch. Photogramm. Remote Sens. Spatial Inf. Sci.* **2015**, *XL-5/W5*, 1–6.

15. Shortis, M.; Harvey, E.; Seager, J. A review of the status and trends in underwater videometric measurement. In Proceedings of the SPIE Conference Videometrics Range Imaging, and Applications IX, San Jose, California, USA, 26–30 August 2007; pp. 1–26.

16. Ishibashi, S. The study of the underwater camera model. In Proceeding of the IEEE OCEANS, Santander, Spain, 6–9 June 2011; pp. 1–6.

17. Bräuer-Burchardt, C.; Heinze, M.; Schmidt, I.; Meng, L.; Ramm, R.; Kühmstedt, P.; Notni, G. Handheld underwater 3D sensor based on fringe projection technique. In Proceedings of the Videometrics, Range Imaging, and Applications XIII, Munich, Germany, 21 June 2015; p. 952809.

18. Kunz, C.; Singh, H. Hemispherical refraction and camera calibration in underwater vision. In Proceeding of the IEEE OCEANS, Quebec City, QC, Canada, 15–18 September 2008; pp. 1–7.

19. Shortis, M.R.; Miller, S.; Harvey, E.S.; Robson, S. An analysis of the calibration stability and measurement accuracy of an underwater stereo-video system used for shellfish surveys. *Geomat. Res. Australas.* **2000**, 1–24.

20. Maas, H.G. On the accuracy potential in underwater/multimedia photogrammetry. *Sensors* **2015**, *15*, 18140–18152.

21. Ray, S.F. *Applied Photographic Optics Lenses and Optical Systems for Photography, Film, Video, Electronic and Digital Imaging*; Focal Press: Oxford, UK, 2002.

22. ShapeGrabber Inc. 3D Scanner, Laser Scanner, 3D Scanners. Available online: http://www.shapegrabber.com/ (accessed on 30 October 2015).

23. Lino Manfrotto + Co., S.p.A. Professional Photo & Video Tripods, Lighting, Bags. Available online: https://www.manfrotto.com/ (accessed on 30 October 2015).

24. Toscani, P. Amateur Photographe En Montagne. Available online: https://www.pierretoscani.com/ (accessed on 30 October 2015).

Camera Calibration for Water-Biota Research: The Projected Area of Vegetation

Rene Wackrow, Edgar Ferreira, Jim Chandler and Koji Shiono

Abstract: Imaging systems have an indisputable role in revealing vegetation posture under diverse flow conditions, image sequences being generated with off the shelf digital cameras. Such sensors are cheap but introduce a range of distortion effects, a trait only marginally tackled in hydraulic studies focusing on water-vegetation dependencies. This paper aims to bridge this gap by presenting a simple calibration method to remove both camera lens distortion and refractive effects of water. The effectiveness of the method is illustrated using the variable projected area, computed for both simple and complex shaped objects. Results demonstrate the significance of correcting images using a combined lens distortion and refraction model, prior to determining projected areas and further data analysis. Use of this technique is expected to increase data reliability for future work on vegetated channels.

Reprinted from *Sensors*. Cite as: Wackrow, R.; Ferreira, E.; Chandler, J.; Shiono, K. Camera Calibration for Water-Biota Research: The Projected Area of Vegetation. *Sensors* **2015**, *15*, 30261–30269.

1. Introduction

The capability of aquatic plants to deform and "reconfigure" is critical to the functioning of lotic ecosystems [1,2]. Specifically, adverse effects imposed by these barriers (in terms of flow resistance) are counterbalanced by a variety of ecosystem services associated with plant motion, namely regulating services [3]. Thus, some authors have sought to quantify plants' morphology as a way to assess the performance of different species [4].

Sagnes [4] describes the technical challenges with quantifying the frontal area of a plant. Specifically, Sagnes identifies that "the projected frontal surface area (A_f) captures flow-induced shape variation and is seemingly the most realistic physical description". Different setups and image perspectives have been adopted to estimate A_f or equivalent descriptors, ranging from: mirrors attached to the bottom part of laboratory facilities or *in situ* environments combined with top view images using regular cameras [4–6]; images acquired in still air [7] and water conditions [8,9]; to submerged digital cameras aligned with the plant mass centre [10,11]. If light absorption or scattering is not dominant in the course of image acquisition, underwater techniques provide the only opportunity to accurately inspect the morphological reconfiguration of vegetation specimens in the field. Nevertheless, non-metric sensors such as consumer grade digital cameras do not possess, as opposed to photogrammetric or metric cameras, a calibration certificate. Basically, this demands deriving a set of parameters which can be used to describe the internal geometry of the imaging system (e.g., focal length, principal point offset, and radial and tangential lens distortion) [12]. This step is crucial, notably if precise spatial information is to be extracted and carried out through a process known as "self-calibrating bundle adjustment" [13,14]. The impact of lens distortion on subsequent measurements have been previously mentioned on vegetated studies, but have been neither thoroughly investigated nor quantified. For instance, Jalonen *et al.* [11] identified that scaling errors can distort the estimated projected area up to 10%, however, it is a plausible conjecture that these results possibly include a combination of errors caused by scale constraints and uncorrected lens distortion. Even in the work conducted by Sagnes [4], possibly the most comprehensive work on the topic that one can find in the literature, overlook this aspect. Our belief is that this is mainly a consequence of user unawareness of imaging geometry or a procedure to appropriately calibrate non-metric imagery. Whittaker [15] states that in the absence of a known focal length, distortion effects cannot be scrutinized and Wunder *et al.* [10] assumed, without apparent reason, that camera distortion effects were minimized in their work. Bearing in mind these considerations, this paper presents a method based on well-established photogrammetric principles to eliminate lens distortion in both dry and wet environments and compares projected areas using non-calibrated and calibrated cameras. Our work proves that a simple methodology, easily adoptable by experimentalists, allows for an effective camera calibration, thus enabling refinement of existing experimental protocols, particularly those prevailing in laboratory-based activities. The present analysis is restricted to the parameter projected area due to its relevance in aquatic studies (e.g., to evaluate the drag coefficient) but conclusions stemming from this work are equally valid for other morphological studies using similar imaging systems.

Tests performed for this work are explained in the next section. Afterwards, the camera calibration procedure is described. Finally, results are presented and some conclusions drawn.

2. Experimental Setup

Three different experimental setups were employed to determine the projected area of an object in dry conditions and in both submerged static and submerged flow conditions (discharge: 0.124 $m^3 \cdot s^{-1}$, water depth: 0.275 m, flume length: 5.24 m, flume width: 0.915 m). Areas evaluated in these practical applications included the use of a simple metal cube, which provided an accurate reference area (0.01055 m^2), and a real plant (bush species: Buxus sempervirens, height: 0.20 m). In all these measurements, distances between photogrammetric targets attached to a wooden frame (Figure 1) were determined using a vernier calliper, and used for scaling purposes in the process of calculating A_f using digital imagery and photogrammetric measurement. The target frame was located in the same plane as the front of the test object. A video sequence of the objects surface area A_f was acquired using an underwater endoscope camera (Figure 2), at an object to camera distance of 0.7 m, and approximately perpendicular to the metal cube and vegetation bodies.

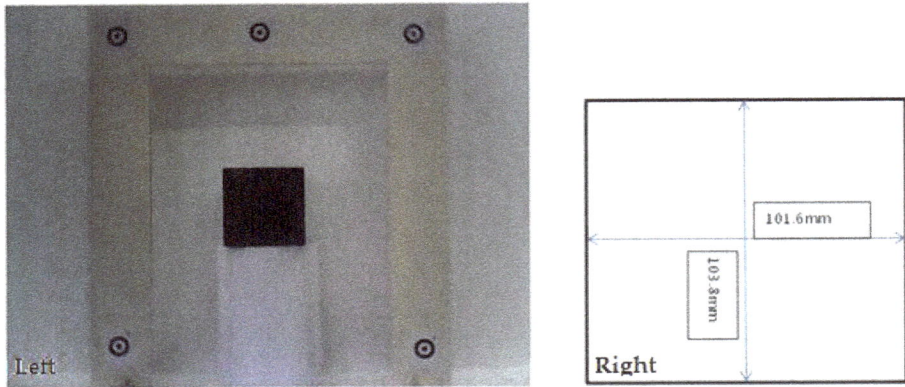

Figure 1. Metal cube and target frame used to provide a reference area (**Left**) and respective dimensions of the cube (**Right**).

Figure 2. Underwater endoscope camera (resolution: 640 × 480 pixels; price July 2013: £25).

The trial was conducted under dry conditions and furthered the opportunity to test the methodology without the additional distorting effect of the light rays passing through water due to refraction. Furthermore, for this attempt, a DSLR camera (Nikon D80, resolution: 3872 × 2592 pixels), shown to be suitable for accurate photogrammetric measurement in the past [16], was also employed for comparing images taken by both cameras. Use of a plastic water tank (submerged static conditions) offered a controlled environment to calibrate the underwater camera and assess if the lens distortion and refractive effects due to the water could be accurately modelled (Figure 3). Results achieved using the plastic water tank encouraged a further test to determine the projected area of both objects, *i.e.*, the cube and the bush, under flow conditions in an open-channel flume (Figure 4).

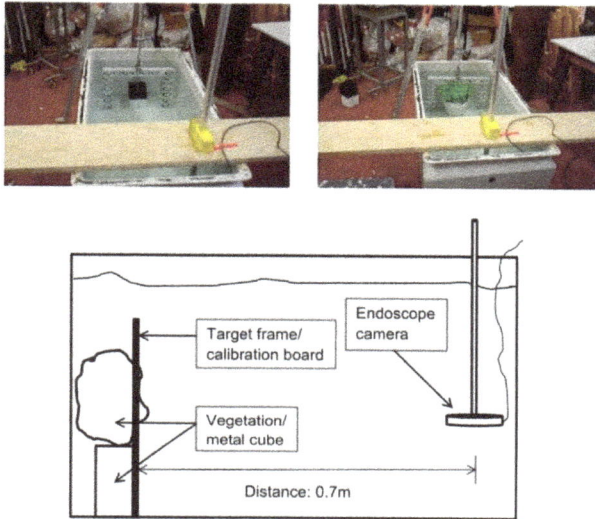

Figure 3. Plastic water tank setup.

Figure 4. Open-channel flume.

For the three cases mentioned above (still air, unstressed and stressed conditions), a Matlab routine was developed to manipulate and measure the images containing the object and the target frame (for stressed flow conditions an image was arbitrarily selected from the video footage). After reading the image file, a Matlab function was used to measure the distance between two photogrammetric targets in the image space. The measured distance in the image space and the distance measured in the object space were used to calculate an image scale factor. In essence, the routine converts an RGB image to a binary image using a simple 2-fold image classification. The pixels in the region of interest (*i.e.*, pixels representing the cube and the bush) are represented by white pixels, whilst all other objects are represented by black pixel values (Figure 5). Pixels representing the cube and the bush were counted automatically and the area was quantified by using the image scale factor. When attempted, image thresholding is almost certainly affected by some degree of uncertainty/imprecision (for example, A_f is slightly overestimated in Figure 5). Hence, in practical terms, each researcher should carry out a systematic modification of the threshold value until the desired classification is reached.

Figure 5. Bush and reference frame used for area determination (**Left**) and binary image obtained from Matlab (**Right**).

It needs to be recognized that most cameras are not designed for accurate photogrammetric measurement [17]. Camera lenses are characterized by significant lens distortion which degrades the achievable accuracy in the object space [18] and additionally, in our study, the distortion effects of the endoscope camera will also change radically when used in an underwater environment as a result of water refraction. Such imaging sensors can be calibrated to minimize the combined effect of these two phenomena, *i.e.*, lens distortions and refraction effects for a specific camera to object distance. Routinely, this is done by assuming that these two components are implicitly considered in the distortion terms of the functional model known as the extended collinearity equations [19,20]. The camera calibration process that has been used prior to computing A_f in both the dry and underwater studies constitutes the core of this work and is portrayed in the subsequent section.

3. Camera Calibration

The extended collinearity equations provide a framework to directly transform the object coordinates into the corresponding photo coordinates [21,22]

$$
\begin{aligned}
x'_a &= x_p - c\frac{[r_{11}(X_0 - X_A) + r_{12}(Y_0 - Y_A) + r_{13}(Z_0 - Z_A)]}{[r_{31}(X_0 - X_A) + r_{32}(Y_0 - Y_A) + r_{33}(Z_0 - Z_A)]} + \Delta x' \\
y'_a &= y_p - c\frac{[r_{21}(X_0 - X_A) + r_{22}(Y_0 - Y_A) + r_{23}(Z_0 - Z_A)]}{[r_{31}(X_0 - X_A) + r_{32}(Y_0 - Y_A) + r_{33}(Z_0 - Z_A)]} + \Delta y'
\end{aligned}
\tag{1}
$$

where (x'_a, y'_a) and (X_A, Y_A, Z_A) represent the coordinates of a generic point A in the image and object space, respectively, (x_p, y_p) are the principal point coordinates, (X_0, Y_0, Z_0) are the coordinates of the perspective centre in the object space, r_{ij} (with $i,j = 1,2,3$) represent the elements of a rotation matrix, c is the principal distance and $\Delta x'$ and $\Delta y'$ are photo coordinate corrections to the combined (radial and decentring) lens distortion. The combined lens distortion terms can be represented by the equations [12]:

$$\Delta x' = \Delta x'_{rad} + \Delta x'_{dec}$$

$$\Delta x' = x' \frac{\Delta r'_{rad}}{r'} + P_1 \left(r'^2 + 2x'^2 \right) + 2P_2 x'y'$$

$$\Delta y' = \Delta y'_{rad} + \Delta y'_{dec} \qquad (2)$$

$$\Delta y' = y' \frac{\Delta r'_{rad}}{r'} + P_2 \left(r'^2 + 2y'^2 \right) + 2P_1 x'y'$$

$$\Delta r'_{rad} = K_1 r'^3 + K_2 r'^5 + K_3 r'^7$$

Both camera exterior orientation (defined by X_0, Y_0, Z_0, and r_{ij}) and interior orientation (comprising x_p, y_p, c, $\Delta x'$, and $\Delta y'$) are typically obtained through a bundle adjustment [23]. Auspiciously, over the past years, continued advances in digital photogrammetry have increased the number of applications of photogrammetry. In particular, automated image-processing algorithms have attenuated competences needed to deal with photogrammetric projects and therefore, this can certainly be a promising solution for hydraulicians studying certain physical processes with the aid of imaging systems [13]. Having these considerations in mind, the PhotoModeler Scanner software (64 bit) [24] was selected to calibrate the two cameras. PhotoModeler models the radial lens distortions and the decentring distortions through Equations (2). As an output, the software provides some quality indicators (average and maximum residuals) which are extremely useful to judge the overall accuracy of the derived calibration data.

Figure 6 represents the image configuration (image frames represented by numbers 1 to 12) and the calibration board used to determine the camera calibration parameters for both the D80 camera and the underwater probe (dry and submerged condition for the underwater probe). The calibration board consisted of 49 coded targets generated by PhotoModeler Scanner. Twelve images of the calibration board were captured, with three image frames rotated by 90 degrees (frames 2, 4, and 6 in Figure 6) to provide the possibility to estimate the principal point offset x_p and y_p of the camera [17,25]. The camera to object distance was set to 0.7 m, the exact same distance used when collecting the metal cube and the bush imagery. The calibration files were subsequently uploaded to a PC, and processed using the camera calibration tool in PhotoModeler Scanner. Finally, camera models determined for the D80 camera and the underwater probe were applied in order to remove the distortion effects of

73

the recorded images. PhotoModeler provides the option to use the estimated camera parameters to produce an undistorted or "idealized" image. In general terms, during idealization, the software re-maps the image pixel by pixel and removes any lens distortion, non-centred principal point and any non-square pixels [24] (Figure 7). The effect of camera calibration on the computation of the surface area will be explored in the following section.

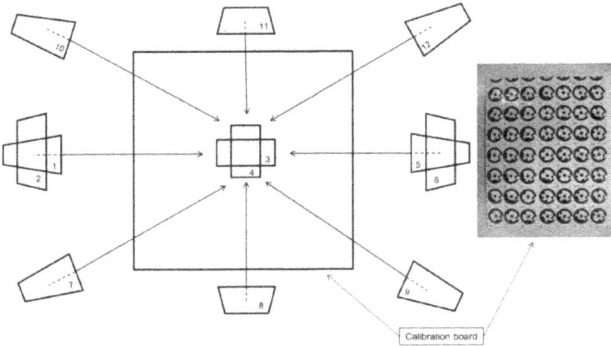

Figure 6. Camera calibration configuration—Note that two distinct environments were considered at this phase: dry and wet (using the plastic water tank) to fully consider the fluid at the camera's interface during area assessment.

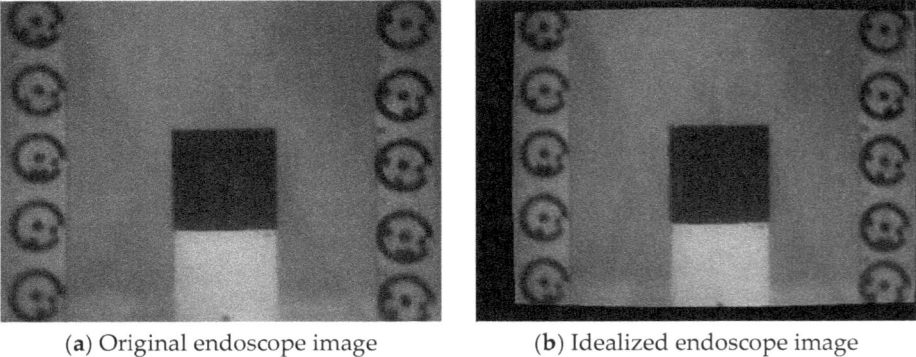

(**a**) Original endoscope image (**b**) Idealized endoscope image

Figure 7. Black metal cube image using the endoscope camera in the plastic water tank.

4. Results and Discussion

4.1. Dry Case

The projected area of a metal cube with known dimensions and a real bush were determined under dry conditions using the endoscope camera and a Nikon D80 DSLR camera normally used for spatial measurement. Table 1 summarizes the estimated cube areas using the two cameras. The first column contains the calibration status of the cameras, whilst the second column tabulates the determined cube area using images acquired with the Nikon camera. The percentage error of the cube area obtained with the Nikon camera is identified in column three and the final two columns represent the cube area and the percentage error determined using the underwater endoscope camera. It should be emphasized that the percentage error was computed as:

$$\left(\frac{|\text{areapredicted} - \text{areaactual}|}{\text{areaactual}} \right) * 100 \tag{3}$$

Both cameras achieved similar results when calibrated (percentage error of 0.1%). The Nikon D80 camera attained a percentage error of 0.8% when camera calibration parameters were not considered, whilst the determined percentage error of the underwater endoscope camera was 1.4%. The performance of both cameras is mainly affected by lens distortion, which evidently is of a different magnitude in these two cases. Nevertheless, results demonstrate that both camera lenses are able to derive an accurate area in dry conditions, if appropriately calibrated.

The metal cube was exchanged for the bush and results are presented in Table 2. Obviously, computed areas at this stage can only be compared in relation to each other, as no "true" area estimation is available. Results reinforce the viability of using this particular endoscope camera to obtain accurate estimates of the projected area, once lens distortion is considered. The areas determined varied by 2.7% using images acquired with non-calibrated cameras. Remarkably, areas of similar orders (discrepancy of 0.3%) have been determined using images where lens distortion was accounted for.

Table 1. Metal cube area.

Camera Calibration	Area D80 Camera [m^2]	Error D80 Camera [%]	Area Endoscope Camera [m^2]	Error Endoscope Camera [%]
Not calibrated dry	0.01063	0.8	0.01040	1.4
Calibrated dry	0.01056	0.1	0.01056	0.1
Not calibrated tank			0.0096	9.0
Calibrated tank			0.0104	1.4
Not calibrated flume			0.0098	7.1
Calibrated flume			0.0107	1.4

Table 2. Bush area dry condition.

Camera Calibration	Area D80 Camera [m^2]	Area Endoscope Camera [m^2]	Difference D80-Endoscope [%]
Not calibrated dry	0.0324	0.0333	2.7
Calibrated dry	0.0318	0.0317	0.3

4.2. Plastic Water Tank

In the presence of static water, distortions are expected to increase since light paths are refracted twice in the vicinity of camera lenses. Again, underwater images of the metal cube were acquired and results are shown in Table 1. Images not corrected for lens distortion exhibit a marked difference to the known metal cube area (error of 9%). However, the percentage error between the computed area and the reference area is reduced to just 1.4% when distortion effects are modelled using the radial lens parameters. This can dramatically reduce the uncertainty of image analysis in these conditions.

The projected area determined for the bush using the endoscope camera image without a lens model diverged from the bush area in dry conditions by 12.3% (Table 3). This error was reduced to just 1.6% when lens distortion was considered. These areas are usually assumed to be coincident since buoyancy effects are taken to be negligible [15]. This is also likely to be true in our case due to the high flexural rigidity of the vegetation stems. Consequently, we hypothesize that this small difference is related to minor experimental errors, e.g., an imperfect alignment of the underwater camera.

Table 3. Bush area using the endoscope camera in the plastic water tank.

Camera Calibration	Area Bush Submerged [m^2]	Area Bush Dry [m^2]	Difference Submerged-Dry [%]
Not calibrated tank	0.0292	0.0333	12.3
Calibrated tank	0.0312	0.0317	1.6

4.3. Open-Channel Flume

This test was conducted under conditions similar to those found in a field environment, especially with respect to water clarity. The water discharge was 0.124 m$^3 \cdot$s^{-1} and use of the underwater camera had the additional advantage of allowing the adoption of reduced object to camera distances with minimal flow disturbance. For the metal cube, a noteworthy discrepancy was found if calibration is ignored (Table 1). This is visible from the substantial departure from the cube reference area (7.1%). Once again, it is clear that the inclusion of a lens model significantly improves area estimations (from one case to the other, area variation was of 8.4%). A similar conclusion was found for the vegetation specimen (Table 4). If distortion effects are compensated, surface area is actually 5% greater in flow conditions. Moreover, area differences between flowing and dry conditions range from 17.3% (not calibrated) to 5.7% (calibrated). This finding is significant since it expresses morphological adjustment of the specimen due to water flowing over and around the plant in "stressed" conditions.

Table 4. Bush area using the endoscope camera in the flume.

Camera Calibration	Area Bush Submerged [m^2]	Area Bush Dry [m^2]	Difference Flowing-Dry [%]
Not calibrated flume	0.0284	0.0333	17.3
Calibrated flume	0.0299	0.0317	5.7
Difference calibrated-not calibrated [%]	5.0	4.9	

This trial demonstrated that image acquisition can be problematic in a real river environment. Due to low illumination of the lower parts of the vegetation specimen, external lighting sources had to be used to improve illumination to a suitable level for image processing. Additionally, a high suspended sediment load in the flume appeared to reduce image quality, although not to a level to affect image processing.

5. Conclusions

Imaging systems are becoming increasingly used by experimentalists, due to their ability to clarify certain aspects of flow-vegetation interactions. This fact together with the notion that calibration of non-metric cameras is vital to extract reliable spatial data [13,14] inspired the present work. The magnitude of lens distortion depends on the combination of several factors, namely the focus settings of the lens, the camera depth of field, the medium of data acquisition, and the lens itself. In other words, lens distortions and/or refraction effects will always be present, to a greater or lesser extent, when image based approaches are used. By assessing the two most demanding arrangements used in this study (*i.e.*, results obtained with the underwater endoscope in the tank and stressed flow conditions) and considering their worst case scenarios, failure to consider camera calibration would lead to errors of 9.0% and 12.3% (cube and bush in the tank, respectively), 7.1% (cube in the open-channel), and 5% (bush in the open-channel). Distortions are clearly case dependent, whereby a sound calibration procedure such as the one presented here can be highly convenient, since simplistic procedures to evaluate lens distortions magnitude (such as the one suggested by Sagnes [4]) are avoided. Our results illustrate the need to consider these distortion effects explicitly, especially in flume and field studies. This will undoubtedly contribute to the refinement of current experimental practices, particularly on vegetated flows research, which is largely focussed on a laboratory scale. This requirement is expected to be even higher in turbid waters, where short focal distances will be needed to attain optimum results, and consequently larger distortions will be created. Although recognizing the existence of other methods to deal with this subject, e.g., the ray tracing approach [20], the author's belief that the approach described above constitutes a valuable starting point for experimentalists whenever environmental conditions (e.g., light and turbidity content) are favourable. This can now be accomplished in a relatively straightforward manner by making use of specialized digital photogrammetry tools.

Acknowledgments: Authors would like to thank the assistance of Nick Rodgers during the experiments. This work was conducted with funding provided by the UK's Engineering and Physical Sciences Research Council (Grant EP/K004891/1).

Author Contributions: All authors have contributed equally in terms of data acquisition, analysis, and writing of this research contribution.

Conflicts of Interest: The authors declare no conflict of interest.

References

1. Marion, A.; Nikora, V.; Puijalon, S.; Koll, K.; Ballio, F.; Tait, S.; Zaramella, M.; Sukhodolov, S.; O'Hare, M.; Wharton, G.; *et al.* Aquatic interfaces: A hydrodynamic and ecological perspective. *J. Hydraul. Res.* **2014**, *52*, 744–758.
2. Puijalon, S.; Bornette, G.; Sagnes, P. Adaptations to increasing hydraulic stress: Morphology, hydrodynamics and fitness of two higher aquatic plant species. *J. Exp. Bot.* **2005**, *56*, 777–786.
3. Luhar, M.; Nepf, H. Flow induced reconfiguration of buoyant and flexible aquatic vegetation. *Limnol. Oceanogr.* **2011**, *56*, 2003–2017.
4. Sagnes, P. Using multiple scales to estimate the projected frontal surface area of complex three-dimensional shapes such as flexible freshwater macrophytes at different flow conditions. *Limnol. Oceanogr. Methods* **2010**, *8*, 474–483.
5. Stazner, B.; Lamouroux, N.; Nikora, V.; Sagnes, P. The debate about drag and reconfiguration of freshwater macrophytes: Comparing results obtained by three recently discussed approaches. *Freshw. Biol.* **2006**, *51*, 2173–2183.
6. Neumeier, U. Quantification of vertical density variations of salt-marsh vegetation. *Estuar. Coast. Shelf Sci.* **2005**, *63*, 489–496.
7. Pavlis, M.; Kane, B.; Harris, J.R.; Seiler, J.R. The effects of pruning on drag and bending moment of shade trees. *Arboric. Urban For.* **2008**, *34*, 207–215.

8. Armanini, A.; Righetti, M.; Grisenti, P. Direct measurement of vegetation resistance in prototype scale. *J. Hydraul. Res.* **2005**, *43*, 481–487.

9. Wilson, C.A.M.E.; Hoyt, J.; Schnauder, I. Impact of foliage on the drag force of vegetation in aquatic flows. *J. Hydraul. Eng.* **2008**, *134*, 885–891.

10. Wunder, S.; Lehmann, B.; Nestmann, F. Determination of the drag coefficients of emergent and just submerged willows. *Int. J. River Basin Manag.* **2011**, *9*, 231–236.

11. Jalonen, J.; Järvelä, J.; Aberle, J. Vegetated flows: Drag force and velocity profiles for foliated plant stands. In *River Flow*, Proceedings of the 6th International Conference on Fluvial Hydraulics, San José, Costa Rica, 5–7 September 2012; Murillo Muñoz, R.E., Ed.; CRC Press: Boca Raton, FL, USA, 2012; pp. 233–239.

12. Brown, D.C. Close-range camera calibration. *Photogramm. Eng.* **1971**, *37*, 855–866.

13. Chandler, J. Effective application of automated digital photogrammetry for geomorphological research. *Earth Surf. Process. Landf.* **1999**, *24*, 51–63.

14. Lane, S.N.; Chandler, J.H.; Porfiri, K. Monitoring river channel and flume surfaces with digital photogrammetry. *J. Hydraul. Eng.* **2001**, *127*, 871–877.

15. Whittaker, P. Modelling the Hydrodynamic Drag Force of Flexible Riparian Woodland. Ph.D. Thesis, Cardiff University, Cardiff, UK, 2014.

16. Wackrow, R.; Chandler, J. A convergent image configuration for DEM extraction that minimizes the systematic effects caused by an inaccurate lens model. *Photogramm. Rec.* **2008**, *23*, 6–18.

17. Fryer, J.G. Camera calibration. In *Close Range Photogrammetry and Machine Vision*; Atkinson, K.B., Ed.; Whittles Publishing: Caithness, UK, 2001.

18. Wackrow, R.; Chandler, J.H.; Bryan, P. Geometric consistency and stability of consumer-grade digital cameras for accurate spatial measurement. *Photogram. Rec.* **2007**, *22*, 121–134.

19. Fryer, J.G.; Fraser, C.S. On the calibration of underwater cameras. *Photogramm. Rec.* **1986**, *12*, 73–85.

20. Harvey, E.S.; Shortis, M.R. Calibration stability of an underwater stereo video system: Implications for measurement accuracy and precision. *Mar. Technol. Soc. J.* **1998**, *32*, 3–17.

21. Cooper, M.A.R.; Robson, S. Theory of close range photogrammetry. In *Close Range Photogrammetry and Machine Vision*; Atkinson, K.B., Ed.; Whittles Publishing: Caithness, UK, 2001.

22. Wackrow, R. Spatial Measurement with Consumer Grade Digital Cameras. Ph.D. Thesis, Loughborough University, Loughborough, UK, 2008.

23. Granshaw, S.I. Bundle adjustment methods in engineering photogrammetry. *Phtogramm. Rec.* **2006**, *10*, 181–207.

24. PhotoModeler Scanner. Available online: http://photomodeler.com/index.html (accessed on 24 September 2005).

25. Fraser, C.S. Digital self-calibration. *ISPRS J. Photogramm. Remote Sens.* **1997**, *52*, 149–159.

Chapter 2:
New 3D Imaging Sensors and Data Processing for Underwater Applications

Optical Sensors and Methods for Underwater 3D Reconstruction

Miquel Massot-Campos and Gabriel Oliver-Codina

Abstract: This paper presents a survey on optical sensors and methods for 3D reconstruction in underwater environments. The techniques to obtain range data have been listed and explained, together with the different sensor hardware that makes them possible. The literature has been reviewed, and a classification has been proposed for the existing solutions. New developments, commercial solutions and previous reviews in this topic have also been gathered and considered.

Reprinted from *Sensors*. Cite as: Massot-Campos, M.; Oliver-Codina, G. Optical Sensors and Methods for Underwater 3D Reconstruction. *Sensors* **2015**, *15*, 31525–31557.

1. Introduction

The exploration of the ocean is far from being complete, and detailed maps of most of the undersea regions are not available, although necessary. These maps are built collecting data from different sensors, coming from one or more vehicles. These gathered three-dimensional data enable further research and applications in many different areas with scientific, cultural or industrial interest, such as marine biology, geology, archeology or off-shore industry, to name but a few.

In recent years, 3D imaging sensors have increased in popularity in fields such as human-machine interaction, mapping and movies. These sensors provide raw 3D data that have to be post-processed to obtain metric 3D information. This workflow is known as 3D reconstruction, and nowadays, it is seen as a tool that can be used for a variety of applications, ranging from medical diagnosis to photogrammetry, heritage reports or machinery design and production [1,2]. Thanks to recent advances in science and technology, large marine areas, including deep sea regions, are becoming accessible to manned and unmanned vehicles; thus, new data are available for underwater 3D reconstruction.

Due to readily-available off-the-shelf underwater camera systems, but also to custom-made systems in deep-sea robotics, an increasing number of images and video are captured underwater. Using the recordings of an underwater excavation site, scientists are now able to obtain accurate 2D or 3D representations and interact with them using standard software. This software allows the scientist to add measurements, annotations or drawings to the model, creating graphic documents. These graphic documents help to understand the site by providing a comprehensive and thematic overview and interface with data entered by experts

83

(pilots, biologists, archaeologists, *etc.*), allowing reasonable access to a set of heterogeneous data [3].

Most 3D sensors developed are designed to operate in air conditions, but the focus of this paper is in the 3D reconstruction of underwater scenes and objects for archeology, seafloor mapping and structural inspection. This data gathering can be performed from a deployed sensor (e.g., from an underwater tripod or a fixed asset), operated by a diver or carried by a towed body, a remotely-operated vehicle (ROV) or an autonomous underwater vehicle (AUV).

Other authors have already reviewed some topics previously mentioned, for example Jaffe *et al.* [4] surveyed in 2001 the different prospects in underwater imaging, foreseeing the introduction of blue-green lasers and multidimensional photomultiplier tube (PMT) arrays. An application of these prospects is shown in Foley and Mildell [5], who covered in 2002 the technologies for precise archaeological surveys in deep water, such as image mosaicking and acoustic three-dimensional bathymetry.

In [6], Kocak *et al.* outlined the advances in the field of underwater imaging from 2005 to 2008, basing their work on a previous survey [7]. Caimi *et al.* [8] wrote their survey in 2008 on underwater imaging, as well, and summarized different extended range imaging techniques, as well as spatial coherency and multi-dimensional image acquisition. Years later, Bonin *et al.* [9] surveyed in 2011 different techniques and methods to build underwater imaging and lighting systems. Finally, in 2013, Bianco *et al.* [10] compared structured light and passive stereo, focusing on close-range 3D reconstruction of objects for the documentation of submerged heritage sites.

Structure from motion and stereoscopy are also studied by Jordt [11], who reported in her PhD thesis (2014) different surveys on 3D reconstruction, image correction calibration and mosaicking.

In this survey, we present a review of optical sensors and associated methods in underwater 3D reconstruction. LiDAR, stereo vision (SV), structure from motion (SfM), structured light (SL), laser stripe (LS) and laser line scanning (LLS) are described in detail, and features, such as range, resolution, accuracy and ease of assembly, are given for all of them, when available. Despite sonar sensors being acoustic, a concise summary is also given due to their extended use in underwater, and figures are presented to be compared to optical systems.

This article is structured as follows: Section 2 presents the underwater environment and its related issues. Section 3 reviews the measuring methods to gather 3D data. Section 4 evaluates the literature and the different types of sensors and technologies. Section 5 shows some commercial solutions, and finally, in Section 6, conclusions are drawn.

2. The Underwater Environment

Underwater imaging [12] has particular characteristics that distinguishes it from conventional systems, which can be summarized as follows:

(1) Limited on-site accessibility, which makes the deployment and operation of the system difficult [13].
(2) Poor data acquisition control, frequently implemented by divers or vehicle operators untrained for this specific task [14].
(3) Insufficient illumination and wavelength-dependent light absorption, producing dim and monotone images [15]. Light absorption also causes darkening on image borders, an effect somewhat similar to vignetting.
(4) Water-glass-air interfaces between the sensor and the scene, modifying the intrinsic parameters of the camera and limiting the performance of the image processing algorithms [16–18], unless specific calibration is carried out [19,20].
(5) Significant scattering and light diffusion that limits the operational distance of the systems.

These distinguishing traits will affect the performance of underwater imaging systems. Particular attention is paid to the typical range, resolution and/or accuracy parameters for the systems discussed in the next sections.

Additionally, images taken in shallow waters (<10 m) can be seriously affected by flickering, which produces strong light fluctuations due to the sunlight refraction on a waving air-water interface. Flickering generates quick changes in the appearance of the scene, making basic image processing functions, like feature extraction and matching, which are frequently used by mapping software [21], more difficult. Although some solutions to this problem can be found in the literature [22], flickering is still a crucial issue in many submarine scenarios.

2.1. Underwater Camera Calibration

Camera calibration was first studied in photogrammetry [23], but it has also been widely studied in computer vision [24–27]. The use of a calibration pattern or set of markers is one of the most reliable ways to estimate a camera's intrinsic parameters [28]. In photogrammetry, it is common to set up a camera in a large field looking at distant calibration patterns or targets whose exact location, size and shape are known.

Camera calibration is a major problem connected with underwater imaging. As mentioned earlier, refraction caused by the air-glass-water interface results in high distortion on images, and it must be taken into consideration during the camera calibration process [29]. This refraction occurs due to the difference in density between two media. As seen in Figure 1, the incident light beam passes through two media changes, modifying the light path.

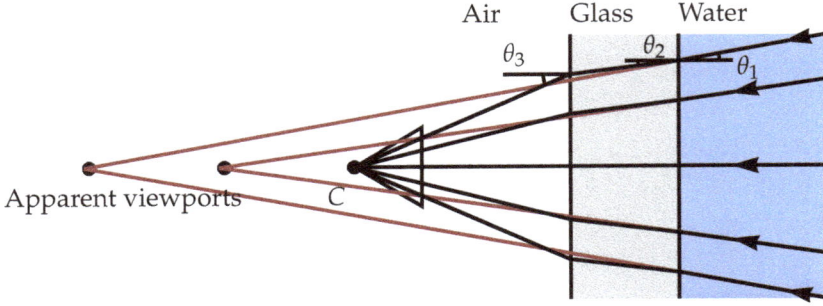

Figure 1. Refraction caused by the air-glass (acrylic)-water interface. The extension of the refracted rays (dashed lines) into air leads to several intersection points, depending on their incidence angles and representing multiple apparent viewpoints. Because of refraction, there is no collinearity between the object point in water, the center of projection of the camera and the image point [20].

According to Figure 1, the incident and emergent angles suffice for Snell's Law, e.g.:

$$r_A^G = \frac{\sin \theta_3}{\sin \theta_2} = \frac{n_G}{n_A} = 1.49, \implies \theta_3 > \theta_2 \tag{1}$$

$$r_G^W = \frac{\sin \theta_2}{\sin \theta_1} = \frac{n_W}{n_G} = 0.89, \implies \theta_2 < \theta_1 \tag{2}$$

where r_A^G is the refractive index between air and glass interfaces and r_G^W is the refractive index between glass and water interfaces (for water $n_W = 1.33$ at $20\,^\circ$C [30], for acrylic glass $n_G = 1.49$ [31]).

If we replace Equation (2) in Equation (1),

$$\frac{\sin \theta_3}{\sin \theta_1} = \frac{n_W}{n_A} = 1.33 \implies \theta_3 > \theta_1 \tag{3}$$

Therefore, the emergent angle θ_3 is bigger than the incident angle θ_1, causing the imaged scene to look wider than it is [14]. For planar interfaces, the deformation increases according to the distance from the center pixel of the camera, called pin-cushion distortion.

Changes in pressure, temperature and salinity alter the refraction index of water and even the camera handling, modifying the calibration parameters [32]. As a result, there is a mismatch between object-plane and image-plane coordinates. This problem has been addressed in two different ways: (1) developing new calibration algorithms that have refraction correction capability [29]; and (2) modifying existing algorithms to reduce the error due to refraction [33]. Other approaches, such as the one reported by Kang *et al.* [34], solve the structure and motion problem taking refraction into consideration.

According to Kwon [29], the refraction error caused by two different media can be reduced by considering radial distortion. Consequently, standard photogrammetric calibration software to calibrate the digital cameras and their housing can be used.

3. Measuring Methods

Sensors for three-dimensional measurement can be classified into three major classes depending on the measuring method: triangulation, time of flight and modulation. A sensor can belong to more than one class, which means that it uses different methods or a combination of them to obtain three-dimensional data, as depicted in Figure 2.

There is also another traditional classification method for sensing devices, active or passive, depending on how they interact with the medium. All of the methods in Figure 2 are active, except for passive imaging.

Active sensors are those that either illuminate, project or cast a signal with respect to the environment to help, enhance or measure the data to gather. An example of an active system is structured light, where a pattern is projected onto the object to reconstruct.

However, according to Bianco [10], those systems using artificial light sources, that are used just to illuminate the scene, but not for the triangulation of the 3D points, are considered passive.

Passive methods sense the environment with no alteration or change of the scene. An example of that is structure from motion, where image features are matched between different camera shots for a post-processed 3D triangulation. Camera-based sensors are the only ones that can be passive for 3D reconstruction, as the others are based on sound or on light projection.

3.1. Time of Flight

Time discrimination methods are based on controlling the travel time of the signal. By knowing the speed of the signal in the medium where it travels, the distance can be drawn. These methods achieve somewhat long distances, especially sonar, but in that case, extra care should be taken to prevent the measures from being affected by alterations in the sound speed, caused by water temperature, salinity and pressure changes.

At short distances, a small inaccuracy in the time measure can cause a great relative error in the result. Furthermore, some sensors require a minimum distance at which they can measure depending on their geometry.

Sonar, LiDAR and pulse gated laser line scanning (PG-LLS) are some examples of sensors using this principle to acquire 3D data.

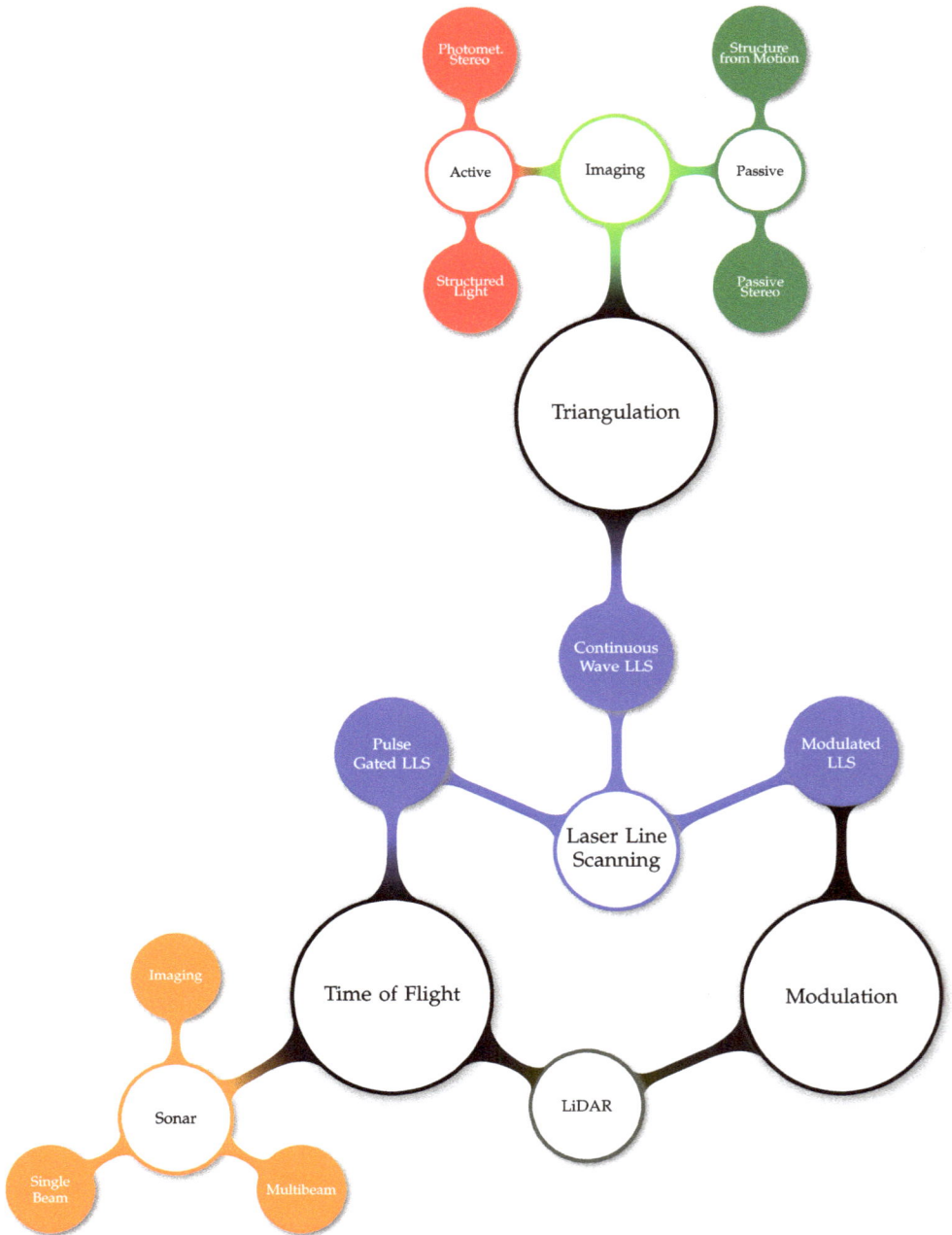

Figure 2. 3D reconstruction sensor classification.

3.2. Triangulation

Triangulation methods are based on measuring the distance from two or more devices (either signal sources or receivers) to a common feature or target with some known parameters.

For example, two cameras can obtain depth (e.g., a stereo rig) by searching in the image gathered by one camera features found in the other one. Once these features have been matched and filtered, the remaining features can be projected on the world as light rays coming from these two cameras. The triangle formed between the feature in the space and the two cameras is the basis for triangulation.

The limitation of triangulation sensors is the need for an overlapping region of the emitter field of view and the receiver one (or the two cameras in the stereo rig case) [17]. Besides, nearby features have a larger parallax, *i.e.*, image disparity, than more distant ones, and as a consequence, the triangulation-based devices have a better z resolution for closer distances than for farther ones. Likewise, the bigger the separation of the cameras (baseline), the better is their z resolution.

Different techniques exist that compute 3D information by triangulation: structured light, laser stripe and photometric stereo (PhS) from active imaging, structure from motion and stereo vision from passive imaging and continuous wave laser line scanning (CW-LLS) from laser line scanning.

3.3. Modulation

While the time domain approach uses amplitude and time to discriminate multiple scattered, diffused photons, the frequency domain uses the differences in the amplitude and phase of a modulated signal to perform this task. The diffused photons that undergo many scattering events produce temporal spreading of the transmitted pulse. Only low frequency components are efficiently transmitted, whilst high frequency components are lost. This method has been reported in the literature both from airborne platforms and from underwater vehicles. They usually modulate the amplitude in frequencies in the order of GHz, thus requiring very sensitive sensors and accurate time scales. The receivers are usually photomultiplier tubes (PMT) or, more recently, photon counters made of avalanche photodiodes (APD). These sensors are generally triggered during a time window, and the incoming light is integrated. After the demodulation step, 3D information can be obtained from the phase difference.

It is known that coherent modulation/demodulation techniques at optical frequencies in underwater environments fall apart due to the high dispersion in the sea water path [6], as well as for the different absorption and scattering coefficients depending on the optical wavelength. Because there is a minimum for these coefficients in the blue-green region of the color spectra, amplitude modulation of the laser carrier of these wavelengths is the most used modulation technique in underwater reconstruction.

4. Sensors and Technologies

This section presents all of the sensors studied in this paper. At the end of each subsection, a table is presented indicating the accuracy and resolution values of the references listed, when available. Furthermore, if a value has been obtained from graphic plots, an approximate ($\approx$) symbol has been used.

4.1. Sonar

The term sonar is an acronym for sound, navigation and ranging. There are two major kinds of sonars, active and passive.

Passive sonar systems usually have large sonic signature databases. A computer system uses these databases to identify classes of ships, actions (*i.e.*, the speed of a ship or the type of weapon released) and even particular ships [35–37]. These sensors are evidently not used for 3D reconstructions; thus, they are discarded in this study.

Active sonars create a pulse of sound, often called a ping, and then listen for reflections of the pulse. The pulse may be at constant frequency or a chirp of changing frequency. If a chirp, the receiver correlates the frequency of the reflections to the known signal. In general, long-distance active sonars use lower frequencies (hundreds of kHz), whilst short-distance high-resolution sonars use high frequencies (a few MHz).

In the active sonar category, we can find three major types of sonars: multibeam sonar (MBS), single beam sonar (SBS) and side scan sonar (SSS). If the across track angle is wide, they are usually called imaging sonars (IS). Otherwise, they are commonly named profiling sonars, as they are mainly used to gather bathymetric data. Moreover, these sonars can be mechanically operated to perform a scan, towed or mounted on a vessel or underwater vehicle.

Sound propagates in water faster than in air, although its speed is also related to water temperature and salinity. One of the main advantages of sonar soundings is their long range, making them a feasible sensor to gather bathymetry data from a surface vessel, even for thousands of meters' depth. At this distance, a resolution of tenths of meters per sounding is a good result, whilst if an AUV is sent to dive at an altitude of 40 m to perform a survey, a resolution of a couple of meters or less can be achieved.

One of the clearest examples of bathymetric data gathering is performed using MBS, as in [38]. This sensor can also be correlated to a color camera to obtain not only 3D, but also color information, as in [39], where its authors scan a pool using this method. However, in this case, its range is lowered to the visual available range.

MBS can also be mounted on pan and tilt systems to perform a complete 3D scan. They are usually deployed using a tripod or mounted on top of an ROV, requiring the ROV to remain static while the scan is done, like in [40].

A scanning SBS can carry out a 3D swath by rotating its head [41], as if it were a one-dimensional range sensor mounted on a pan and tilt head. The data retrieval is not as fast as with an MBS.

Profiling can also be done with SSS, which is normally towed or mounted in an AUV to perform a gridded survey. The SSS is mainly used on-board a constant speed vehicle describing straight transects. Even though SSS can be considered as a 2D imaging sonar, 3D information can be inferred from it, as depicted by Coiras et al. in [42].

Imaging sonars (IS) differ from MBS or SBS by a broadened beam angle (e.g., they capture a sonic image of the sea bottom instead of a thin profile). For instance, in [43], Brahim et al. use an imaging sonar with a field of view of 29 (azimuth) × 10.8 (elevation) to produce either 48 × 512 or 96 × 512 azimuth by-range images where each pixel contains the backscattered energy for all of the points in the scene located at the same distance with the same azimuth from the camera.

Other exotic systems have been researched, combining IS with conventional cameras to enhance the 3D output and to better correlate the sonar correspondences. In [44], Negahdaripour uses a stereo system formed by a camera and an imaging sonar. Correspondences between the two images are described in terms of conic sections. In [45], a forward looking sonar and a camera are used, and feature correspondences between the IS and the camera image are provided manually to perform reconstructions. Furthermore, in [46], an SfM approach from a set of images taken from an imaging sonar is used to recover 3D data.

The object shadows in a sonic image can also be used to recover 3D data, as in [47], where Aykin et al. are capable of reconstructing simple geometric forms on simple backgrounds. Its main requirement is that the shadow is distinguishable and that it lays on a known flat surface.

Beamforming (BF) is a technique aimed at estimating signals coming from a fixed steering direction, while attenuating those coming from other directions. When a scene is insonified by a coherent pulse, the signals representing the echoes backscattered from possible objects contain attenuated and degraded replicas of the transmitted pulse. It is a spatial filter that combines linearly temporal signals spatially sampled by a discrete antenna. This technique is used to build a range image from the backscattered echoes, associated point by point with another type of information representing the reliability (or confidence) of such an image. Modeling acoustic imaging systems with BF has also been reported by Murino in [48,49], where an IS of 128 × 128 pixels achieves a range resolution of ±3.5 cm. One pulse of this sonar system covers a footprint of 3.2 × 3.2 m^2.

In [50], Castellani *et al.* register multiple MBS range measurements using global registration (ICP) with an average error of 15 cm.

Kunz *et al.* [51] fuse acoustic and visual information from a single camera, so that the imagery can be texture-mapped onto the MBS bathymetry (binned at 5 cm from 3 m), obtaining three-dimensional and color information.

Table 1 shows a comparison of the 3D reconstruction techniques using sonar.

Table 1. Summary of sonar 3D reconstruction solutions.

References	Sonar Type	Scope	Accuracy	Resolution
Pathak [38]	MBS	Rough map for path planning	$\approx$1 m	2.5 cm
Rosenblum [52]	MBS	Small object reconstruction	-	$\approx$8 cm
Hurtos [39]	MBS + Camera	Projects images on 3D surfaces	2.34 cm	-
Guo [41]	SBS	Small target 3D reconstruction	2.62 cm	-
Coiras [42]	SSS	Seabed elevation with UW pipe	19 cm	5.8 cm
Brahim [43]	IS	Sparse scene geometry	0.5 m	-
Aykin [47]	IS	Smooth surfaces 3D reconstruction	$\approx$15 cm	1 cm
Negahdaripour [44–46]	IS + Camera	Alternative to stereo systems	$\approx$5 cm	-

4.2. Light Detection and Ranging

Airborne scanning light detection and ranging (LiDAR) is widely used as a mapping tool for coastal and near shore ocean surveys. Similar to LLS, but surveyed from an aircraft, a laser line is scanned throughout the landscape and the ocean. Depending on the laser wavelength, LiDAR is capable of recovering both the ocean surface and the sea bottom. In this particular case, a green 532-nm laser that penetrates the ocean water over 30 m [53] is used in combination with a red or infrared laser. Both lasers return the echo from the sea surface, but only one reaches the underwater domain.

LiDAR has been used for underwater target detection (UWTD), usually mines, as well as for coastal bathymetry [54,55]. It is normally surveyed at heights of hundreds of meters (Pellen *et al.* survey mostly uniformly at 300 m [53]) with a swath of 100 to 250 m with a typical resolution in the order of decimeters. In [53], a resolution of 0.7 m is achieved. Moreover, the LiDAR signal can be modulated, enhancing its range capabilities and rejecting underwater backscatter [56,57].

Although this paper focuses on underwater sensors, LiDAR has been briefly mentioned, as it is capable of reconstructing certain coastal regions from the air. In Table **??**, two 3D reconstruction references using LiDAR are compared.

Table 2. Summary of LiDAR 3D reconstruction solutions.

References	Class	Wavelength	LiDAR Model	Combination	Accuracy	Resolution
Reineman [53]	ToF	905 nm	Riegl LMS-Q240i	Camera, GPS	0.42 m	0.5 m
Cadalli [54]	ToF	532 nm	U.S. Navy prototype	PMT + 64 × 64 CCD	-	≈10 m
Pellen [55]	UWTD[1]	532 nm	ND:YAG laser	PMT	-	-
Mullen [56,57]	UWTD[1]	532 nm	ND:YAG laser	PMT + Microwave	-	-

[1] Underwater target detection. No 3D reconstruction.

4.3. Laser Line Scanning

To increase the resolution of the systems exposed above, laser combined with imaging devices can be used. Green lasers working at 532 nm are a common solution as a light source because of their good trade-off between price, availability and low absorption and scattering coefficients in seawater. At the reception side, photomultiplier tubes (PMT) or photon counters can be used, although many approaches also use photodiodes or cameras.

For a larger operational range, preventing the effects of light scattering in the water, some LLS systems send out narrow laser pulses that will be gathered by range gated receivers.

There are three main categories of LLS: continuous wave LLS (CW-LLS), pulse gated LLS (PG-LLS) and modulated LLS. In Table 3, the reader can find a summary of the different LLS 3D reconstruction solutions. In addition to reconstruction, LLS are also used for long-range imaging (from ≈ 7 m). Some additional references are listed in Table 3, as well.

Table 3. Summary of laser line scanning 3D reconstruction solutions.

References	Aim	Type	Wavelength	Receiver	Accuracy	Resolution
Moore [58]	3D	CW-LLS	532 nm	Linescan CCD	-	1 mm
Moore [59]	3D	CW-LLS	532 nm	Linescan CCD	-	3 mm
McLeod [60]	3D	PG-LLS	-	-	7 mm	1 mm
Cochenour [61]	3D	Mod-LLS	532 nm	PMT	-	-
Rumbaugh [62]	3D	Mod-LLS	532 nm	APD	4.5 cm	1 cm
Dominicis [63]	3D	Mod-LLS	405 nm	PMT	5 mm	1 mm
Dalgleish [64]	Img.[1]	CW-LLS	532 nm	PMT	-	-
Dalgleish [64]	Img.[1]	PG-LLS	532 nm	PMT	-	-
Gordon [65]	Img.[1]	PG-LLS	488-514.5 nm	PMT	-	-
Mullen [66]	Img.[1]	Mod-LLS	532 nm	PMT	-	-

[1] The technique is aimed at extended range imaging.

4.3.1. Continuous Wave LLS

This subcategory uses a triangulation method to recover the depth. A camera-based triangulation device using a laser scan concept can be built using a moving laser pointer made of a mirror galvanometer and a line-scan camera, as shown in [58,59].

The geometric relationship between the camera, the laser scanner and the illuminated target spot is shown in Figure 3. The depth D of a target can be calculated from Equation (4).

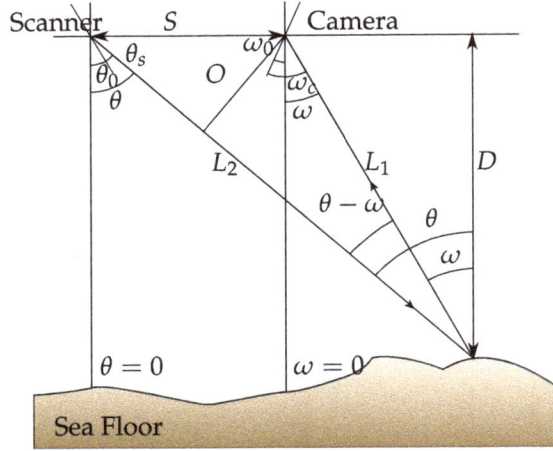

Figure 3. Triangulation geometry principle for a laser scanning system.

$$D = L_1 \cos(\omega) \tag{4}$$

as:

$$L_1 = \frac{S \cos(\theta)}{\sin(\theta - \omega)} \tag{5}$$

since:

$$\sin(\theta - \omega) = \frac{O}{L_1}, \text{ and } O = S \cos(\theta) \tag{6}$$

therefore:

$$D = \frac{S}{\tan(\theta) - \tan(\omega)} \tag{7}$$

where S is the separation (e.g., baseline) between the center of the scanning mirror and the center of the primary receiving lens of the camera (e.g., the center of perspective). Here, θ and ω are the scanning and camera pixel viewing angles, respectively.

The angles ω_0 and θ_0 are the offset mounting angles of the scanner and camera, and θ_s and ω_c are the laser beam angle known from a galvanometer or an encoder and the pixel viewing angle (with respect to the camera housing). Thus,

$$\theta = \theta_0 + \theta_s \tag{8}$$

$$\omega = \omega_0 + \omega_c \tag{9}$$

and:

$$D = \frac{S}{\tan(\theta_0 + \theta_s) - \tan(\omega_0 + \omega_c)} \tag{10}$$

Both θ_0 and ω_0 have to be computed by calibration, so that afterwards, the distance to the target can be computed.

4.3.2. Pulse Gated LLS

This ToF sensor has a simple principle: it illuminates a narrow area with a laser light pulse while keeping the receivers shutter closed. Then, it waits for the return of the light from the object by estimating its distance from the sensor and then opens the shutter so that only the light returning from the target is captured. For instance, in Figure 4, the shutter should have been opened from 80 to 120 ns to get rid of the unwanted backscatter.

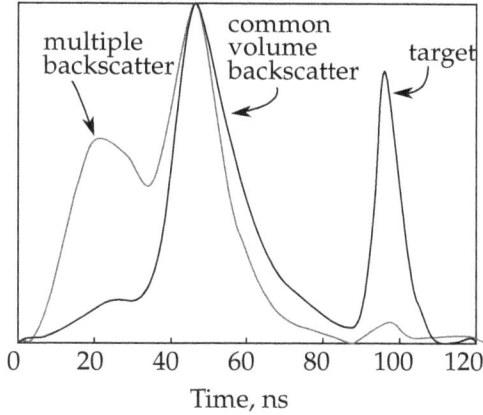

Figure 4. Representative normalized returning signal from an LLS. At higher turbidity (dashed gray line), the backscatter peak is stronger and the target return is weaker. The common volume backscatter is light that has been deflected once, whilst the multiple backscatter has been deflected twice or more times.

This setup has been highly used in extended range imagery. In the early 1990s, the LLS system in [65] was used on the USS Dolphin research submarine and as a towed body to perform high resolution imagery at an extended range. This prototype used an argon ion gas laser, with a high power budget not available for most unmanned vehicles (ROVs or AUVs).

Dalgleish *et al.* [64] compared PG-LLS with CW-LLS as imaging systems. The experimental results demonstrate that the PG imager improved contrast and SNR (signal-to-noise ratio). Their sensor becomes limited by forward backscatter at seven attenuation lengths, whilst CW at six.

In true ToF 3D reconstruction, McLeod *et al.* [60] published a paper about a commercial sensor [67] mounted on the Marlin AUV. Their setup achieves an accuracy of 7 mm in a good visibility scenario, when measuring a point at 30 m.

4.3.3. Modulated LLS

A modulated LLS characterizes the use of the frequency domain, instead of the spatial or time domain, to discern a change in the sent signal. In sonar chirps (radar as well), the modulation and posterior de-modulation of the signal give insight into the distance from the sensor to the target.

As stated before, amplitude modulation is the only realizable modulation in underwater scenarios. The original and the returned signal are subtracted, and the distance is obtained by demodulation of the remainder.

The same approach can be used for extended range imaging, as well, as seen in [66], where Mullen *et al.* have developed a modulated LLS that uses frequency modulation in the laser source in order to identify the distance at which the target has been illuminated. The optical modulation is used to discriminate scattered light. Different frequencies are compared experimentally, finding that a high frequency (90 MHz) reaches further than a lower one (50 MHz or 10 MHz). The setup used by the authors can be seen in Figure 5.

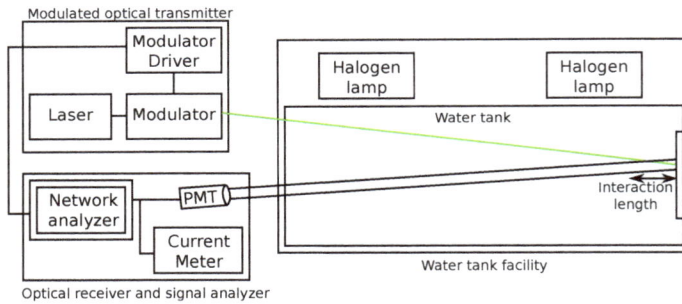

Figure 5. Laser line scanning setup including a modulated optical transmitter, an optical receiver and signal analyzer and a water tank facility. The interaction length is the distance over which the transmitted beam and the receiver field of view overlap. Reproduced from [66].

In [61], different modulation techniques based on ST-MP (single-tone modulated pulse) and PN-MP (pseudorandom coded modulated pulse) are compared for one-dimensional ranging. The results show that in clear water, the PN-MP stands as an improvement over the ST-MP due to the excellent correlation properties of pseudorandom codes.

In [62], a one-axis ranging solution is proposed. Although the authors characterize the solution as LiDAR, their setup is more similar to LLS, and the measurements are not taken from a plane. In the paper, a resolution of 1 cm from a distance of 60 cm is reported. This system could then be swept for a 3D reconstruction and work as a true LLS.

In [63], a simpler approach using an amplitude modulated blue laser (405 nm) at 80 MHz was used, called the MODEM-based 3D laser scanning system, that can reconstruct objects 8.5 meters away within a 5% of error. The system is similar to those described before, but this study focuses on the 3D reconstruction of the object, showing the potential of this technique for long-range underwater reconstruction.

4.4. Structured Light

These systems consist of a camera and a color (or white light) projector. The triangulation principle is used between these two elements and the projected object.

The projector casts a known pattern on the scene, normally a set of light planes, as shown in Figure 6, where both the planes and the camera rays are known. The intersection between them is unknown and can be calculated as follows.

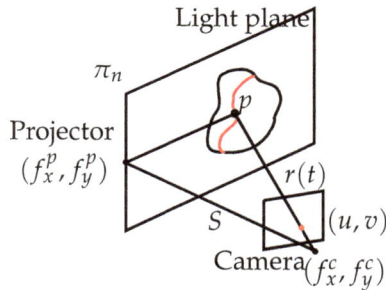

Figure 6. Triangulation geometry principle for a structured light system.

Mathematically, a line can be represented in parametric form as:

$$r(t) = \begin{cases} x &= \frac{u-c_x}{f_x}\,t \\ y &= \frac{v-c_y}{f_y}\,t \\ z &= t \end{cases} \tag{11}$$

where (f_x, f_y) is the camera focal length in the x and y axes, (c_x, c_y) is the central pixel in the image and (u, v) is one of the detected pixels in the image. Supposing a calibrated camera and the origin in the camera frame, the light plane can be represented as in Equation (12).

$$\pi_n : \; Ax + By + Cz + D = 0 \qquad (12)$$

To find the intersection point, Equation (11) is substituted into Equation (12), giving Equation (13).

$$t = \frac{-D}{A\frac{u-c_x}{f_x} + B\frac{v-c_y}{f_y} + C} \qquad (13)$$

Different patterns have been used in the literature [68], even though it is a fact that binary patterns are the most used ones, because they are easy to achieve with a projector and simple to process. Binary patterns use only two states of light stripes in the scene, usually white light. At the beginning, there is only one division (black-to-white) in the pattern. In the following pattern projections, a subdivision of the previous pattern is projected until the software cannot segment two consecutive stripes. The correspondence of consecutive light planes is solved using time multiplexing. The number of light planes achievable with this method is fixed, normally to the resolution of the projector.

Time multiplexing methods are based on the codeword created by the successive projection of patterns onto the object surface (see Figure 7). Therefore, the codeword associated to a position in the image is not completely formed until all patterns have been projected. Usually, the first projected pattern corresponds to the most significant bit, following a coarse-to-fine paradigm. Accuracy directly depends on the number of projections, as every pattern introduces finer resolution in the image. In addition, the codeword basis tends to be small, providing resistance against noise [68].

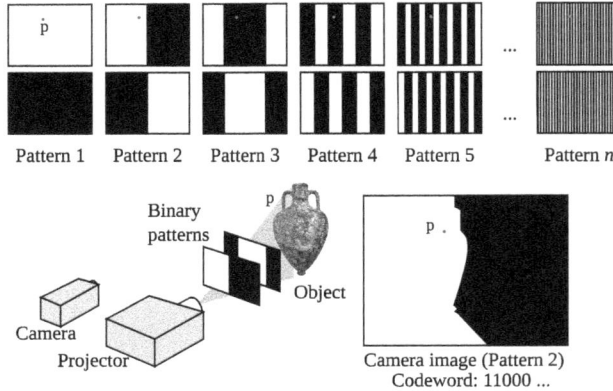

Figure 7. Binary structured light patterns. The codeword of a point p is created by the successive projection of patterns.

On the other hand, phase shifting patterns use sinusoidal projections in the same operating mode to cover wider values in gray scale. By unraveling the phase value, different light planes can be obtained for just one state in the equivalent binary pattern. Phase shifting patterns are also time multiplexing patterns. Frequency multiplexing methods provide dense reconstruction for moving scenarios, but present high sensitivity to the non-linearities of the camera, reducing the accuracy and sensitivity to details on the surface of the target.

These methods use more than one projection pattern to obtain range information. De Bruijn sequences can achieve one-shot reconstructions by using pseudo-random sequences formed by alphabets of symbols in a circular string. If this theory is brought to matrices instead of vectors (e.g., strings), then those patterns are called M-arrays. These can be constructed by following a pseudo-random sequence [69]. Usually, these patterns use color to better distinguish the symbols in the alphabet. However, not all kinds of surface finishes and colors reflect correctly the incoming color spectra back to the camera [70,71]. One-shot coded patterns have also been used in air. However, to the best knowledge of the authors, there are no reports of these codification strategies in underwater scenarios.

In the literature, Zhang *et al.* project a grey scale four-step sinusoidal fringe [72]. Therefore, the pattern is a time multiplexing method using four different patterns. In their article, SL is compared to SV showing better behavior in SL on textureless objects. Similar results were obtained by Törnblom, projecting 20 different grey coded patterns in a pool [73]. An accuracy in the z direction of 2% was achieved with this system.

Bruno *et al.* [70] also project gray coded patterns with a final code shift of four pixel-wide bands. With these last shifts, better accuracy can be obtained compared to narrowing the pattern to only one pixel-wide patterns, where finding all of the thin black and white lines is more difficult. In this setup, a total of 48 patterns were used. However, this particular setup calculates the 3D points using the positions of two cameras determined during the calibration phase. The projector is used to illuminate the scene, whilst depth is obtained from the stereo rig. Thus, no lens calibration is needed for the projector, and any commercially-available projector can be used without compromising the accuracy of the measurements. This system would be a hybrid between SL and SV.

Another way to triangulate information using structured light is to sweep a light plane. This light plane can be swept either using the available pixels in the projector or by moving the projector. Narasimhan and Nayar [74] sweep a light plane into a tank with diluted milk and recover 3D information even in high turbidity scenarios where it is impossible to see anything but backscattering when using conventional floodlights. By narrowing the illuminated area to a light plane, the shapes of the objects in the distance can be picked out and therefore triangulated.

The use of infrared projectors, such as Kinect, has also been tested underwater [75]. The attempt confirmed that the absorption of the infrared spectrum is too strong to reach distances greater than a few centimeters.

Laser-Based Structured Light Systems

The systems presented in this section project laser light into the environment. Laser stripe (LS) systems are a subgroup of laser-based structured light systems (LbSLS), where although the pattern is fixed to be a line (a laser plane), the projector is swept across the field of view of the camera. Thus, for this setting, a motorized element is needed in addition to the laser if the system holding the camera and the laser is not moving. The relative position and orientation of the laser and camera system must be known in order to perform the triangulation process. The resolution of these systems is usually higher than stereoscopy, but they are still limited by absorption and scattering. The range of LS does not normally go over 3 m in clear waters [76], as will be seen later in the commercial solutions.

According to Bodenmann [77], the attenuation of light is significantly more pronounced in water than in air or in space, and so in order to obtain underwater images in color, it is typically necessary to be within 2 to 3 m of the seafloor or the object of interest. Moreover, these are some of the reported ranges for LS: 3 m for Inglis [76], 250 mm for Jakas [78] and 2 m for Roman [79].

Using an underwater stripe scanning system was initially proposed by Jaffe and Dunn in [80] to reduce backscattering. Tetlow and Spours [81] show in their article a laser stripe system with an automatic threshold setup for the camera, making this sensor robust to pixel saturation if the laser reflection is too strong. To that end, they programmed a table with the calibrated thickness of the laser stripe depending on the distance to the target, achieving resolutions of up to five millimeters at a distance of three meters.

Kondo et al. [82] tested an LS system in the Tri-Dog I AUV. Apart from using it for 3D reconstruction, they also track the image in real time to govern the robot. To keep a safe distance from the seabed, they center the laser line in the camera image by changing the depth of the vehicle. They report a resolution of 40 mm at three meters.

Hildebrandt et al. [83] mount a laser line onto a servomotor that can be rotated $45°$ with an accuracy of $0.15°$. The camera is a 640×480 CMOS shooting at 200 frames per second (fps) with a $90°$ HFOV (horizontal field of view). The system returns 300k points in 2.4 seconds. Calibration is made in his article with a novel rig consisting of a standard checkerboard next to a grey surface on one side. The laser is better detected on a grey surface. On a white surface, light is strongly reflected, and the camera has to compensate for the vast amount of light by shortening the exposure time. The detection of the laser in the same plane of the calibration pattern is used to calculate the position of the laser sheet projector with respect to the camera.

In [84], a system consisting of a camera, a laser line and an LED light are mounted on the AUV Tuna Sand to gather 3D information, as well as imagery. The laser is pointed at the upper part of the image, whilst the lighting is illuminating the lower part. Therefore, there is enough contrast to detect the laser line. In [77,85,86], a similar system, called SeaXerocks (3D mapping device), is mounted on the ROV Hyper-Dolphin. With this system, the authors perform 3D reconstructions in real intervention scenarios, such as in hydrothermal sites and shipwrecks.

In [87], the Tuna Sand AUV is used with a different sensor. In this case, a camera and a motorized laser stripe are mounted in two independent watertight housings. By keeping the robot as static as possible, the laser is projected onto the scene whilst rotating it. Then, the camera captures the line deformation, from which the 3D information is recovered. In this paper, multiple laser scans from sea experiments at Kagoshima Bay are combined using the iterative closest point (ICP) algorithm. The reconstructed chimney is three meters tall at a 200-meter depth.

In [63,78], Jakas and Dominicis use a dual laser scanner to increase the field of view of a single laser stripe. The reported horizontal field of view is 180°. The system is very similar to the commercial sensor in [88]. They approximate the detected laser lines to be Gaussian and explain an optimization method to calibrate the camera-to-laser transformation. The authors claim that the achieved measuring error is below 4%.

Prats *et al.* [89–91] mount a camera fixed to the AUV Girona 500 frame and a laser stripe on an underwater manipulator carried by the vehicle. The stripe sweeps the scene by means of the robot arm, and the resulting point cloud is used to determine the target grasping points. The sea bottom is tracked to estimate the robot motion during the scanning process, so small misalignments between the data can be compensated.

Different approaches to the common laser stripe scanning have also been reported. In [92], two almost-parallel laser stripes are projected to compute the distance between these lines captured from a camera, to know the distance to the target. These values are used as an underwater rangefinder. However, 3D reconstruction was not the aim of the research.

In [93], Caccia mounts four laser pointers lined with a camera in an ROV. The four imaged pointers are used to calculate the altitude and the heading of the vehicle, assuming the seabed is flat.

Yang *et al.* mount a camera and a vertical laser stripe in a translation stage [94]. They recover 3D data interpolating from a data table previously acquired from calibration. Whenever a laser pixel is detected in the image, its depth value is calculated from the four closest points in the calibration data.

Massot and Oliver [95–97], designed a laser-based structured light system that enhances simpler laser stripe approaches by using a diffractive optical element

(DOE) to enhance a simple laser pointer, shaping the beam into 25 parallel lines, called a laser-based structured light (LbSL) system. The pattern is projected on the environment and recovered by a color camera. In one camera shot, this solution is capable of recovering sparse 3D information, as seen in Figure 8, whilst with two or more shots, denser information can be obtained. The system is targeted at underwater autonomous manipulation stages where a high density point cloud of a small area is needed, and during the manipulation, a one-shot and fast reconstruction aids the intervention.

In Table 4, the different SL references are compared. For the solutions with no clear results, the resolution has been deduced from the graphics in their respective articles.

Table 4. Summary of structured light 3D reconstruction solutions.

References	Type	Color/Wavelength	Pattern	Accuracy	Resolution
Zhang [72]	SL	Grayscale	Sinusoidal Fringe	≈1 mm	-
Tornblom [73]	SL	White	Binary pattern	4 mm	0.22 mm
Bruno [70]	SL	White	Binary pattern	0.4 mm	0.3 mm
Narasimhan [74]	SL	White	Light plane sweep	9.6 mm	-
Bodenmann [84,85]	LS	532 nm	Laser line	-	-
Yang [94]	LS	532 nm	Laser line	-	
Kondo [82]	LS	532 nm	Laser line	-	≈1 cm
Tetlow [81]	Mot. LS	532 nm	Laser line	1 cm	5 mm
Hildebrandt [83]	Mot. LS	532 nm	Laser line	-	-
Prats [89]	Mot. LS	532 nm	Laser line	≈1 cm	-
Nakatani [87]	Mot. LS	532 nm	Laser line	≈1 cm	-
Jakas [63,78]	Dual LS	405 nm	Laser line	See [88]	≈1 cm
Massot [96]	LbSL	532 nm	25 laser lines	3.5 mm	-

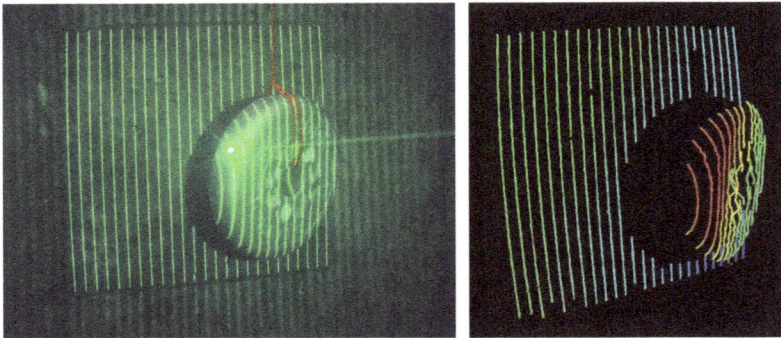

(a) Captured image with laser pattern. (b) One shot 3D reconstruction, lateral view.

Figure 8. 3D reconstruction of a 1-kg plate using LbSLS from [95].

4.5. Photometric Stereo

In situations where light stripe scanning takes too long to be practical, photometric stereo provides an attractive alternative. This technique for scene reconstruction requires a small number of images captured under different lighting conditions. In Figure 9, there is a representation of a typical PhS setup with four lights.

3D information can be obtained by changing the location of the light source whilst keeping the camera and the object in a fixed position. Narasimhan and Nayar present a novel method to recover albedo, normals and depth maps from scattering media [74]. Usually, this method requires a minimum of five images. In special conditions, such as the ones presented in [74], four different light conditions can be enough.

In [98], Tsiotsios *et al.* show that three lights are enough to compute tridimensional information. They also compensate the backscatter component by fitting a backscatter model for each pixel.

Like in time multiplexing SL techniques, PhS also suffers from long acquisition times; hence, these techniques are not suitable for moving objects. However, the cited references report them to be effective in clear waters for close range static objects.

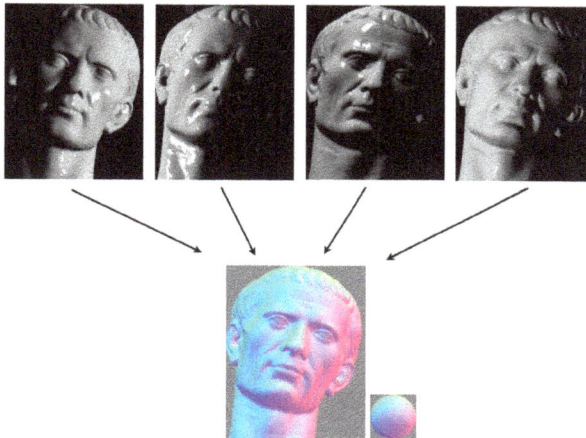

Figure 9. Photometric stereo setup: four lights are used to illuminate an underwater scene. The same scene with lighting from different sources results in the images used to recover three-dimensional information [99].

4.6. Structure from Motion

SfM is a triangulation method that consists of taking images of an object or scene using a monocular camera. From these camera shots, image features are detected and matched between consecutive frames to know the relative camera motion and, thus, its 3D trajectory.

First, suppose a calibrated camera, where the principal point and calibration are known, as well as lens distortion and refractive elements to ensure an accurate 3D result.

Given m images of n fixed 3D points, then m projection matrices $\mathbf{P}_i$ and n 3D points $\mathbf{X}_j$ from the $m \cdot n$ correspondences $\mathbf{x}_{ij}$ are to be estimated.

$$\mathbf{x}_{ij} = \mathbf{P}_i \mathbf{X}_j, \ i = 1, \ldots, m, \ j = 1, \ldots, n \tag{14}$$

Therefore, if the entire scene is scaled by some factor k and, at the same time, the projection matrices by a factor of $1/k$, the projection of the scene points remain the same. Thus, only with SfM, the scale is not available, although there are methods that compute it from known objects or by knowing the constraints of the robot carrying the camera.

$$\mathbf{x} = \mathbf{PX} = \left(\frac{1}{k}\mathbf{P}\right)(k\mathbf{X}) \tag{15}$$

The one-parameter family of solutions parametrized by λ is:

$$\mathbf{X}(\lambda) = \mathbf{P}^+ \mathbf{x} + \lambda \mathbf{c} \tag{16}$$

where $\mathbf{P}^+$ is the pseudo-inverse of $\mathbf{P}$ (*i.e.*, $\mathbf{PP}^+ = \mathbf{I}$) and $\mathbf{c}$ is its null-vector, namely the camera center, defined by $\mathbf{Pc} = 0$.

The approach of SfM is the least expensive in terms of hardware and the easiest to install in a real robot. Only a still camera or a video recorder is needed, with enough storage to keep a full dive in memory. Later, the images can be processed to obtain the required 3D models.

In the underwater medium, both feature detection and matching suffer from diffusion, non-uniform light and, eventually, sun flickering, making the detection of the same feature more difficult from different viewpoints. Depending on the distance from the camera to the 3D point, the absorption and scattering components vary, changing the colors and the sharpness of that particular feature in the image. More difficulties arise if images are taken from the air to the ocean [100].

Sedlazeck *et al.* show in [101] a real 3D scenario reconstructed from the ROV Kiel 6000 using an HD color camera. Features are selected using a corner detector based on image gradients. Later, the RANSAC [102] procedure is used to filter outliers after the features have been matched.

Pizarro *et al.* [103] use the SeaBED AUV to perform optical surveys, equipped with a 1280 × 1024 px CCD camera. The feature detector used is a modified Harris corner detector, and its descriptor is a generalized color moment.

In [104], Meline *et al.* compare Harris and SIFT features using a 1280 × 720 px camera in shallow water. In the article, the authors reconstruct a statue bust. They conclude that SIFT is not robust to speckle noise, contrary to Harris. Furthermore, Harris presented a better inlier count in the different scenarios.

McKinnon et al. [105] use GPU SURF features and a high resolution camera of 2272×1704 px to reconstruct a piece of coral. This setup presents several challenges in terms of occlusions of the different views. With their SfM approach, they achieve 0.7 mm accuracy at 1 to 1.5 m.

Jordt-Sedlazeck and Koch develop a novel refractive structure from motion algorithm that takes into account the refraction of glass ports in water [106]. By considering the refraction coefficient between the air-glass-water interface, their so-called refractive SfM improves the results of generic SfM.

Cocito et al. [107] use images captured by divers that always contain a scaling cube to recover scaled 3D data. The processing pipeline requires an operator to outline silhouettes of the area of interest of the images. In the case of the application in that paper, they were measuring bryozoan colonies' volume.

In [108], the documentation of an archaeological site where experimental cleaning operations were conducted is shown. A commercial software, Photoscan by Agisoft, was used to perform a multi-view 3D reconstruction.

Nicosevici et al. [109] use SIFT features in a robotics approach, with an average error of 11 mm.

Ozog et al. [110] reconstruct a ship hull from an underwater camera that also acts as a periscope when the vehicle navigates on surface. Using SLAM and a particle filter, they achieve faster execution times (compared to FabMap). The error distribution achieved has a mean of 1.31 m and a standard deviation of 1.38 m. However, using planar constraints, they reduced the mean and standard deviation to 0.45 and 0.19 m, respectively.

The solutions presented are summarized in Table 5. Known reference distances must be visible in the images to recover the correct scale. In the solutions where a result is given, the authors have manually scaled the resulting point cloud to match a particular feature or human-made object.

Table 5. Summary of structure from motion 3D reconstruction solutions.

References	Feature	Matching Method	Accuracy	Resolution
Sedlazeck [101]	Corner	KTL Tracker	-	-
Pizarro [103]	Harris	Affine invariant region	3.6 cm	-
Meline [104]	Harris	SIFT	-	-
McKinnon [105]	SURF	SURF	0.7 mm	-
Jordt-Sedlazeck [106]	-	KLT Tracker	-	-
Cocito [107]	Silhouettes	Manually	$\approx$1 cm	-
Bruno [108]	SIFT	SIFT	4.5 mm	-
Nicosevici [109]	SIFT	SIFT	11 mm	-
Ozog [110]	SIFT	SIFT	0.45 m	-

4.7. Stereo Vision

Stereoscopy follows the same working principle as SfM, but features are matched between left and right frames of a stereo camera to compute 3D correspondences. Once a stereo rig is calibrated, the relative position of one camera with respect the other is known, and therefore, the scale ambiguity is solved.

The earliest stereo matching algorithms were developed in the field of photogrammetry for automatically constructing topographic elevation maps from overlapping aerial images. In computer vision, the topic of stereo matching has been widely studied [111–115], and it is still one of the most active research areas.

Suppose two cameras C_L and C_R and two similar features F_L and F_R in each camera image. To compute the 3D coordinates of the feature F, whose projection in C_L is F_L and in C_R is F_R, we trace a line L_L that crosses C_L focal point and F_L and another line L_R that crosses C_R focal point and F_R. If both cameras' calibration are perfect, $F = L_L \cap L_R$. However, as camera calibration is usually solved by least squares, the solution is not always perfect. Therefore, the approximate solution is taken as the closest point between L_L and L_R [116].

By knowing the relative position of the cameras and the location of the same feature in both images, the 3D coordinates of the feature in the world can be computed by triangulation. In Figure 10, the corresponding 3D point of the image coordinates $x = (u_L, v_L)$ and $x' = (u_R, v_R)$ is the point $p = (x^W, y^W, z^W)$, which can also be written as $x'Fx = 0$ where F is the fundamental matrix [116].

Once the camera rig is calibrated (known baseline, relative pose of the cameras and no distortion in the images), 3D imaging can be obtained calculating the disparity for each pixel, e.g., perform a 1D search for each pixel in the left and right images, where block matching is normally used. The disparity is the difference in pixels from the left to the right image, where the same patch has been found; so, the depth z is given by:

$$z = \frac{f \cdot b}{d} \tag{17}$$

where d is the disparity in pixels, f is the focal distance in pixels, b is the baseline in meters and z is the depth or distance of the pixel perpendicular to the image plane, in meters.

Once these 3D data have been gathered, the registration between consecutive frames can be done using 2D or 3D features or even 3D registration methods, such as ICP.

Fairly different feature descriptors and matchers have been used in the literature. SIFT [117–122] is one of the most used, as well as SURF [123], or even direct 3D registration with SIFT 3D [118] or ICP [117]. For instance in [124], Servos *et al.* perform refractive projection correction on depth images generated from a Bumblebee2 camera (12-cm baseline). The results obtained with this correction have better accuracy and more pixel correspondences, compared to standard methods. The registration is directly done in the generated point cloud using ICP.

Schmidt *et al.* [120] use commercial GoPro cameras to set a 35-mm baseline stereo rig and perform micro bathymetry using SIFT features. They achieve a resolution of 3 mm in their reconstructions.

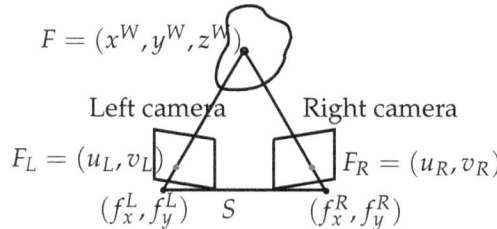

Figure 10. Triangulation geometry principle for a stereo system.

In [122], the stereo system IRIS is hung from the tip of the arm of the Victor6000 ROV. The system uses SIFT combined with RANSAC to discard outliers. After that, a sparse bundle adjustment is performed to correct the navigation to survey natural underwater objects.

In [125], Hogue *et al.* combine a Bumblebee stereo and a inertial unit housed in a watertight case, called Aquasensor. This system is used to reconstruct and register dense stereo scenes. The reconstruction shows high drift if the IMU is not used; thus, an erroneous camera model is assumed to be the cause of this inaccuracy. The system is used by the authors to perform a reconstruction of a sunken barge.

Beall *et al.* [123] use a wide baseline stereo rig and extract SURF features from left and right image pairs. They track these features to recover the structure of the environment after a SAM (smoothing and mapping) step. Then, the 3D points are triangulated using Delaunay triangulation, and the image texture is mapped to the mesh. This setup is applied to reconstruct coral reefs in the Bahamas.

Negre *et al.* [126,127] perform 3D reconstruction of underwater environments using a graph SLAM approach in a micro AUV equipped with two stereo rigs. In Figure 11, a 3D reconstruction of Santa Ponça Bay is displayed, covering an area of 25 × 10 m.

Johnson-Roberson *et al.* [128] studied the generation and visualization of large-scale reconstructions using stereo cameras. In their manuscript, image blending techniques and mesh generation are discussed to improve visualization by reducing the complexity of the scene in proportion to the viewing distance or relative size in screen space.

Fused stereoscopy and MBS have been reported in [129]. There, Galceran *et al.* provide a simultaneous reconstruction of the frontal stereo camera and the downwards-looking MBS.

Another example of this set of sensors is shown by Gonzalez-Rivero [130], where its output is used to monitor a coral reef ecosystem and to classify the different types of corals.

Figure 11. 3D reconstruction from SV using graph SLAM (25 × 10 m, Mallorca) [127,131].

Nurtantio *et al.* [119] use three cameras and extract SIFT features. The reconstruction of the multi-view system is triangulated using Delaunay triangulation. However, they manually preprocess the images to select whether they are suitable for an accurate reconstruction. The outlier removal stage is also manual.

Inglis and Roman constrain stereo correspondences using multibeam sonar [132]. From the Hercules ROV, navigation data, multibeam and stereo are preprocessed to reduce the error, and then, the sonar and optical data are mapped into a common coordinate system. They back project the range data coming from the sonar to the camera image and limit the available z correspondence range for the algorithm. To simplify this approach, they tile the sonar back projections into the image and generate tiled minimum and maximum disparity values for an image region (e.g., a tile). The number of inliers obtained with this setup increases significantly compared to an unconstrained system.

In Table 6, the different solutions are presented and compared.

Table 6. Summary of stereoscopy 3D reconstruction solutions.

References	Feature	Matching Method	Baseline	Accuracy	Resolution
Kumar [117]	SIFT	RANSAC and ICP	-	-	-
Jasiobedzki [118]	SIFT	SIFT3D and SLAM	-	-	-
Nurtantio [119]	SIFT	SIFT	8 and 16 cm	-	-
Schmidt [120]	SIFT	SIFT	35 mm	-	3 mm
Brandou [122]	SIFT	SIFT	-	-	-
Beall [123]	SURF	SURF and SAM	60 cm	-	-
Servos [124]	-	ICP	12 cm	26.4 cm	-
Hogue [125]	Corners	KLT tracker	12 cm	2 cm	-
Inglis [132]	SIFT	SIFT	42.5 cm	-	-

4.8. Underwater Photogrammetry

It is commonly accepted that photogrammetry is defined as the science or art of obtaining reliable measurements by means of photographs [133]. Therefore, any practical 3D reconstruction method that uses photographs (e.g., imaging-based methods) to obtain measurements are photogrammetric methods. Photogrammetry comprises methods of image measurement and interpretation often shared with other scientific areas in order to derive the shape and location of an object or target from a set of photographs. Hence, techniques such as structure from motion and stereo vision belong to both photogrammetric and computer vision communities.

In photogrammetry, it is common to set up a camera in a large field looking at distant calibration targets whose exact location has been precomputed using surveying equipment. There are different categories for photogrammetric applications depending on the camera position and object distance. For example, aerial photogrammetry is normally surveyed at a height of 300 m [134].

On the other hand, close-range photogrammetry applies to objects ranging from 0.5 to 200 m in size, with accuracies under 0.1 mm and around 1 cm at each end. In a close-range setup, the cameras observe a specific volume where the object or area to reconstruct is totally or partially in view and has been covered with calibration targets. The location of these targets can be known as before or calculated after the images have been captured if their shape and dimensions are known [134].

Image quality is a very important topic in photogrammetry. One of the main important fields of this community is camera calibration, a topic that has already been introduced in Section 2.1. If absolute metric accuracy is required, it is imperative to pre-calibrate the cameras using one of the techniques previously mentioned and to use ground control points to pin down the reconstruction. This is particularly true for classic photogrammetry applications, where the reporting of precision is almost always considered mandatory [135].

Underwater reconstructions can also be referred to as underwater photogrammetric reconstructions when they have a scale or dimension associated with the objects or pixels of the scene (e.g., if the resulting 3D model is metric) and if the data were gathered using cameras.

According to Abdo *et al.* [136], an underwater photogrammetric system for obtaining accurate measurements of complex biological objects needs to: (1) be suitable for working in restrictive spaces; (2) allow one to investigate relatively large areas carried out on one or numerous organisms; (3) admit the acquisition of data easily, performed *in situ* and efficiently; and (4) provide a measurement process that is easy to perform, precise, accurate and accomplished in a reasonable time lapse.

The most accurate way to recover structure and motion [137] is to perform robust non-linear minimization of the measurement (re-projection) errors, which is commonly known in the photogrammetry communities as bundle adjustment [28]. Bundle adjustment is now the standard method of choice for most structure-from-motion problems and is commonly applied to problems with hundreds of weakly calibrated images and tens of thousands of points. In computer vision, it was first applied to the general structure from motion problem and then later specialized for panoramic image stitching [28].

Image stitching originated in the photogrammetry community, where more manually-intensive methods based on surveyed ground control points or manually registered tie points have long been used to register aerial photos into large-scale photo-mosaics [23]. The literature on image stitching dates back to work in the photogrammetry community in the 1970s [138,139].

Underwater photogrammetry can also be associated with other types of measures, such as the measure of biological organisms' volumes with 3D reconstruction using an stereo pair [136], the sustainability of fishing stocks [140], examining spatial biodiversity, counting fish in aquaculture [141], continuous monitoring of sediment beds [142] or to map and understand seabed habitats [13,21].

Zhukovsky *et al.* [143] reconstruct an antique ship, similar to [144]. In [32], Menna *et al.* reconstruct the sunken vessel Costa Concordia using photogrammetric targets to reconstruct and assess the damaged hull.

Photogrammetry is also performed by fusing data from diverse sensors, such as in [145], where chemical sensors, a monocular camera and an MBS are fused in an archaeological investigation, and in [146], where a multimodal topographic model of Panarea Island is obtained using a LiDAR, an MBS and a monocular camera.

Planning a photogrammetric network with the aim of obtaining a highly-accurate 3D object reconstruction is considered as a challenging design problem in vision metrology [147]. The design of a photogrammetric network is the process of determining an imaging geometry that allows accurate 3D reconstruction. There are very few examples of the use of a static deployment of cameras working as underwater photogrammetric networks [148] because this type of approach is not readily adapted to such a dynamic and non-uniform environment [149].

In [150], de Jesus *et al.* show an application of photogrammetry for swimming movement analysis with four cameras, two underwater and two aerial. They use a calibration prism composed of 236 markers.

Leurs *et al.* [151] estimate the size of white sharks using a camera and two laser pointers, with an accuracy of ± 3 cm from a distance of 12 m.

Different configurations to monocular or stereo camera systems have also been reported. In [152], Brauer *et al.* use a stereo rig and a projector (SL). Using fringe projection, they achieve a measurement field of 200×250 mm and a resolution of 150 μm.

In [153], Ekkel *et al.* use a stereo laser profiler (four cameras, two for positioning with targets and two for laser triangulation) using a 640-nm laser. They report an accuracy of 0.05 mm in the object plane.

5. Commercial Solutions

There exist different commercial solutions for gathering 3D data or to help with calculating it. In Table 7, a selection of alternatives is shown.

Teledyne sells an underwater LLS called INSCAN [154]. This system must be deployed underwater or fixed to a structure. The device samples 1 m^2 in 5 s at a 5-m range.

SL1 is a similar device from 3D at Depth [67]. In fact, this company worked with Teledyne in this design [155], and the specifications of these two pieces of equipment are quite close.

3DLS is a triangulation sensor formed by an underwater dual laser projector and a camera. It is produced by Smart light devices and uses a 15-W green laser.

2G Robotics has three models of triangulation-based laser scanners fitting different ranges [156–158]. These are motorized solutions, so they must be deployed and static during their scan.

Savante provides three products. Cerberus [159] is a triangulation sensor formed by a laser pointer and a receiver, capable of recovering 3D information. SLV-50 [160] is another triangulation sensor formed by a laser stripe and a high sensitivity camera, and finally, Lumeneye [161] is a laser stripe that only casts laser light on the scene.

Tritech provides (similar to Savante) a green laser sheet projector called SeaStripe [162]. The 3D reconstruction must be performed by the end-user camera and software.

111

Table 7. Available commercial solutions to perform 3D reconstruction.

Commercial Solutions Name	Company	Range (m) Min	Max	Depth (m)	Resolution (mm)	Field of view (deg)	Motorized	Method
INSCAN [154]	Teledyne CDL	2	25	3000	5	30 × 30 × 360	yes	TOF
SL1 [67]	3D at Depth	2	30	3000	4	30 × 30 × 360	yes	TOF
3DLS [88]	Smart Light Devices	0.3	2	4000	0.1	-	-	Triangulation
ULS-100 [156]	2g Robotics	0.1	1	350	1	50 × 360	yes	Triangulation
ULS-200 [157]	2g Robotics	0.25	2.5	350	1	50 × 360	yes	Triangulation
ULS-500 [158]	2g Robotics	1	10	3000	3	50 × 360	yes	Triangulation
Cerberus [159]	Savante	-	10	6000	-	-	-	Triangulation
SLV-50 [160]	Savante	-	2.5	6000	1	60	no	Triangulation
Lumeneye [161]	Savante	-	-	6500	-	65	no	Laser only
SeaStripe [162]	Tritech	-	-	4000	-	64	no	Laser only

6. Conclusions and Prospects

The selection of a 3D sensing system to be used in underwater applications is non-trivial. Basic aspects that should be considered are: (1) the payload volume, weight and power available, in case the system is an on-board platform, (2) the measurement time, (3) the budget and (4) the expected quality of the data gathered. Regarding the quality, optical sensors are very sensitive to water turbidity and surface texture. Consequently, factors, such as the target dimensions, surface, shape or accessibility, may influence the choice and adaptiveness of the sensor to the reconstruction problem. Table 8 presents a comparison of the solutions surveyed in this article according to its typical operative range, resolution, ease of use, relative price and its suitability to be used on different platforms.

Underwater 3D mapping has been historically carried out by means of acoustic multibeam sensors. In that case, the information is normally gathered as an elevation map, and more recently, color and texture can be added afterwards from photo-mosaics, if available.

Color or texture information must be acquired using cameras operating at relatively short distances (<5 m, typically) and with a low cruise speed. In general, mono-propeller AUVs are not appropriate for optical imaging applications, because they cannot slow down their speed as required by the optical equipments. On the other hand, hovering vehicles are suitable for imaging-based sensors, as they can adjust their velocity to the sensors' needs. In some particular cases, even divers can be a choice.

Optical mapping can also be accomplished with only SfM and, as industrial ROVs most often incorporate a video camera, it is feasible to record the needed images and reconstruct an entire scene (see Campos *et al.* [163], for example). However, these reconstructions lack a correct scale, and they are computationally demanding. If, instead, a stereo rig is used, SV techniques can be applied and can solve the scale problem.

According to Bruno, SV is the easiest way to obtain the depth of a submarine scene [70]. These passive sensors are widely used because of their low cost and

112

simplicity. Similarly to SfM, SV needs textured scenes to achieve satisfactory result, giving rise to missing parts corresponding to untextured regions in the final reconstruction.

Table 8. Strengths and weaknesses of the sensors and techniques for 3D reconstruction.

3D technique	Range	Platform	Resolution	Ease of assembly	Price
MBS	<11,000 m	V,[1] T,[2] ROV, AUV	Low	Intermediate	High
SBS	< 6000 m	V, ROV, AUV	Low	Intermediate	High
SSS	< 150 m	T, AUV	Low	Intermediate	High
IS	< 150 m	V, T, ROV, AUV	Low	Intermediate	High
LiDAR	<20 m	Aerial	Low		High
CW-LLS	< 10 m	ROV	Intermediate	Low	High
PG-LLS	< 10 m	ROV	Intermediate	Low	High
Mod. LLS	< 10 m	ROV	Intermediate	Low	High
SfM	< 3 m	ROV, AUV	Intermediate	High	Low
SV	< 3 m	ROV, AUV	Intermediate	Intermediate	Low
PhS	< 3 m	ROV	Intermediate	Intermediate	Low
VW-SL	< 3 m	ROV, AUV	High	Intermediate	Intermediate
CW-SL	< 10 m	ROV, AUV	High	Intermediate	Intermediate

[1] Vessel; [2] Towed.

To overcome the above-mentioned problems of SfM and SV and trying to increase the resulting resolution, SL uses light projection to cast features on the environment. These sensors are capable of working at short distances with high resolution, even for objects without texture. The drawback, compared to SV, is a slower acquisition time caused by the need to move the projection atop the scene or even to use different patterns. The acquisition time is a relevant problem that limits the use of SL systems in real conditions where the relative movement between the sensor and the scene can give rise to reconstruction errors.

In addition, acquiring data from dark objects using SL is, in general, strongly influenced by illumination and contrast conditions [70]. Shiny objects are also challenging for SL, because the reflected light may mislead the pattern decoder. Moreover, due to the large illuminated water volume, this technique is strongly affected by scattering, reducing its range.

To minimize absorption, as well as common volume scattering, LbSL systems take advantage of selected wavelength sources in the green-blue region of the spectrum, extending their capable range. For an improved reduction of the scattering effects, the receiver window can be narrowed as in LLS sensors; even more, the emitter and the receiver can also be pulse gated [64], even though this strategy can be limited by a contrast decline.

On the other hand, when a precise and closer look at an object or structure is needed, LLS technology is not always suitable, as it has a large minimum measuring distance.

Amongst optical solutions, laser-based sensors present a good trade-off between cost and accuracy, as well as an acceptable operational range. Accordingly, regarding the foreseeable future, more research on laser-based structured light and on laser line scanning underwater is needed. These new devices should be able to scan while the sensor is moving, just like MBS, so software development and enhanced drivers are also required.

Another challenge for the future is to develop imaging systems that can eliminate or reduce scattering while imaging. Solutions such as pulse gated cameras and laser emitters are effective [164], but still expensive.

Overall, it is quite clear that no single optical imaging system fits all of the 3D reconstruction needs, covering very different ranges and resolutions. Besides, it is important to point out the lack of systematic studies to compare, with as much precision as possible, the performance of different sensors on the same scenario and conditions. One of these studies is authored by Roman *et al.* [79], who compared laser-based SL to SV and MBS, mapping a small area of a real underwater scenario using an ROV. In that case, the stereo data showed less definition than the sonar and the SL. The comparison was made during a survey where laser images were collected at 3 Hz, at a speed of 2 to 5 cm/s from 3 m above the bottom, whilst stereo imagery was captured on a separate survey at 0.15 Hz at a speed of 15 cm/s and a distance of 1.5 to 3 m, giving a minimum overlap of 50%. MBS was captured during the laser survey at 5 Hz. As seen in these numbers, a different data rate induces less or more spatial resolution. Nonetheless, Roman *et al.* concluded that SL offers a high resolution mapping capability, better than SV and MBS for close-range reconstructions, such as the investigation of archaeological sites.

Massot *et al.* in [96] provide a systematic analysis comparing SV and LbSL in a controlled environment. To that end, a robot arm is used to move the sensors describing a precise path, surveying a 3×2 m underwater scene created in a water tank containing different objects of known dimensions. Apart from other numerical details, the authors conclude that for survey missions, stereo data may be enough to recover the overall shape of the underwater environment, whenever there is enough texture and visibility. In contrast, when the mission is aimed at manipulation and precise measurements of reduced areas are needed, LbSL is a better option.

It would be advisable to work on similar approaches to the aforementioned for the near future, contributing to a better knowledge of each individual sensor behavior when used in diverse situations and applications and also to the progress in multisensor data integration methodologies.

Table 9. Strengths and weaknesses of the sensors and techniques for 3D reconstruction.

Technology	Strength	Weakness
MBS	Early adopted Long range and coverage Independent of water turbidity	High cost High minimum distance Low resolution
SBS	Early adopted Long range Independent of water turbidity	High cost Echoes Low resolution
SSS	Good data acquisition rate Independent of water turbidity Long range	High cost Needs constant speed Unknown dimension
IS	Medium to large range Independent of water turbidity	High cost Unknown dimension
LiDAR	Not underwater	Limited to first 15 meters Safety constraint
LLS	Medium data acquisition rate Medium range Good performance in scattering waters	High cost Safety constraint
SfM	Simple and inexpensive High accuracy on well-defined targets Close range	Computation demanding Sparse data covering Needs textured scenes Unknown scale
SV	Simple and inexpensive High accuracy on well-defined targets Close range	Computation demanding Sparse data covering Low data acquisition rate
PhS	Simple and inexpensive Close range	Limited to smooth surfaces Needs fixed position
VW-SL	High data acquisition rate Close range	Computation demanding Missing data in occlusions and shadows Needs fixed position
CW-SL	High data acquisition rate Medium range	Computation demanding Missing data in occlusions and shadows Safety constraint if laser source

Table 9 summarizes the main strengths and weaknesses of the solutions surveyed in this article. The comments in the table are quite general, and a number of exceptions may exist. Furthermore, these pros and cons may also be mitigated or increased depending on the application and/or the platform used.

With regard to the use of standard robots as data-gathering platforms, at present, scientists can mount their systems in the payload area, but in general, these systems are independent from the control architecture of the vehicle. As a consequence, the payload and robot work independently; thus, the generation and control of surveys for data sampling missions is still an issue. An adaptive data sampling mission should allow scientists to program the data density in a required area or volume. Then, the controlled robot would only proceed from one mission waypoint to the next only if the data sampling requirement were met. In this way, the resulting data would not lack spatial or temporal resolution. However, work class ROVs or commercial AUVs do not normally have this type of control interface available.

Finally, as was mentioned earlier, to overcome the limitations of each individual sensor type, advanced reconstruction systems can combine various sensors of the same or different natures. This solution can be suited to an underwater robot or to a fleet of them, as using several sensing modalities often requires different speeds and distances from the sea bottom. To make these solutions really functional, much more research effort has to be focused on underwater localization, so that data can be consistently registered and finally integrated in a unique framework.

Acknowledgments: Acknowledgments: This work has been partially supported by grant BES-2012-054352(FPI), contract DPI2014-57746-C3-2-R by the Spanish Ministry of Economy and Competitiveness, Illes Balears Local Government AAEE60/2014 and FEDER funding.

Author Contributions: Author Contributions: Miquel Massot carried out a literature survey. Miquel Massot and Gabriel Oliver proposed the methods and techniques to be summarized and compared. Miquel Massot wrote the whole paper and Gabriel Oliver supervised the drafts and the bibliography. All authors revised and approved the final submission.

Conflicts of Interest: Conflicts of Interest: The authors declare no conflict of interest.

Abbreviations

APD: Avalanche photodiode
AUV: Autonomous underwater vehicle
BF: Beam forming
CCD: Charged-couped device
CMOS: Complementary metal-oxide semiconductor
CW-LLS: Continuous wave LLS
DOE: Diffractive optical element
fps; Frames per second
GPS: Global positioning system

GPU: Graphics processor unit

HD: High definition

ICP: Iterative closest point

IMU: Inertial measurement unit

IS: Imaging sonar

KLT: Kanade-Lucas-Tomasi feature tracker

LbSL: Laser-based SL

LED: Light-emitting diode

LiDAR: Light detection and ranging

LLS; Laser line scanning

LS: Laser stripe

MBS: Multibeam sonar

NA: Not available

PG-LLS: Pulse gated LLS

PMT: Photomultiplier tube

PN-MP: Pseudorandom coded modulation pulse

RANSAC: Random sample and consensus

ROV: Remoted operated vehicle

SAM: Smoothing and mapping

SBS: Single beam sonar

SIFT: Scale invariant feature transform

SL: Structured light

SNR: Signal to noise ratio

SONAR: Sound navigation and ranging

SSS: Sidescan sonar

ST-MP: Single tone modulated pulse[SURF] Speeded up robust features

SV: Stereo vision

TOF: Time of flight

UW: Underwater

UWTD: Underwater target detection

References

1. Blais, F. Review of 20 years of range sensor development. *J. Electron. Imaging* **2004**, *13*, 231–243.
2. Malamas, E.N.; Petrakis, E.G.; Zervakis, M.; Petit, L.; Legat, J.D. A survey on industrial vision systems, applications and tools. *Image Vis. Comput.* **2003**, *21*, 171–188.
3. Drap, P. Underwater Photogrammetry for Archaeology. In *Special Applications of Photogrammetry*; InTech: Rijeka, Croatia, 2012; pp. 111–136.
4. Jaffe, J.S.; Moore, K.; McLean, J.; Strand, M. Underwater Optical Imaging: Status and Prospects. *Oceanography* **2001**, *14*, 64–75.

5.   Foley, B.; Mindell, D. Precision Survey and Archaeological Methodology in Deep Water. *MTS J.* **2002**, *36*, 13–20.

6.   Kocak, D.M.; Dalgleish, F.R.; Caimi, F.M. A focus on recent developments and trends in underwater imaging. *Mar. Technol. Soc. J.* **2008**, *42*, 52–67.

7.   Kocak, D.M.; Caimi, F.M. The Current Art of Underwater Imaging With a Glimpse of the Past and Vision of the Future. *Mar. Technol. Soc. J.* **2005**, *39*, 5–26.

8.   Caimi, F.M.; Kocak, D.M.; Dalgleish, F.; Watson, J. Underwater imaging and optics: Recent advances. In Proceedings of the MTS/IEEE Oceans, Quebec City, QC, Canada, 15–18 September 2008; pp. 1–9.

9.   Bonin-Font, F.; Burguera, A. Imaging systems for advanced underwater vehicles. *J. Marit. Res.* **2011**, *8*, 65–86.

10. Bianco, G.; Gallo, A.; Bruno, F.; Muzzupappa, M. A comparative analysis between active and passive techniques for underwater 3D reconstruction of close-range objects. *Sensors* **2013**, *13*, 11007–11031.

11. Jordt-Sedlazeck, A. Underwater 3D Reconstruction Based on Physical Models for Refraction and Underwater Light Propagation. Ph.D. Thesis, Kiel University, Kiel, Germany, 2014. Available online: http://www.inf.uni-kiel.de/de/forschung/publikationen/kcss (accessed on 18 September 2015).

12. Höhle, J. Reconstruction of the Underwater Object. *Photogramm. Eng. Remote Sens.* **1971**, *37*, 948–954.

13. Drap, P.; Seinturier, J.; Scaradozzi, D. Photogrammetry for virtual exploration of underwater archeological sites. In Proceedings of the 21st International Symposium, CIPA 2007: AntiCIPAting the Future of the Cultural Past, Athens, Greece, 1–6 October 2007; pp. 1–6.

14. Gawlik, N. 3D modelling of underwater archaeological artefacts Natalia Gawlik. Master's Thesis, Norwegian University of Science of Technology, Trondheim, Norway, 2014. Available online: http://hdl.handle.net/11250/233084 (accessed on 18 September 2015).

15. Pope, R.M.; Fry, E.S. Absorption spectrum (380–700 nm) of pure water. *Appl. Opt.* **1997**, *36*, 8710–8723.

16. McGlamery, B.L. *Computer Analysis and Simulation of Underwater Camera System Performance*; Technical Report; Visibility Laboratory, University of California, San Diego and Scripps Insitution of Oceanography: Oakland, CA, USA, 1975.

17. Jaffe, J.S. Computer modeling and the design of optimal underwater imaging systems. *IEEE J. Ocean. Eng.* **1990**, *15*, 101–111.

18. Schechner, Y.; Karpel, N. Clear underwater vision. In Proceedings of the 2004 IEEE Computer Society Conference on Computer Vision and Pattern Recognition (CVPR), Washington, DC, USA, 27 June–2 July 2004; pp. 536–543.

19. Jordt, A.; Koch, R. Refractive calibration of underwater cameras. In *Computer Vision–ECCV 2012*; Springer Berlin Heidelberg: Berlin, Germany, 2012; pp. 1–14.

20. Treibitz, T.; Schechner, Y.Y.; Kunz, C.; Singh, H. Flat refractive geometry. *IEEE Trans. Pattern Anal. Mach. Intell.* **2012**, *34*, 51–65.

21. Henderson, J.; Pizarro, O.; Johnson-Roberson, M.; Mahon, I. Mapping Submerged Archaeological Sites using Stereo-Vision Photogrammetry. *Int. J. Naut. Archaeol.* **2013**, *42*, 243–256.

22. Gracias, N.; Negahdaripour, S.; Neumann, L.; Prados, R.; Garcia, R. A motion compensated filtering approach to remove sunlight flicker in shallow water images. In Proceedings of the Oceans 2008, Quebec City, QC, Canada, 15–18 September 2008; pp. 1–7.

23. Slama, C.C.; Theurer, C.; Henriksen, S.W. *Manual of Photogrammetry*; Va. American Society of Photogrammetry: Bethesda, MD, USA, 1980.

24. Grossberg, M.D.; Nayar, S.K. A general imaging model and a method for finding its parameters. In Proceedings of the Eighth IEEE International Conference on Computer Vision (ICCV), Vancouver, BC, Canada, 7–14 July 2001; pp. 108–115.

25. Zhang, Z. A flexible new technique for camera calibration. *IEEE Trans. Pattern Anal. Mach. Intell.* **2000**, *22*, 1330–1334.

26. Tsai, R.Y. A versatile camera calibration technique for high-accuracy 3D machine vision metrology using off-the-shelf TV cameras and lenses. *IEEE J. Robot. Autom.* **1987**, *3*, 323–344.

27. Brown, L.G. A Survey of Image Registration Techniques. *ACM Comput. Surv.* **1992**, *24*, 325–376.

28. Szelisky, R. *Computer Vision: Algorithms and Applications*; Springer: London, UK, 2011.

29. Kwon, Y.H. Object Plane Deformation Due to Refraction in Two-Dimensional Underwater Motion Analysis. *J. App. Biomech.* **1999**, *15*, 396–403.

30. Zajac, A.; Hecht, E. *Optics*, 4th ed.; Pearson Higher Education: San Francisco, CA, USA; Massachusetts Institute of Technology: Cambridge, MA, USA; 2003.

31. Refractiveindex.info. Refractive Index and Related Constants—Poly(methyl methacrylate) (PMMA, Acrylic glass). Available online: http://refractiveindex.info/?shelf=organic&book=poly%28 methyl_methacrylate% 29&page=Szczurowski (accessed on 5 October 2015).

32. Menna, F.; Nocerino, E.; Troisi, S.; Remondino, F. A photogrammetric approach to survey floating and semi-submerged objects. In Proceedings of the Videometrics, Range Imaging and Applications XII, Munich, Germany, 14–16 May 2013; pp. 1–15.

33. Kwon, Y.h.; Lindley, S.L. Applicability of the localized-calibration methods in underwater motion analysis. In Proceedings of the Sixteenth International Conference on Biomechanics in Sports, Vienna, Austria, 12–16 July 2000; pp. 1–8.

34. Kang, L.; Wu, L.; Yang, Y.H. Two-view underwater structure and motion for cameras under flat refractive interfaces. *Lect. Notes Comput. Sci.* **2012**, *7575*, 303–316.

35. Yang, T.C. Data-based matched-mode source localization for a moving source. *J. Acoust. Soc. Am.* **2014**, *135*, 1218–1230.

36. Candy, J.; Sullivan, E.J. Model-based identification: An adaptive approach to ocean-acoustic processing. *IEEE J. Ocean. Eng.* **1996**, *21*, 273–289.

37.   Buchanan, J.L.; Gilbert, R.P.; Wirgin, A.; Xu, Y. Identification, by the intersecting canonical domain method, of the size, shape and depth of a soft body of revolution located within an acoustic waveguide. *Inverse Probl.* **2000**, *16*, 1709–1926.

38.   Pathak, K.; Birk, A.; Vaskevicius, N. Plane-based registration of sonar data for underwater 3D mapping. In Proceedings of the 2010 IEEE/RSJ International Conference on Intelligent Robots and Systems, Taipei, Taiwan, 18–22 October 2010; pp. 4880–4885.

39.   Hurtos, N.; Cufi, X.; Salvi, J. Calibration of optical camera coupled to acoustic multibeam for underwater 3D scene reconstruction. In Proceedings of the MTS/IEEE Oceans, Seattle, WA, USA, 20–23 September 2010; pp. 1–7.

40.   Blueview, T. 3D Mechanical Scanning Sonar. Available online: http://www.blueview.com/ products/3d-multibeam-scanning-sonar (accessed on 30 September 2015).

41.   Guo, Y. 3D underwater topography rebuilding based on single beam sonar. In Proceedings of the 2013 IEEE International Conference on Signal Processing, Communication and Computing (ICSPCC 2013), Kunming, China, 5–8 August 2013; pp. 1–5.

42.   Coiras, E.; Petillot, Y.; Lane, D.M. Multiresolution 3-D reconstruction from side-scan sonar images. *IEEE Trans. Image Process.* **2007**, *16*, 382–390.

43.   Brahim, N.; Gueriot, D.; Daniel, S.; Solaiman, B. 3D reconstruction of underwater scenes using DIDSON acoustic sonar image sequences through evolutionary algorithms. In Proceedings of the MTS/IEEE Oceans, Santander, Spain, 6–9 June 2011; pp. 1–6.

44.   Negahdaripour, S.; Sekkati, H.; Pirsiavash, H. Opti-acoustic stereo imaging: On system calibration and 3-D target reconstruction. *IEEE Trans. Image Process.* **2009**, *18*, 1203–1214.

45.   Babaee, M.; Negahdaripour, S. 3-D Object Modeling from Occluding Contours in Opti-Acoustic Stereo Images. In Proceedings of the MTS/IEEE Oceans, San Diego, CA, USA, 23–27 September 2013; pp. 1–8.

46.   Negahdaripour, S. On 3-D reconstruction from stereo FS sonar imaging. In Proceedings of the MTS/IEEE Oceans, Seattle, WA, USA, 20–23 September 2010; pp. 1–6.

47.   Aykin, M.; Negahdaripour, S. Forward-Look 2-D Sonar Image Formation and 3-D Reconstruction. In Proceedings of the MTS/IEEE Oceans, San Diego, CA , USA, 23–27 September 2013; pp. 1–4.

48.   Murino, V.; Trucco, A.; Regazzoni, C.S. A probabilistic approach to the coupled reconstruction and restoration of underwater acoustic images. *IEEE Trans. Pattern Anal. Mach. Intell.* **1998**, *20*, 9–22.

49.   Murino, V. Reconstruction and segmentation of underwater acoustic images combining confidence information in MRF models. *Pattern Recognit.* **2001**, *34*, 981–997.

50.   Castellani, U.; Fusiello, A.; Murino, V. Registration of Multiple Acoustic Range Views for Underwater Scene Reconstruction. *Comput. Vis. Image Underst.* **2002**, *87*, 78–89.

51.   Kunz, C.; Singh, H. Map Building Fusing Acoustic and Visual Information using Autonomous Underwater Vehicles. *J. Field Robot.* **2013**, *30*, 763–783.

52. Rosenblum, L.; Kamgar-Parsi, B. 3D reconstruction of small underwater objects using high-resolution sonar data. In Proceedings of the 1992 Symposium on Autonomous Underwater Vehicle Technology, Washington, DC, USA, 2–3 June 1992; pp. 228–235.

53. Reineman, B.D.; Lenain, L.; Castel, D.; Melville, W.K. A Portable Airborne Scanning Lidar System for Ocean and Coastal Applications. *J. Atmos. Ocean. Technol.* **2009**, *26*, 2626–2641.

54. Cadalli, N.; Shargo, P.J.; Munson, D.C., Jr.; Singer, A.C. Three-dimensional tomographic imaging of ocean mines from real and simulated lidar returns. In Proceedings of the SPIE 4488, Ocean Optics: Remote Sensing and Underwater Imaging, San Diego, CA, USA, 29 July 2001; pp. 155–166.

55. Pellen, F.; Jezequel, V.; Zion, G.; Jeune, B.L. Detection of an underwater target through modulated lidar experiments at grazing incidence in a deep wave basin. *Appl. Opt.* **2012**, *51*, 7690–7700.

56. Mullen, L.; Vieira, A.; Herezfeld, P.; Contarino, V. Application of RADAR technology to aerial LIDAR systems for enhancement of shallow underwater target detection. *IEEE Trans. Microw. Theory Tech.* **1995**, *43*, 2370–2377.

57. Mullen, L.J.; Contarino, V.M. Hybrid LIDAR-radar: Seeing through the scatter. *IEEE Microw. Mag.* **2000**, *1*, 42–48.

58. Moore, K.; Jaffe, J.S.; Ochoa, B. Development of a new underwater bathymetric laser imaging system: L-bath. *J. Atmos. Ocean. Technol.* **2000**, *17*, 1106–1117.

59. Moore, K.; Jaffe, J.S. Time-evolution of high-resolution topographic measurements of the sea floor using a 3-D laser line scan mapping system. *IEEE J. Ocean. Eng.* **2002**, *27*, 525–545.

60. McLeod, D.; Jacobson, J. Autonomous Inspection using an Underwater 3D LiDAR. In Proceedings of the MTS/IEEE Oceans, San Diego, CA, USA, 23–27 September 2013.

61. Cochenour, B.; Mullen, L.J.; Muth, J. Modulated pulse laser with pseudorandom coding capabilities for underwater ranging, detection, and imaging. *Appl. Opt.* **2011**, *50*, 6168–6178.

62. Rumbaugh, L.; Li, Y.; Bollt, E.; Jemison, W. A 532 nm Chaotic Lidar Transmitter for High Resolution Underwater Ranging and Imaging. In Proceedings of the MTS/IEEE Oceans, San Diego, CA, USA, 23–27 September 2013; pp. 1–6.

63. De Dominicis, L.; Fornetti, G.; Guarneri, M.; de Collibus, M.F.; Francucci, M.; Nuvoli, M.; Al-Obaidi, A.; Mcstay, D. Structural Monitoring Of Offshore Platforms By 3d Subsea Laser Profilers. In Proceedings of the Offshore Mediterranean Conference, Ravenna, Italy, 20–22 March 2013.

64. Dalgleish, F.R.; Caimi, F.M.; Britton, W.B.; Andren, C.F. Improved LLS imaging performance in scattering-dominant waters. In Proceedings of the SPIE 7317, Ocean Sensing and Monitoring, Orlando, FL, USA, 29 April 2009.

65. Gordon, A. Use of Lases Scanning Systems on Mobile Underwater Platforms. In Proceedings of the 1992 Symposium on Autonomous Underwater Vehicle Technology, Washington, DC, USA, 2–3 June 1992; pp. 202–205.

66. Mullen, L.J.; Contarino, V.M.; Laux, A.; Concannon, B.M.; Davis, J.P.; Strand, M.P.; Coles, B.W. Modulated laser line scanner for enhanced underwater imaging. In Proceedings of the SPIE. 3761, Airborne and In-Water Underwater Imaging, Denver, CO, USA, 28 October 1999; Volume 3761, pp. 2–9.

67. 3D at Depth. SL1 High Resolution Subsea Laser Scanner. Available online: http://www.3datdepth.com/sl1overview/ (accessed on 30 September 2015).

68. Salvi, J.; Fernandez, S.; Pribanic, T.; Llado, X. A state of the art in structured light patterns for surface profilometry. *Pattern Recognit.* **2010**, *43*, 2666–2680.

69. Salvi, J.; Pagès, J.; Batlle, J. Pattern codification strategies in structured light systems. *Pattern Recognit.* **2004**, *37*, 827–849.

70. Bruno, F.; Bianco, G.; Muzzupappa, M.; Barone, S.; Razionale, A. Experimentation of structured light and stereo vision for underwater 3D reconstruction. *ISPRS J. Photogramm. Remote Sens.* **2011**, *66*, 508–518.

71. Zhang, S. Recent progresses on real-time 3D shape measurement using digital fringe projection techniques. *Opt. Lasers Eng.* **2010**, *48*, 149–158.

72. Zhang, Q.; Wang, Q.; Hou, Z.; Liu, Y.; Su, X. Three-dimensional shape measurement for an underwater object based on two-dimensional grating pattern projection. *Opt. Laser Technol.* **2011**, *43*, 801–805.

73. Törnblom, N. Underwater 3D Surface Scanning Using Structured Light. Ph.D. Thesis, Uppsala Universitet, Uppsala, Sweden, 2010. Available online: http://www.diva-portal.org/smash/get/diva2:378911/FULLTEXT01.pdf (accessed on 18 September 2015).

74. Narasimhan, S.; Nayar, S. Structured Light Methods for Underwater Imaging: Light Stripe Scanning and Photometric Stereo. In Proceedings of the MTS/IEEE Oceans, Washington, DC, USA, 18–23 September 2005; pp. 1–8.

75. Dancu, A.; Fourgeaud, M.; Franjcic, Z.; Avetisyan, R. Underwater reconstruction using depth sensors. In *SIGGRAPH Asia 2014 Technical Briefs on—SIGGRAPH ASIA'14*; ACM Press: New York, NY, USA, 2014; pp. 1–4.

76. Inglis, G.; Smart, C.; Vaughn, I.; Roman, C. A pipeline for structured light bathymetric mapping. In Proceedings of the 2012 IEEE/RSJ International Conference on Intelligent Robots and Systems, Algarve, Portugal, 7–12 October 2012; pp. 4425–4432.

77. Bodenmann, A.; Thornton, B.; Nakajima, R.; Yamamoto, H.; Ura, T. Wide Area 3D Seafloor Reconstruction and its Application to Sea Fauna Density Mapping. In Proceedings of the MTS/IEEE Oceans, San Diego, CA, USA, 23–27 September 2013; pp. 4–8.

78. Liu, J.; Jakas, A.; Al-Obaidi, A.; Liu, Y. Practical issues and development of underwater 3D laser scanners. In Proceedings of the 2010 IEEE 15th Conference on Emerging Technologies & Factory Automation (ETFA 2010), Bilbao, Spain, 13–16 September 2010; pp. 1–8.

79. Roman, C.; Inglis, G.; Rutter, J. Application of structured light imaging for high resolution mapping of underwater archaeological sites. In Proceedings of the MTS/IEEE Oceans, Sydney, Australia, 24–27 May 2010; pp. 1–9.

80. Jaffe, J.S.; Dunn, C. A Model-Based Comparison Of Underwater Imaging Systems. In Proceedings of the Ocean Optics IX, Orlando, FL, USA, 4 April 1988; pp. 344–350.

81. Tetlow, S.; Spours, J. Three-dimensional measurement of underwater work sites using structured laser light. *Meas. Sci. Technol.* **1999**, *10*, 1162–1167.

82. Kondo, H.; Maki, T.; Ura, T.; Nose, Y.; Sakamaki, T.; Inaishi, M. Structure tracing with a ranging system using a sheet laser beam. In Proceedings of the 2004 International Symposium on Underwater Technology (IEEE Cat. No.04EX869), Taipei, Taiwan, 20–23 April 2004; pp. 83–88.

83. Hildebrandt, M.; Kerdels, J.; Albiez, J.; Kirchner, F. A practical underwater 3D-Laserscanner. In Proceedings of the MTS/IEEE Oceans, Quebec City, QC, Canada, 15–18 September 2008; pp. 1–5.

84. Bodenmann, A.; Thornton, B.; Nakatani, T.; Ura, T. 3D colour reconstruction of a hydrothermally active area using an underwater robot. In Proceedings of the OCEANS 2011, Waikoloa, HI, USA, 19–22 September 2011; pp. 1–6.

85. Bodenmann, A.; Thornton, B.; Hara, S.; Hioki, K.; Kojima, M.; Ura, T.; Kawato, M.; Fujiwara, Y. Development of 8 m long range imaging technology for generation of wide area colour 3D seafloor reconstructions. In Proceedings of the MTS/IEEE Oceans, Hampton Roads, VA, USA, 14–19 October 2012; pp. 1–4.

86. Bodenmann, A.; Thornton, B.; Ura, T. Development of long range color imaging for wide area 3D reconstructions of the seafloor. In Proceedings of the 2013 IEEE International Underwater Technology Symposium (UT), Tokyo, Japan, 5–8 March 2013; pp. 1–5.

87. Nakatani, T.; Li, S.; Ura, T.; Bodenmann, A.; Sakamaki, T. 3D visual modeling of hydrothermal chimneys using a rotary laser scanning system. In Proceedings of the 2011 IEEE Symposium on Underwater Technology and Workshop on Scientific Use of Submarine Cables and Related Technologies, Tokyo, JApan, 5–8 April 2011; pp. 1–5.

88. Smart Light Devices, S. 3DLS Underwater 3D Laser Imaging Scanner. Available online: http://www.smartlightdevices.co.uk/products/3dlaser-imaging/ (accessed on 30 September 2015).

89. Prats, M.; Fernandez, J.J.; Sanz, P.J. An approach for semi-autonomous recovery of unknown objects in underwater environments. In Proceedings of the 13th International Conference on Optimization of Electrical and Electronic Equipment (OPTIM), Brasov, Romania, 24–26 May 2012; pp. 1452–1457.

90. Prats, M.; Fernandez, J.J.; Sanz, P.J. Combining template tracking and laser peak detection for 3D reconstruction and grasping in underwater environments. In Proceedings of the IEEE/RSJ International Conference on Intelligent Robots and Systems, Algarve, Portugal, 7–12 October 2012; pp. 106–112.

91. Sanz, P.J.; Penalver, A.; Sales, J.; Fornas, D.; Fernandez, J.J.; Perez, J.; Bernabe, J. GRASPER: A Multisensory Based Manipulation System for Underwater Operations. In Proceedings of the IEEE International Conference on Systems, Man, and Cybernetics, Manchester, UK, 13–16 October 2013; pp. 4036–4041.

92. Cain, C.; Leonessa, A. Laser based rangefinder for underwater applications. In Proceedings of the 2012 American Control Conference (ACC), Montreal, QC, USA, 27–29 June 2012; pp. 6190–6195.

93. Caccia, M. Laser-Triangulation Optical-Correlation Sensor for ROV Slow Motion Estimation. *IEEE J. Ocean. Eng.* **2006**, *31*, 711–727.

94. Yang, Y.; Zheng, B.; Zheng, H. 3D reconstruction for underwater laser line scanning. In Proceedings of the MTS/IEEE Oceans, Bergen, Norway, 10–14 June 2013; pp. 2008–2010.

95. Massot-Campos, M.; Oliver-Codina, G. One-shot underwater 3D reconstruction. In Proceedings of the 2014 IEEE Emerging Technology and Factory Automation (ETFA), Barcelona, Spain, 16–19 September 2014; pp. 1–4.

96. Massot, M.; Oliver, G.; Kemal, H.; Petillot, Y.; Bonin-Font, F. Structured light and stereo vision for underwater 3D reconstruction. In Proceedings of the MTS/IEEE Oceans, Genoa, Italy, 18–21 May 2015.

97. Massot-Campos, M.; Oliver-Codina, G. Underwater Laser-based Structured Light System for one-shot 3D reconstruction. In Proceedings of the 2014 IEEE Sensors, Valencia, Spain, 2–5 November 2014; pp. 1138–1141.

98. Tsiotsios, C.; Angelopoulou, M.; Kim, T.K.; Davison, A. Backscatter Compensated Photometric Stereo with 3 Sources. In Proceedings of the 2014 IEEE Conference on Computer Vision and Pattern Recognition (CVPR), Columbus, OH, USA, 23–28 June 2014; pp. 2259–2266.

99. Robust Photometric Stereo via Low-Rank Matrix Completion and Recovery. Available online: http://perception.csl.illinois.edu/matrix-rank/stereo.html (accessed on 18 September 2015).

100. Wen, Z.Y.; Fraser, D.; Lambert, A.; Li, H.D. Reconstruction of underwater image by bispectrum. In Proceedings of the International Conference on Image Processing (ICIP), San Antonio, TX, USA, 16 September–19 October 2007; pp. 545–548.

101. Jordt-Sedlazeck, A.; Koser, K.; Koch, R. 3D reconstruction based on underwater video from ROV Kiel 6000 considering underwater imaging conditions. In Proceedings of the Proceedings of MTS/IEEE Oceans, Bremen, Germany, 11–14 May 2009; pp. 1–10.

102. Fischler, M.A.; Bolles, R.C. Random Sample Consensus: A Paradigm for Model Fitting with Applications to Image Analysis and Automated Cartography. *Commun. ACM* **1981**, *24*, 381–395.

103. Pizarro, O.; Eustice, R.M.; Singh, H. Large Area 3-D Reconstructions From Underwater Optical Surveys. *IEEE J. Ocean. Eng.* **2009**, *34*, 150–169.

104. Meline, A.; Triboulet, J.; Jouvencel, B. Comparative study of two 3D reconstruction methods for underwater archaeology. In Proceedings of the 2012 IEEE/RSJ International Conference on Intelligent Robots and Systems, Algarve, Portugal, 7–12 October 2012; pp. 740–745.

105. McKinnon, D.; He, H.; Upcroft, B.; Smith, R. Towards automated and in-situ, near-real time 3-D reconstruction of coral reef environments. In Proceedings of the OCEANS 2011, Waikoloa, HI, USA, 19–22 September 2011; pp. 1–10.

106. Jordt-Sedlazeck, A.; Koch, R. Refractive Structure-from-Motion on Underwater Images. In Proceedings of the 2013 IEEE International Conference on Computer Vision, Sydney, Australia, 1–8 December 2013; pp. 57–64.

107. Cocito, S.; Sgorbini, S.; Peirano, A.; Valle, M. 3-D reconstruction of biological objects using underwater video technique and image processing. *J. Exp. Mar. Biol. Ecol.* **2003**, *297*, 57–70.

108. Bruno, F.; Gallo, A.; Muzzupappa, M.; Davidde Petriaggi, B.; Caputo, P. 3D documentation and monitoring of the experimental cleaning operations in the underwater archaeological site of Baia (Italy). In Proceedings of the Digital Heritage International Congress (DigitalHeritage), Marseille, France, 28 October–1 November 2013; pp. 105–112.

109. Nicosevici, T.; Gracias, N.; Negahdaripour, S.; Garcia, R. Efficient three-dimensional scene modeling and mosaicing. *J. Field Robot.* **2009**, *26*, 759–788.

110. Ozog, P.; Carlevaris-Bianco, N.; Kim, A.; Eustice, R.M. Long-term Mapping Techniques for Ship Hull Inspection and Surveillance using an Autonomous Underwater Vehicle. *J. Field Robot.* **2015**, *24*, 1–25.

111. Seitz, S.M.; Curless, B.; Diebel, J.; Scharstein, D.; Szeliski, R. A comparison and evaluation of multi-view stereo reconstruction algorithms. In Proceedings of the 2006 IEEE Computer Society Conference on Computer vision and pattern recognition, New York, NY, USA, 17–22 June 2006; pp. 519–528.

112. Brown, M.Z.; Burschka, D.; Hager, G.D. Advances in computational stereo. *IEEE Trans. Pattern Anal. Mach. Intell.* **2003**, *25*, 993–1008.

113. Barnard, S.T.; Fischler, M.A. Computational stereo. *ACM Comput. Surv.* **1982**, *14*, 553–572.

114. Dhond, U.R.; Aggarwal, J.K. Structure from stereo-a review. *IEEE Trans. Syst. Man Cybern.* **1989**, *19*, 1489–1510.

115. Scharstein, D.; Szeliski, R. A taxonomy and evaluation of dense two-frame stereo correspondence algorithms. *Int. J. Comput. Vis.* **2002**, *47*, 7–42.

116. Hartley, R.I.; Zisserman, A. *Multiple View Geometry in Computer Vision*, 2nd ed.; ISBN: 0521540518; Cambridge University Press: Cambridge, UK, 2004.

117. Ku3, N.S.; Ramakanth Kumar, R. Design & development of autonomous system to build 3D model for underwater objects using stereo vision technique. In Proceedings of the 2011 Annual IEEE India Conference, Hyderabad, India, 16–18 December 2011; pp. 1–4.

118. Jasiobedzki, P.; Se, S.; Bondy, M.; Jakola, R. Underwater 3D mapping and pose estimation for ROV operations. In Proceedings of the MTS/IEEE Oceans, Quebec City, QC, Canada, 15–18 September 2008; pp. 1–6.

119. Nurtantio Andono, P.; Mulyanto Yuniarno, E.; Hariadi, M.; Venus, V. 3D reconstruction of under water coral reef images using low cost multi-view cameras. In Proceedings of the 2012 International Conference on Multimedia Computing and Systems, Tangier, Morocco, 10–12 May 2012; pp. 803–808.

120. Schmidt, V.; Rzhanov, Y. Measurement of micro-bathymetry with a GOPRO underwater stereo camera pair. In Proceedings of the Oceans 2012, Hampton Roads, VA, USA, 14–19 October 2012; pp. 1–6.

121. Dao, T.D. Underwater 3D Reconstruction from Stereo Images. Msc Erasmus Mundus in Vision and Robotics, University of Girona (Spain), University of Burgundy (France), Heriot Watt University (UK), 2008. Available online: https://fb.docs.com/VA95 (accessed on 18 September 2015).

122. Brandou, V.; Allais, A.G.; Perrier, M.; Malis, E.; Rives, P.; Sarrazin, J.; Sarradin, P.M. 3D Reconstruction of Natural Underwater Scenes Using the Stereovision System IRIS. In Proceedings of the MTS/IEEE Oceans, Aberdeen, UK, 18–21 June 2007; pp. 1–6.

123. Beall, C.; Lawrence, B.J.; Ila, V.; Dellaert, F. 3D reconstruction of underwater structures. In Proceedings of the 2010 IEEE/RSJ International Conference on Intelligent Robots and Systems, Taipei, Taiwan, 18–22 October 2010; pp. 4418–4423.

124. Servos, J.; Smart, M.; Waslander, S.L. Underwater stereo SLAM with refraction correction. In Proceedings of the 2013 IEEE/RSJ International Conference on Intelligent Robots and Systems, Tokyo, Japan, 3–7 November 2013; pp. 3350–3355.

125. Hogue, A.; German, A.; Jenkin, M. Underwater environment reconstruction using stereo and inertial data. In Proceedings of the 2007. ISIC. IEEE International Conference on Systems, Man and Cybernetics, Montreal, QC, Canada, 7–10 October 2007; pp. 2372–2377.

126. Negre Carrasco, P.L.; Bonin-Font, F.; Oliver-Codina, G. Stereo Graph-SLAM for Autonomous Underwater Vehicles. In Proceedings of the 13th International Conference on Intelligent Autonomous Systems (IAS 2014), Padova, Italy, 15–19 July 2014; pp. 351–360.

127. Bonin-Font, F.; Cosic, A.; Negre, P.L.; Solbach, M.; Oliver, G. Stereo SLAM for robust dense 3D reconstruction of underwater environments. OCEANS 2015, Genova, Italy, 18–21 May 2015; pp. 1–6.

128. Johnson-Roberson, M.; Pizarro, O.; Williams, S.B.; Mahon, I. Generation and visualization of large-scale three-dimensional reconstructions from underwater robotic surveys. *J. Field Robot.* **2010**, *27*, 21–51.

129. Galceran, E.; Campos, R.; Palomeras, N.; Ribas, D.; Carreras, M.; Ridao, P. Coverage Path Planning with Real-time Replanning and Surface Reconstruction for Inspection of Three-dimensional Underwater Structures using Autonomous Underwater Vehicles. *J. Field Robot.* **2014**, *24*, 952–983.

130. González-Rivero, M.; Bongaerts, P.; Beijbom, O.; Pizarro, O.; Friedman, A.; Rodriguez-Ramirez, A.; Upcroft, B.; Laffoley, D.; Kline, D.; Bailhache, C.; *et al.* The Catlin Seaview Survey—Kilometre-scale seascape assessment, and monitoring of coral reef ecosystems. *Aquat. Conserv. Mar. Freshw. Ecosyst.* **2014**, *24*, 184–198.

131. Pointclouds. System, Robotics and Vision, University of the Balearic Islands. Available online: http://srv.uib.es/point clouds (accessed on 2 December 2015).

132. Inglis, G.; Roman, C. Sonar constrained stereo correspondence for three-dimensional seafloor reconstruction. In Proceedings of the MTS/IEEE Oceans, Sydney, Australia, 24–27 May 2010; pp. 1–10.

133. Zhizhou, W. A discussion about the terminology "photogrammetry and remote sensing". *ISPRS J. Photogramm. Remote Sens.* **1989**, *44*, 169–174.

134. Luhmann, T.; Robson, S.; Kyle, S.; Boehm, J. *Close-Range Photogrammetry and 3D Imaging*, 2nd ed.; De Gruyter Textbook, ISBN: 9783110302783; John Wiley & Sons Ltd.: Berlin, Germany, 2013.

135. Förstner, W. Uncertainty and projective geometry. In *Handbook of Geometric Computing*; Springer Berlin Heidelberg: Berlin, Germany, 2005; pp. 493–534.

136. Abdo, D.A.; Seager, J.W.; Harvey, E.S.; McDonald, J.I.; Kendrick, G.A.; Shortis, M.R. Efficiently measuring complex sessile epibenthic organisms using a novel photogrammetric technique. *J. Exp. Mar. Biol. Ecol.* **2006**, *339*, 120–133.

137. Westoby, M.; Brasington, J.; Glasser, N.; Hambrey, M.; Reynolds, J. "Structure-from-Motion" photogrammetry: A low-cost, effective tool for geoscience applications. *Geomorphology* **2012**, *179*, 300–314.

138. Milgram, D.L. Computer methods for creating photomosaics. *IEEE Trans. Comput.* **1975**, 1113–1119, doi:10.1109/T-C.1975.224142.

139. Milgram, D.L. Adaptive techniques for photomosaicking. *IEEE Trans. Comput.* **1977**, *100*, 1175–1180.

140. Kocak, D.; Jagielo, T.; Wallace, F.; Kloske, J. Remote sensing using laser projection photogrammetry for underwater surveys. In Processings of the IEEE International Geoscience and Remote Sensing Symposium, IGARSS'04, Anchorage, AK, USA, 20–24 September 2004; pp. 1451–1454.

141. Schewe, H.; Monchrieff, E.; Gründig, L. Improvement of fishfarm pen design using computational structural modelling and large-scale underwater photogrammetry (cosmolup). *Int. Arch. Photogramm. Remote Sens.* **1996**, *31*, 524–529.

142. Bouratsis, P.P.; Diplas, P.; Dancey, C.L.; Apsilidis, N. High-resolution 3-D monitoring of evolving sediment beds. *Water Resour. Res.* **2013**, *49*, 977–992.

143. Zhukovsky, M.O.; Kuznetsov, V.D.; Olkhovsky, S.V. Photogrammetric Techniques for 3-D Underwater Record of the Antique Time Ship From Phanagoria. *Int. Arch. Photogramm. Remote Sens. Spat. Inf. Sci.* **2013**, *40*, 717–721.

144. Nornes, S.M.; Ludvigsen, M.; Odegard, O.; Sorensen, A.J. Underwater Photogrammetric Mapping of an Intact Standing Steel Wreck with ROV. In Processings of the 4th IFAC Workshop onNavigation, Guidance and Control of Underwater VehiclesNGCUV, Copenhagen, Denmark, 24–26 August 2015; pp. 206–211.

145. Bingham, B.; Foley, B.; Singh, H.; Camilli, R.; Delaporta, K.; Eustice, R.; Mallios, A.; Mindell, D.; Roman, C.; Sakellariou, D. Robotic tools for deep water archaeology: Surveying an ancient shipwreck with an autonomous underwater vehicle. *J. Field Robot.* **2010**, *27*, 702–717.

146. Fabris, M.; Baldi, P.; Anzidei, M.; Pesci, A.; Bortoluzzi, G.; Aliani, S. High resolution topographic model of Panarea Island by fusion of photogrammetric, lidar and bathymetric digital terrain models. *Photogramm. Rec.* **2010**, *25*, 382–401.

147. Atkinson, K. *Close Range Photogrammetry and Machine Vision*; Whittles Publishing: London, UK, 1996.

148. Lavest, J.M.; Guichard, F.; Rousseau, C. Multi-view reconstruction combining underwater and air sensors. In Processings of the International Conference on Image, Rochester, NY, USA, 24–28 June 2002; pp. 813–816.

149. Shortis, M.; Harvey, E.; Seager, J. A Review of the Status and Trends in Underwater Videometric Measurement. Invited paper, SPIE Conference, 2007. Available online: http://www.geomsoft.com/markss/papers/Shortis_etal_paper_Vid_IX.pdf (accessed on 18 September 2015)

150. De Jesus, K.; de Jesus, K.; Figueiredo, P.; Vilas-Boas, J.A.P.; Fernandes, R.J.; Machado, L.J. Reconstruction Accuracy Assessment of Surface and Underwater 3D Motion Analysis: A New Approach. *Comput. Math. Methods Med.* **2015**, *2015*, 1–8.

151. Leurs, G.; O'Connell, C.P.; Andreotti, S.; Rutzen, M.; Vonk Noordegraaf, H. Risks and advantages of using surface laser photogrammetry on free-ranging marine organisms: A case study on white sharks *"Carcharodon carcharias"*. *J. Fish Biol.* **2015**, *86*, 1713–1728.

152. Bräuer-Burchardt, C.; Heinze, M.; Schmidt, I.; Kühmstedt, P.; Notni, G. Compact handheld fringe projection based underwater 3D scanner. *Int. Arch. Photogramm. Remote Sens. Spat. Inf. Sci.* **2015**, *XL-5/W5*, 33–39.

153. Ekkel, T.; Schmik, J.; Luhmann, T.; Hastedt, H. Precise laser-based optical 3D measurement of welding seams under water. *ISPRS Int. Arch. Photogramm. Remote Sens. Spat. Inf. Sci.* **2015**, *XL-5/W5*, 117–122.

154. Teledyne CDL. INSCAN 3D Scanning Subsea Laser. Available online: http://teledyne-cdl.com/events/inscan-demonstration-post-press-release (accessed on 30 September 2015).

155. Hannon, S. Underwater Mapping. *LiDAR Mag.* **2013**, *3*, 1–4.

156. 2G Robotics. ULS-100 Underwater Laser Scanner for Short Range Scans. Available online: http://www.2grobotics.com/products/underwater-laser-scanner-uls-100/ (accessed on 30 September 2015).

157. 2G Robotics. ULS-200 Underwater Laser Scanner for Mid Range Scans. Available online: http://www.2grobotics.com/products/underwater-laser-scanner-uls-200/ (accessed on 30 September 2015).

158. 2G Robotics. ULS-500 Underwater Laser Scanner for Long Range Scans. Available online: http://www.2grobotics.com/products/underwater-laser-scanner-uls-500/ (accessed on 30 September 2015).

159. Savante. Cerberus Subsea Laser Pipeline Profiler. Available online: http://www.savante.co.uk/subsea-laser-scanner/cerberus-subsea-laser-pipeline-profiler/ (accessed on 21 September 2015).

160. Savante. SLV-50 Laser Vernier Caliper. Available online: http://www.savante.co.uk/wp-content/uploads/2015/02/Savante-SLV-50.pdf (accessed on 21 September 2015).

161. Savante. Lumeneye Subsea Laser Module. Available online: http://www.savante.co.uk/portfolio-items/lumeneye-subsea-line-laser-module/?portfolioID=5142 (accessed on 21 September 2015).

162. Tritech. Sea Stripe ROV/AUV Laser Line Generator. Available online: http://www.tritech.co.uk/product/rov-auv-laser-line-generator-seastripe (accessed on 30 September 2015).

163. Campos, R.; Garcia, R.; Alliez, P.; Yvinec, M. A Surface Reconstruction Method for In-Detail Underwater 3D Optical Mapping. *Int. J. Robot. Res.* **2015**, *34*, 64–89.

164. Tan, C.S.; Sluzek, A.; Seet, G.L.; Jiang, T.Y. Range Gated Imaging System for Underwater Robotic Vehicle. In Proceedings of the MTS/IEEE Oceans, Singapore, 16–19 May 2007; pp. 1–6.

The Bubble Box: Towards an Automated Visual Sensor for 3D Analysis and Characterization of Marine Gas Release Sites

Anne Jordt, Claudius Zelenka, Jens Schneider von Deimling, Reinhard Koch and Kevin Köser

Abstract: Several acoustic and optical techniques have been used for characterizing natural and anthropogenic gas leaks (carbon dioxide, methane) from the ocean floor. Here, single-camera based methods for bubble stream observation have become an important tool, as they help estimating flux and bubble sizes under certain assumptions. However, they record only a projection of a bubble into the camera and therefore cannot capture the full 3D shape, which is particularly important for larger, non-spherical bubbles. The unknown distance of the bubble to the camera (making it appear larger or smaller than expected) as well as refraction at the camera interface introduce extra uncertainties. In this article, we introduce our wide baseline stereo-camera deep-sea sensor bubble box that overcomes these limitations, as it observes bubbles from two orthogonal directions using calibrated cameras. Besides the setup and the hardware of the system, we discuss appropriate calibration and the different automated processing steps deblurring, detection, tracking, and 3D fitting that are crucial to arrive at a 3D ellipsoidal shape and rise speed of each bubble. The obtained values for single bubbles can be aggregated into statistical bubble size distributions or fluxes for extrapolation based on diffusion and dissolution models and large scale acoustic surveys. We demonstrate and evaluate the wide baseline stereo measurement model using a controlled test setup with ground truth information.

Reprinted from *Sensors*. Cite as: Jordt, A.; Zelenka, C.; von Deimling, J.S.; Koch, R.; Köser, K. The Bubble Box: Towards an Automated Visual Sensor for 3D Analysis and Characterization of Marine Gas Release Sites. *Sensors* **2015**, *15*, 30716–30735.

1. Introduction

The oceans' seafloors host significant amounts of the greenhouse gas methane in the form of solid gas hydrate, dissolved in sedimentary pore water, or as free gas. The occurrence of free methane gas in the seabed and respective release into the water column in the form of rising gas bubbles represents a global phenomenon termed *gas seepage* (cf. to [1,2], see also Figure 1).

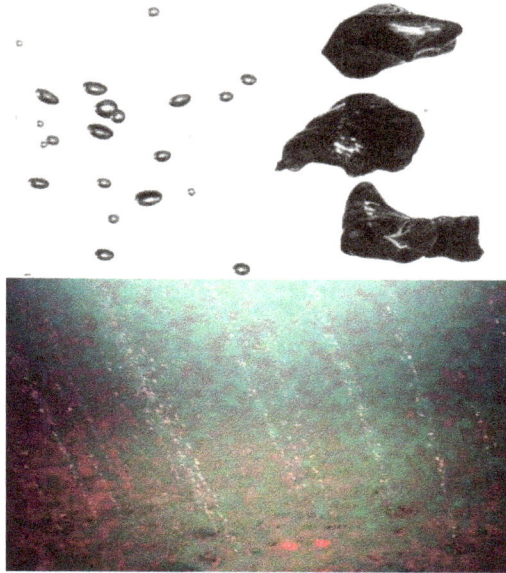

Figure 1. The shape and speed of gas bubbles rising in a liquid depends on the bubbles' volume and surface area and other (physical and chemical) properties of gas and liquid [3]. For instance, air bubbles in water are almost spherical up to millimeter size (small bubbles to the **left**), become approximately ellipsoidal in the several millimeter range and can be irregularly shaped when much larger. The image to the **right** shows an area in the North Sea where methane is released from the ocean floor and forms distinct bubble streams. Measuring the flux, the rise speed, and the bubble size distribution is of crucial importance for larger scale ocean and atmosphere models.

Methane, however, represents a strong greenhouse gas on Earth. Consequently, if methane release from the seabed migrates towards the atmosphere, it then represents a threat in terms of global warming [2,4], but even very rough estimates for the oceans' contributions to the atmosphere's methane budget are very challenging. The majority of known marine gas seepages originates from microbial degradation of organic matter into methane. Subsequently, the major fraction of methane is filtered *in situ* by a complex microbial community on the seabed [5]. However, gas bubbles bypass this benthic filter enabling a highly efficient transport of CH_4 from the sediment into the water column and ultimately to the atmosphere. To a lesser degree, the release of gas from the seafloor is associated with sub-seabed volcanic activity and emission of abiotic CO_2, methane, and H_2S gas [6]. Recent work also discusses anthropogenic contributions of gas leakage from the seafloor by marine oil or gas exploitation activities, carbon storage and enhanced oil recovery facilities, gas pipelines, and abandoned wells [7,8]. After gas bubbles release from the seafloor into the water column, the bubbles undergo diffusion with ambient seawater. During their rise with decreasing hydrostatic pressure, gas bubbles may lose large amounts of their initial amounts of moles, e.g., driven by undersaturation of seawater with methane. Numerical models are available to predict the diffusion and dissolution of gas bubbles in seawater and allow for calculation of bubble lifetime, mole fractions, and rise height of seepage gas bubbles [9,10]. The input parameters initial gas bubble size, water depth, gas mole fraction, and environmental properties of the ambient seawater are required for modeling the chemical bubble evolution and lifetime. However, only a very limited number of studies investigated the crucial parameter of bubble size distribution so far [11,12]. So called gas hydrate skin effects at water depths greater than approximately 500 m further complicate the physicochemical behavior and lifetime modeling of gas bubbles [13]. Shipborn acoustic echosounding represents the state-of-the-art technique to remotely sense marine seepage. Until recently, singlebeam echosounders were operated to reliably sense gas seepage sites even beyond 2000 m water depth [14]. Today, wide coverage multibeam echosounder systems can be used for more efficient *in situ* and shipborn bubble detection [15–17] and even allow for bubble rise velocimetry adapted from visual computer vision [18]. However, the acoustic absolute quantification of seepage flux remains a very challenging task, and without the knowledge of bubble size distributions and rise velocity, the inversion of single frequency echosounder data into gas flux estimates remains ambiguous [19]. Moreover, acoustic analyses of individual bubble shape, small-scale bubble trajectory, rising speed, surface characterization, exact bubble size determination, and upwelling phenomena of ambient seawater are limited due to wavelength and resolution. This shortcoming can be overcome by high resolution *in situ* visual sensors that allow for measuring the bubble size distribution of a stream and the rise velocity of bubbles (see e.g., [20–22]). These numbers can then again be used to enhance acoustic inversion and bubble lifetime modeling and support regional acoustic inversions. In this contribution, we present first investigations towards a novel 3D sensor called bubble box to quantitatively investigate natural gas seepage bubbles in the oceans. It can be deployed on or towed across the seafloor using a TV-guided frame or an ROV (remotely operated vehicle) and is depth rated up to 6000 m. The goal of this work is automated determination of bubble size distribution and rise velocity as well as overall gas flux. The bubble box in its current design (see Figure 2 for sample images) allows for flux validation by capturing a reference volume at the top. It is equipped with a wide baseline stereo camera system and background illumination to capture the 3D shape of rising bubbles in order to obtain bubble characteristics for a certain vent, or ultimately, the statistics of vents in an entire area.

Figure 2. Three views of the bubble box, from **left** to **right**: box without cameras during development, box on ROV Phoca before being deployed (research cruise on RV Poseidon) and box in shallow water near Panarea, an area with heavy carbon dioxide release.

State of the Art in Visual Characterization of Bubble Streams

Mechanical installations such as inverted funnels can help estimating an average flux but cannot provide the rise speed or bubble size distribution required by diffusion and dissolution models as discussed above. This information is also an input parameter of inverse acoustical methods to compute flux from large scale sonar surveys. On the other hand, in visual data it is possible to distinguish and track bubbles for rise speed and size distribution analyses, so that we discuss the state of the art in visual characterization of bubble streams in the next paragraphs.

Several visual systems have been presented for *in situ* measurement of bubbles in the ocean. The systems differ in whether they are intended for shallow water and waves [20], for moderate depth [12,23] or for the deep sea [21]. Besides dedicated sensors, bubbles have also been analyzed from ROV cameras [24]. These systems have been primarily designed for capturing good data and require a lot of interactive manual work to process the data afterwards. All systems referenced above rely on a monocular camera. Monocular cameras are well suited for observing spherical bubbles, when the distance of the bubble to the camera is exactly known. However, even if the distance to the seepage source is measured carefully, the camera distance to individual bubbles in a stream may vary, as bubbles can show zig-zag motion or oscillation (*cf.* to [3]) and several bubbles can appear next to each other. By the intercept theorem from elementary geometry, it can be understood that an error of 25% in assumed distance results also in an error of 25% in estimated sphere radius and a volume error of almost 100% (*cf.* Figure 3). In order to remedy this problem, recently Wang and Socolofsky [22] have presented a stereo-camera system. The use of two cameras allows for measuring the distance of each bubble to the camera more precisely, even if it is not exactly in the center of an assumed corridor. Then, for spherical bubbles, the 3D sphere size can be computed from the observed circle in the image. As we will outline in the next section, the stereo configuration can however be improved for analysis of ellipsoidal bubbles (see Figure 1 for different shapes).

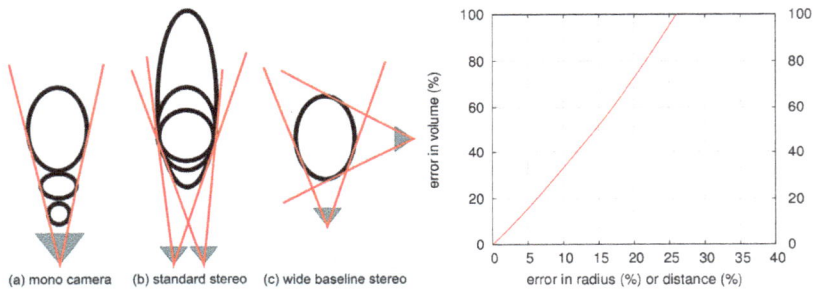

(a) mono camera   (b) standard stereo   (c) wide baseline stereo   error in radius (%) or distance (%)

Figure 3. Different settings to observe bubbles: (**a**) For a monocular camera a large distant bubble looks the same as a small bubble closeby; (**b**) A small baseline stereo camera system can measure the distance of spherical bubbles, but the extent of ellipsoidal bubbles in viewing direction is very uncertain; (**c**) A wide baseline stereo system can determine the ellipsoidal shape well.

The problem is also related to image processing based on particle tracking in lab environments. Here, [25] compute the rise speed of the center of mass of bubbles in an aquarium using a mirror but do not perform volume or sizes estimates. Bian *et al.* [26] estimate a restricted ellipsoid model for a single bubble. Both studies do not report on how they handle the refraction at the air-glass-water interfaces, which has to be considered properly in multi-media photogrammetry (*cf.* e.g., to [27]). Towards this end, recently, novel techniques for modeling and calibration of distance, thickness, and normal of the interface of the camera housing have been proposed [28,29]. We carefully analyze the calibration issues for the proposed bubble box sensor. In terms of pure 2D image processing, [30] presents automated detection and shape estimation as well as tracking using a monocular camera. In the next section, we will present the details of our system and outline the design considerations.

2. Bubble Box Design and Calibration

The visual sensor that we entitle bubble box can be seen in Figure 2. The goal of the sensor is to measure rise speed, flux, shape, and bubble size distribution at the source position of methane or carbon dioxide seeps at the ocean floor. For this, the system must be both mobile and robust, such that it can be deployed by a remotely operated vehicle (ROV) from a research vessel. The body therefore consists of a steel frame of a 30 cm by 30 cm footprint on the ground (plus surrounding plate for carrying batteries), with approximately 1 m height.

2.1. Design Considerations

The box provides a vertical corridor where a bubble stream can rise through. To avoid horizontal drift of bubbles due to currents, the corridor is protected to the sides by acrylic glass. The corridor width of 20 cm was chosen to minimize boundary effects for thin bubble streams with a distinct source, when the box is centered on the source. Two of the four outer vertical faces of the corridor are made of white acrylic glass with strong scattering and little attenuation properties (6 mm polymethylmethacrylate, "Perspex"). These are illuminated from the back and function as bright, white background plates when the box is in operation (different illumination setting tests, e.g., from the front, from the top and from the back can be seen in Figure 4, and compare also to [21]). The two other vertical faces use transparent acrylic glass and contain a camera each. The system has been designed to carry robust deep sea cameras (1024 × 1024 pixel resolution machine vision cameras with 70° field of view, controlled by a mini computer, in titanium housings with 10 cm diameter dome ports). Both cameras are triggered using a separate micro controller and send the images via a gigabit Ethernet connection to a small computer inside the pressure housing. Images are finally stored on a shock resistant 1 TB solid state disk. The system can be powered by the ROV that deploys it or as a standalone system and requires approximately 60 W (including lighting) during operation and negligible power during standby. Storage and power are sufficient for single day missions of many deployments on several seeps but are currently a limiting factor for long term monitoring. We have also built a light-weight version using (1280 × 720 pixel resolution) GoPro Hero3 cameras in Polyoxymethylene housings for shallow water). The cameras are arranged in a way that they observe the bubble corridor from a distance of 10 cm and from two approximately orthogonal perspectives. This way, each dimension of a bubble is observed by at least one of the cameras. In contrast, in standard stereo settings, where both cameras are next to each other and share almost the same perspective, noisy observations of bubble rims lead to high uncertainty of the bubble extent in viewing direction (see also Figure 3).

The target capture rate of the system is 120 Hz. The observation corridor is approximately 20 cm high, which means that a bubble rising with 25 cm/s (compare [9] for a thorough discussion on rise speeds) in the center of the corridor will be photographed approximately 100 times. Faster bubbles will be photographed slightly less frequently but can still be evaluated. The exposure time has to be selected such that a bubble moves less than half a pixel. In the scenario above, this is approximately 0.1 mm and leads to an exposure time of less than 5 ms. We are generally interested in observing bubbles larger than 0.5 mm in diameter (that have enough buoyancy to detach from the sediment and rise towards the surface). The minimum bubble size both camera systems can observe is therefore in the order of 0.5 mm in diameter. If smaller bubbles need to be observed (e.g., created from exploding bigger bubbles), different lenses or cameras with higher resolutions could be used, but this is out of scope of this article.

The acrylic glass walls are each illuminated by a high-powered LED from the back, which is synchronized to output all energy during the exposure time of the cameras. To make the system more lightweight/mobile, it is also possible to illuminate from the back using e.g., ROV lights or diver-provided illumination.

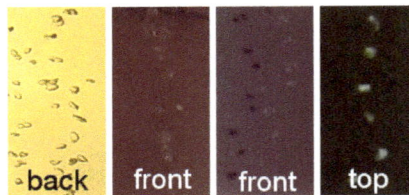

Figure 4. Different lighting configurations. By far the best results are achieved by back lighting, *i.e.*, using a diffuse light source behind the bubbles. This causes the bubbles to appear as dark contours. When using a diffuse light source from the same direction as the camera, the bubbles are less well distinguishable from the background. Using non-diffuse front lighting causes the bubbles to have shadows, which complicates the detection process. Another interesting option can be top lighting, which can be seen in the right image.

2.2. Calibration

In this contribution, the term camera calibration stands for geometric calibration, *i.e.*, determining which pixel corresponds to which 3D ray in water. In the case of a stereo camera rig, where the cameras are rigidly mounted, this means that the cameras need to be synchronized in order to gain corresponding images of the bubble stream at the same time. This, in turn, allows for determining the extrinsic parameters, *i.e.*, the position and orientation of one camera with respect to the other. Additionally, intrinsic parameters like focal length, but also of the underwater housing need to be calibrated.

2.2.1. Synchronization

The deep sea camera system is hardware triggered and will capture images at the same time. In order to temporally calibrate the lightweight stereo system of GoPros, and later to match bubbles across a stereo pair, the two video streams need to be synchronized with an accuracy down to a single frame even at frame rates of 120 frames per second. The GoPro Hero 3 black edition does not allow synchronized capture with that accuracy. To allow temporal alignment of the videos in postprocessing, a short blinking light (flash) is presented such that both cameras can see it. Then, the streams are synchronized manually by importing both streams into the open source video editing software cinelerra (http://cinelerra.org) and aligning the light on/off frames. The corresponding left and right images were put together into a single (synchronized) stereo image that was then later used for further processing for calibration and bubble measurement. This method allows to synchronize the streams but requires some manual work.

2.2.2. Stereo Calibration

The perspective camera model in air describes how a camera projects 3D points into the image plane, after intersecting the center of projection (single-view-point camera). In this work, a standard model with focal length, principal point, and radial distortion [31] is used and calibration is performed according to the method described in [32]. It relies on checkerboard images that are captured from different points of view, including different distances from the camera and different angles relative to the camera's optical axis.

When using a camera underwater, it needs to be enclosed in an underwater housing viewing the scene through a glass port. In case of flat ports, the different media (water, glass, air) cause the light to be refracted, and hence the perspective camera model to become invalid, introducing a systematic measurement error. The magnitude of this systematic measurement error depends on the camera-glass configuration and the distance of objects. Dome ports, semi-spherical glass housings, do not suffer from refraction, when the camera center is aligned with the sphere center. On the other hand, cameras behind flat ports that are centimeters away from the glass, possibly even with an inclination angle between camera and glass, produce a large measurement error if refraction is ignored. Each such underwater camera system that is to be used for measurements needs to be carefully evaluated to see if refraction needs to be modeled explicitly (refer for example to [28,33,34] for calibration methods). Treibitz *et al.* [33] showed that the caustic size is a measure for deviation from the single view point camera model. In particular, for the center bubble area in a low resolution GoPro mounted very close to the interface, the caustic is negligible and perspective calibration can be used. This however depends on the actual camera and housing used as well as on the setup and the introduced error has to be analyzed carefully. In any case, the cameras need to be calibrated by capturing a set of underwater checkerboard images as in Figure 5. Then, depending on the expected caustic size, either the perspective calibration method is used (not explicitly modeling refraction [35]) or a refractive method, e.g., [28] can be used to determine housing parameters like glass distance and inclination.

Figure 5. Calibration in water, sample input images from different points of view.

Afterwards, the relative translation and rotation between the two cameras need to be determined by using checkerboard images that are completely visible in both cameras.

3. Bubble Stream Characterization

The previous section described the hardware and the calibration routines for the bubble box. Once a set of bubble images has been captured, automated methods for image processing are required that estimate properties like overall volume, rise velocity, and size distribution. In a preprocessing step, the images' regions of interest (ROI) are determined and in an optional step, possible blur is removed. Then, the bubbles are detected and matched across the stereo image pairs. This allows for computing ellipsoids for each bubble. Finally, it is of interest to track bubble streams over time to determine 3D bubble paths and to compute rise velocities. Refer also to Figure 6 for an overview of the method.

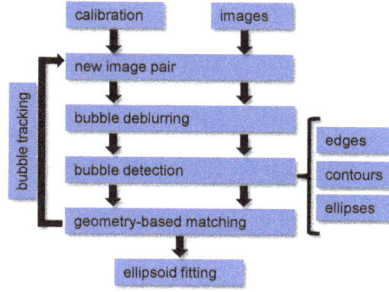

Figure 6. Flow diagram of the stereo bubble box data processing steps.

3.1. Preprocessing—Automatic Region of Interest Selection

It is assumed that only the bubbles move within the images and that they occupy a fairly constant region of interest (ROI). However, bubble detection is based on gradients within the images and will also fire on other possible structures like, for example, markers for measurement as can be seen on the diffuse acrylic plates in Figure 2. Therefore, it is important to have a region of interest in both images that contain the bubble plume only. This can either be determined manually or by utilizing the assumption that the only moving objects in the images are the bubbles. This movement can then be detected using optical flow methods [36]. The areas of the images, where movement is measured over time are then accumulated in a heat map and the region of interest showing the bubbles to be measured is the resulting surrounding rectangle (refer to Figure 7). For the evaluation presented in this article, a static ROI has been computed for each sequence and was not changed during the sequence. In addition to determining the ROIs, computing the heat map by using optical flow methods allows to initially predict bubble motion between consecutive frames, which will be used for tracking bubbles across time. However, the optical flow estimation results contain outliers due to sporadic large bubble displacements, such that the results usually cannot be utilized directly for bubble tracking over time.

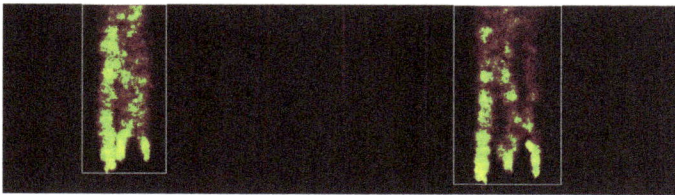

Figure 7. Heat map showing the automatically determined bounding boxes (**white**) for bubble detection in the left and right stereo images.

139

3.2. Preprocessing—Bubble Deblurring

As already stated in the introduction, the frame rates for capturing gas bubbles need to be high and exposure times need to be short in order to avoid motion blur in the images. Consequently, the capture of bubble streams requires strong illumination. For long term observations with limited energy at the sea floor, or for lightweight deployment of the system without integrated lighting and battery (*i.e.*, mobile lights from divers), too little light might be available for good quality images. To some extent, this can be compensated by longer exposure times that lead to moderate motion blur or a larger aperture that leads to a smaller depth of field [37]. However, larger apertures can cause the bubble to be outside the depth of field and will hence cause defocus blur.

Both blur effects skew the measurements of the bubble's shape and size, which are important measures in bubble box applications. The fidelity of images acquired under less-than-perfect conditions can be improved by blind deconvolution techniques with a gradient sparsity prior. This will compensate for the influence of motion and defocus blur to some extent. This section focuses on the practical application of the algorithm, for more details, see results already published in [38].

Blind deconvolution is based on the following image model. The observed image O is formed from the undisturbed image S, convolved with the blur kernel or point spread function (PSF) B and additional noise n:

$$O = S \otimes B + n \tag{1}$$

Blind deconvolution means that both S and B have to be recovered from a single measurement O. This problem of recovering these two unknowns (S, B) from one measurement (O) is ill-posed in general as shown in [39]. However, by using additional constraints or exploiting special properties, solutions can be found [39,40]. For instance, [41] uses a heavy tailed gradient sparsity prior, designed for the restoration of natural images [39]. Bubble box images typically show a uniform background with low gradient values and bubbles with a stronger gradient rim. Consequently, also for bubble sharpening, a MAP (Maximum *a posteriori*)-estimation gradient sparsity blind deconvolution algorithm (*cf.* to [41]) is suitable. The formulation is based on the Bayesian framework and uses the Maximum *a posteriori* (MAP) principle as its foundation. For the MAP principle, the latent distribution of the observed image O is denoted as $P(O)$. $P(S)$ and $P(B)$ denote the *a priori* distributions of the undisturbed image and of the blur kernel, both with applied priors. The algorithm recovers the undisturbed image S and the blur kernel B, by finding their MAP, dependent on O:

$$P(S, B|O) = \frac{P(O|S, B)P(S, B)}{P(O)}$$
$$\propto P(O|S, B)P(S, B) = P(O|S, B)P(S)P(B) \tag{2}$$

while $P(O|S, B)$ follows a Gaussian distribution with parameter γ that encapsulates the Gaussian's standard deviation:

$$P(O|(S, B)) \propto e^{-\frac{\gamma}{2}\|S \otimes B - O\|^2} \tag{3}$$

On the recovered signal S and the blur kernel B, the regularizers $Q(S)$ and $R(B)$ are employed. Q controls the gradient sparsity prior on the image and R the assumptions on the blur kernel. This allows the recovery of the blur kernel and sharp image from the blurred image. For a more detailed description of the algorithm, see [41].

3.3. Bubble Detection

Within the region of interest in an image, the bubbles are to be detected and tracked automatically (refer to [30] for a more detailed description of the bubble detector and tracker). For this, the images are interpreted as a function mapping a gray value to each pixel. In this function, gradients can be computed and used for line-detection with the Canny edge-detector as in [21]. The edges are then used to determine the bubble contours by tracing the convex hull of the connected components. Extra contours inside the bubbles are rejected, by checking for overlapping detections. The final ellipses are fit into the remaining contours using a method by [42]. Figure 8 shows intermediate and final results of bubble detection.

At this point, the image ellipses around the detected bubbles would allow the calculation of the volume by assuming that each bubble is a rotationally symmetric, 3D ellipsoid. This method based on a monocular camera only is often used in the literature for bubble volume estimation. An additional assumption required in this case is that the distance between camera and bubble is constant and known, thus allowing to convert the measured volume to metric units, *i.e.*, computing a factor, which determines how many pixels correspond to one millimeter. This method works well if the bubble box contains only one bubble plume at a precisely known distance to the camera. However, deviations from the assumed distance affect the estimated volume with an error in the third power of the error of the distance, resp. conversion factor (see Figure 3).

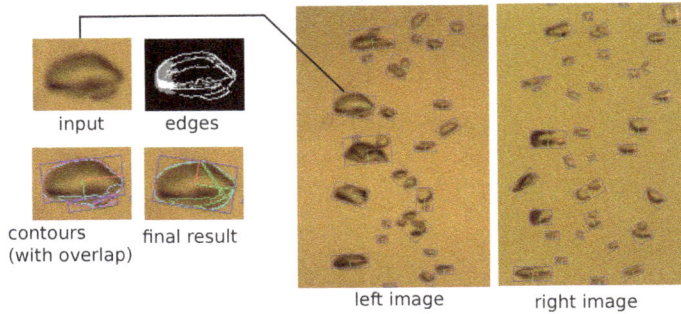

Figure 8. Detected bubbles in image ROIs. In **cyan**: contour around the bubble rim. In **blue**: rectangles describing the ellipse.

3.4. Stereo Matching and Ellipsoid Triangulation

In the case of more than one bubble plume being measured in the bubble box or in case the distance between bubbles and camera is not known with high accuracy, a wide-baseline stereo camera rig is used to accurately triangulate 3D bubble positions and reconstruct their ellipsoidal shape approximation. This is achieved by utilizing the knowledge about the calibrated rig to compute 3D cones of sampled rays for each bubble rim, which can then be intersected. Consequently, the exact distance of each bubble from the two cameras is computed. This allows not only precise volume measurements, but also to reconstruct the 3D information of single bubbles in the plume and track its 3D movement upwards.

The method then works as shown in Figure 6. The stereo rig calibration and a series of input images serve as input. In each image pair, the bubbles are detected with the method described in Section 3.3. Due to the wide baseline, it is not possible to match the bubbles according to their shape (refer to Figure 8), so matching has to rely on two-view geometric constraints. For perspective cameras, this can be done using epipolar geometry [43], based on the idea that a point in one image has to lie on a corresponding line in the second image. However, in presence of refraction, epipolar lines become curves. In this case, we use piecewise, linear approximations of the epipolar curve. Note that for this kind of geometric matching, the rig calibration needs to be very accurate, otherwise, the computed epipolar lines will not intersect the matching bubbles. The distances to the epipolar lines are used as weights. Then, the two sets of bubbles form the two sets of a weighted, bipartite matching problem (see Figure 9 for results).

With the stereo correspondences, the 3D position and ellipsoidal shape of the bubbles are triangulated, by utilizing the calibration information about the camera configuration. For the purpose of shape reconstruction, different approaches are feasible. [26] uses an analytic approach of calculating the parameters, which delivers

142

exact results and also requires exact synchronization. Note that shape-from-silhouette approaches [44] are also an interesting alternative in general; however, they usually require a large number of projections of 3D points into the images. When using a refractive camera model, such approaches are prohibitively expensive. Therefore, the described system intersects the bubbles' viewing cones: for each bubble correspondence in a stereo image pair, the outer contour is sampled discretely and for each sample, the corresponding ray in space is computed. The set of 3D rays for each bubble contour forms a characteristic cone (see Figure 10). The bubble is then reconstructed by fitting an ellipsoid on the inside of the cone intersection. Note that in the presence of noisy observations, this kind of cone intersection is more ambiguous in small baseline stereo camera setups than in our wide-baseline setting (compare to Figure 3).

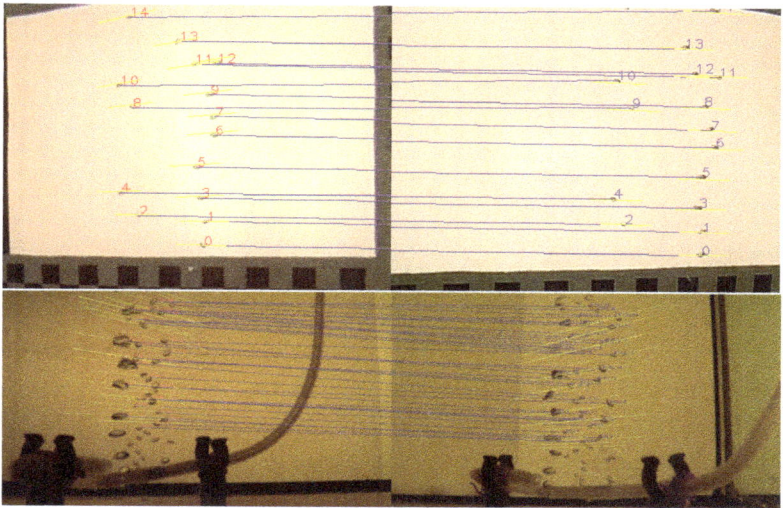

Figure 9. Exemplary results for stereo matching. **Top**: few bubbles, matching works very well. **Bottom**: many bubbles with overlap, where the matching procedure will make some errors.

Compared to estimating the bubble volume by assuming certain fixed distance for each bubble from the camera, the approach described in this section triangulates the 3D bubble position and determines its 3D ellipsoidal shape. From the ellipsoid, the bubble's volume can be computed directly. In case of failed matching, the bubble volume can still be approximated from one of the stereo images by assuming a rotationally symmetrical bubble at an average distance (traditional monocular assumption).

143

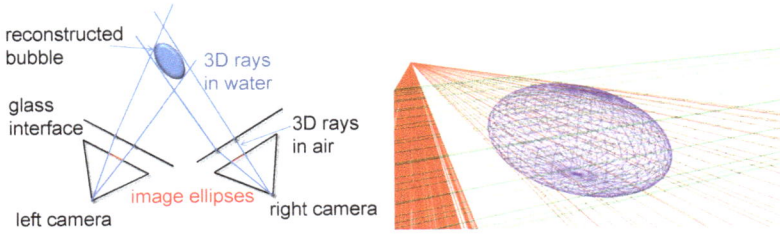

Figure 10. Left: ellipsoids are computed by determining the boundary rays corresponding to the contours in the images. **Right**: when viewed in 3D, so-called boundary cones are defined when discretely sampling the points on the contours in the images. The red cone is from the first camera and the green cone from the second camera. The blue ellipsoid is fitted to lie inside the cone intersection. Note that the computation of 3D rays does not require the projection of 3D points into the images, which is infeasible when using the refractive camera model.

3.5. Bubble Tracking

To reconstruct the motion of the bubbles, they are tracked over the image sequence, *i.e.*, over time. Due to the bubble usually following a constant upwards motion with certain deviations to the left and right, it is possible to specify an upwards and a sideways motion constraint, thus allowing the determination of a set of possible matching candidates between two consecutive images. The matching candidates are assigned a weighting factor based on the motion constraints, which leads to a minimum-weighted, bipartite matching problem, which is a classical graph problem and can be solved by the Hungarian algorithm [45]. When combining the correspondences over time, the entire trajectory of the individual bubbles can be inferred.

4. Assessment

Before evaluating the performance of the presented system, results of the optional deblurring step will be shown. Then, the described method will be evaluated on synthetic and real data.

4.1. Deblurring

Figure 11 shows results of the gradient sparsity algorithm as suggested in [41], which produces good results on the input image with only slight defocus (Figure 11a).

For motion-blurred images, the deblurring of [41] also shows a good improvement in sharpness, but remains incomplete, see Figure 12b. Clearer contours can be achieved as described in [38], *i.e.*, by using 2.5 times the weight for the sparsity prior of the blur kernel (compare Figure 12b,c). In addition, notice the strong halo artifacts around the bubbles caused by this parameter setting. However, they do not

144

disturb the bubble detection, due to the gradient on the bubble rim being increased, while the gradient between background and halo is low in comparison. Therefore, to compensate motion blur, these specialized parameters should be employed for bubble images.

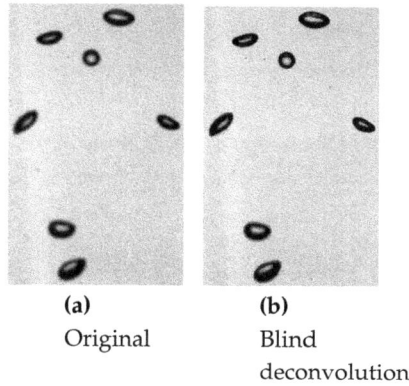

(a)
Original

(b)
Blind
deconvolution

Figure 11. Experimental results of blind deconvolution on an input image with defocus: (**a**) shows the input image and (**b**) the result of gradient sparsity blind deconvolution.

The correct restoration is confirmed by the recovered PSFs in Figure 13 with its characteristic shapes for defocus (cf. to Figure 13a) and upwards motion blur (cf. to Figure 13b)

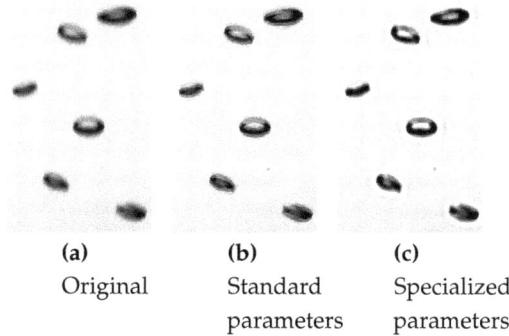

(a)
Original

(b)
Standard
parameters

(c)
Specialized
parameters

Figure 12. Cropped results of blind deconvolution on an input image with motion blur: (**a**) shows the cropped original image; (**b**) gradient sparsity blind deconvolution with standard; (**c**) gradient sparsity blind deconvolution, with a specialized parametrization for bubble box images, see Section 4.1 for details.

145

The results of deconvolution indicate that the fattening of bubble contours caused by motion and defocus blur, which leads to an overestimation of the bubble size, can be reduced. This has also been confirmed in a detailed analysis in [38]. Blind deconvolution is computationally expensive and the run-time for the gradient sparsity algorithm in a Matlab implementation is up to 10 s on an image with 261 pixels width and 612 pixels height. A hybrid approach that uses blind deconvolution to estimate the PSF once and uses a non-blind deconvolution with a lower run-time for the following images, is a feasible and practical approach.

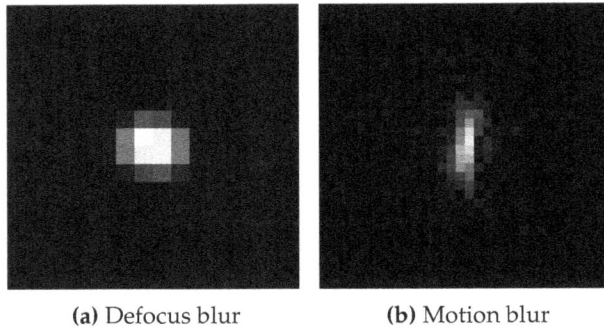

(a) Defocus blur **(b)** Motion blur

Figure 13. Blur kernels recovered by blind deconvolution, **(a)** shows the 15×15 blur kernel recovered from the defocused image in Figure 11a, while restoring Figure 11b; **(b)** shows the 32×32 blur kernel recovered from image with motion blur in Figure 12a, while restoring Figure 12b. Values between 0 (**black**) 1 (**white**), values scaled for better visibility.

4.2. Bubble Simulator

In order to verify the developed algorithms on ground truth data, a bubble simulator was used that rendered synthetic images with known ground truth. For this, a simplified model for the bubbles is implemented. Randomized 3D ellipsoids are generated that move upwards on a randomized path. Additionally, the length of the ellipsoids' main axes change from frame to frame without changing the volume. Thus, for each frame to be rendered, a known 3D representation with ellipsoids is known in addition to the exact 3D path a bubble took.

For rendering the images, we aim to simulate the back-lighting of the bubbles as shown in Figure 4. Bubbles do not follow the standard computer graphics Blinn–Phong shading model [46], but instead feature a darker rim and a bright core (see Figure 1). The reason for this darker rim is that while light reaching a bubble surface is refracted according to Snell's law, if the angle of incidence is small, total internal reflection occurs. Our approximation of this process is to model the color of the bubbles with the angle of incidence of the cast viewing ray. This allows rendering reasonable test images using a simple raycaster, without requiring the complexities of a full, non-sequential raytracer.

For comparison, the bubble stream is rendered from three points of view. The first two views are set in the wide baseline scenario with a 90° angle, while the first and the third camera feature the small-baseline scenario with a 20° angle. All three cameras view the same set of bubbles, and hence the simulator allows for the comparison of performances of the proposed wide-baseline scenario, the small-baseline scenario, and the use of a monocular camera. Each camera observes the bubbles using the refractive camera model and during bubble size computation, refraction at the underwater housing is modeled explicitly.

Using the simulator, we generated a sequence of 100 images with three bubble plumes. For each bubble, the ground truth position, volume, and size are known, therefore, the performance of the proposed system can be evaluated on this data. Figure 14 shows an exemplary input image of the simulator on the left, followed by a bubble detection result, a matching result, and a 3D view of the bubble plumes.

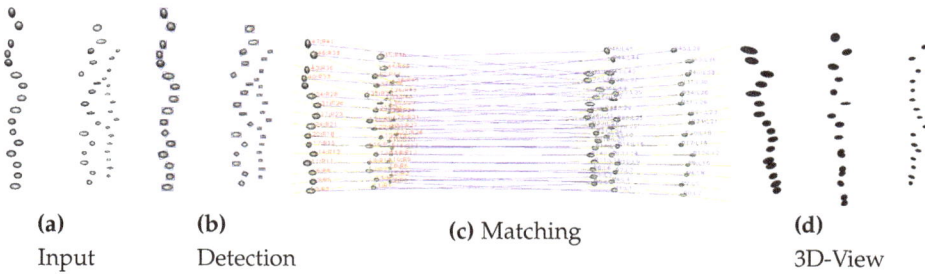

| (a) | (b) | (c) Matching | (d) |
| Input | Detection | | 3D-View |

Figure 14. Exemplary simulator images. From **left** to **right**: input image, bubble detection result, matching result, and 3D view of the original 3D bubbles. In case of the matching image (c), the blue line indicates matching bubbles and the yellow lines indicate the local approximations of the epipolar lines.

Table 1 shows a summary of the results over all images for the wide-baseline scenario (WB), the small-baseline scenario (SB), and using the monocular camera with an assumed pixel-to-mm conversion ratio (Mono). The ground truth volume of the bubble stream is higher than the estimate for the wide-baseline scenario by our algorithm due to some mismatches and therefore missed bubbles. Note that the average bubble volume was estimated accurately, showing that the ellipsoid computation at the 90° configuration yields good results. The velocity of the generated bubbles was higher than in real bubble scenarios, but was also estimated accurately. The results of the small-baseline scenario demonstrate the major problem of this method. Matching accuracy was comparable in both scenarios, but when computing the ellipsoid, bubble size is usually over-estimated due to ambiguities in bubble shape. This is demonstrated on an exemplary bubble in Figure 15, where the left image shows the ellipsoid in the wide-baseline scenario and the right image shows the ellipsoid computed in the small-baseline scenario (compare also to Figure 3). The histogram of bubble volumes in Figure 16 confirms this observation. While the results of the wide-baseline scenario match the ground truth histogram well, in the small-baseline setting the bubble volume in general is overestimated in our experiment. Table 1 also gives some results using a monocular camera. In this case, a fixed pixel-to-mm conversion ratio is assumed, to estimate bubble volume with the assumption that all bubbles have the same distance to the camera and are rotationally symmetric.

Table 1. Results on synthetic data. The first column shows the ground truth results. WB stands for wide-baseline scenario, SB for small-baseline scenario and Mono shows an estimate for using one camera only based on a pixel-to-mm conversion ratio.

	Ground Truth	WB	SB	Mono
Synthetic Data				
volume	127,85.72 mm^3	11,391.93 mm^3	18,823.21 mm^3	15,028.2 mm^3
average volume	42.62 mm^3	39.83 mm^3	66.04 mm^3	52.18 mm^3
average velocity	35.94 cm·s^{-1}	36.00 cm·s^{-1}	36.16 cm·s^{-1}	-
# bubbles	300	286	285	288

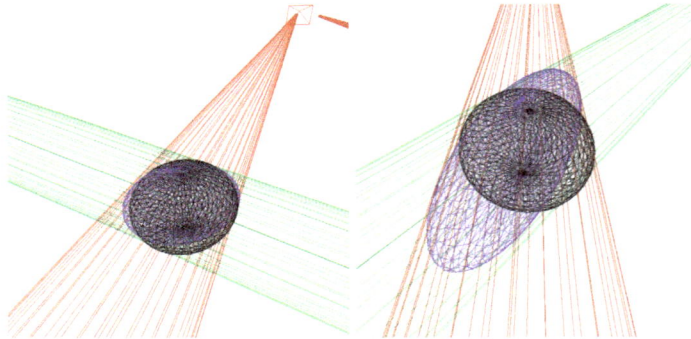

Figure 15. Ellipsoid triangulation from wide and small baseline. **Left**: wide baseline scenario, where blue shows the computed ellipsoid and black the ground truth ellipsoid. The green and red lines show the 3D cone limiting the ellipsoid. **Right**: small baseline scenario. Note how the blue ellipsoid cannot be determined uniquely due to the elongated intersection of the red and green viewing cone.

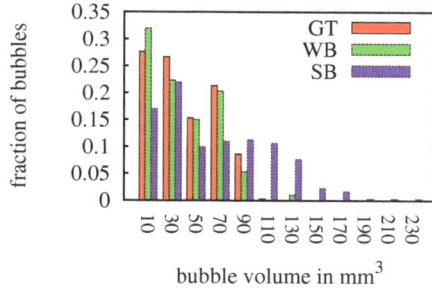

Figure 16. Histogram showing the bubble volume distribution on synthetic data computed in the wide-baseline scenario and the small-baseline scenario compared to ground truth.

4.3. Test Setup with Air Bubbles in Water

The camera setup of the bubble box has been tested prior to building the actual box using lab experiments in a fish tank filled with water (refer to Figure 17). Inside of the water, three bubble streams were produced. Note that in order to test the automated data processing algorithms, a simple stream of air coming out of a flexible tube is sufficient for which the actual overall volume was measured using a measuring cup. In order to illuminate the bubble streams, two adjoining glass walls of the fish tank were equipped with white acrylic glass that strongly diffuses the light. Both acrylic planes were lit with 1000 W halogen lamps. Inside the fish tank, two GoPro 3 cameras were set at a 90° angle at a distance of about 25 cm to the bubble streams. The GoPro cameras have a very small distance between glass and camera and the cutouts used in this example only utilize a small opening angle of the overall image. For this cutout, the caustic and therefore the deviation from the single-view-point camera model is very small and the cameras were calibrated perspectively using checkerboard images captured underwater.

By manipulating the tube, it is possible to create many or few, small or large bubbles (refer to Figure 17). However, for our real-world applications, we are mainly interested in bubble sizes of several millimeters, which maintain approximately ellipsoidal shapes.

The system was evaluated on a sequence of 600 stereo images and the summarized results can be seen in Table 2. The proposed method underestimates the flux, which can be explained by bubbles that could not be detected properly and are therefore very difficult to match with the geometric matching procedure and need to be discarded. In the last column, we therefore present an approximate estimation of the flux of the missed bubbles. This flux was estimated using the monocular approach, *i.e.*, by assuming a constant distance between camera and bubbles and a

149

known pixel-to-mm conversion ratio. Note, however, that, in this case, the distance between camera and bubbles is unknown and differs between the bubbles inside a plume but also between the three different outlets, and we therefore can only roughly estimate the volume. Additionally, bubbles that were not detected individually, but were overlapping are not compensated for.

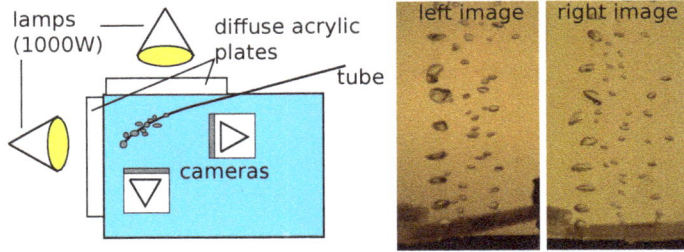

Figure 17. Left: test setup for evaluation of the system. Two back-illuminated planes provide a bright background behind an air bubble stream generated by a small tube. The cameras are arranged in a wide baseline setting viewing the scene from 90° different perspectives. **Middle** and **right**: exemplary images showing the produced bubble streams with small, ellipsoidal bubbles and large bubbles with arbitrary shapes.

Figure 18 shows histograms of the bubble volume distribution and the bubble velocity. Throughout the sequence, there were three bubble streams one of which consisted of larger bubbles, while the other two contained smaller bubbles. This is reflected in the one-sided volume histogram. The velocity is normal distributed for all bubbles and Figure 19 shows three different views of the reconstructed 3D bubbles.

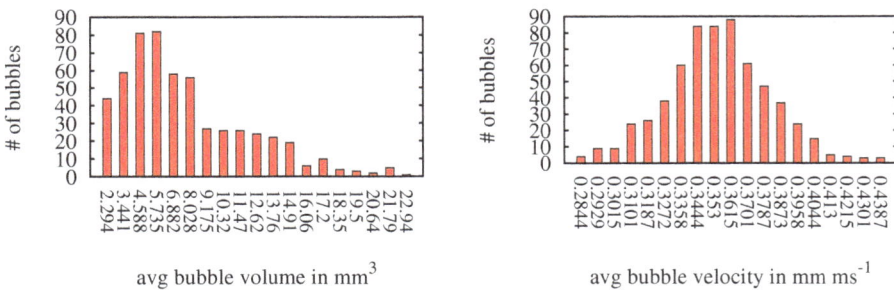

Figure 18. Histogram showing volume and velocity distribution of the bubbles.

Table 2. Results on real data image sequence.

	Measured	Results	Comment
Real Data			
flux	$4.177\ \mathrm{mL \cdot s^{-1}}$	$2.504\ \mathrm{mL \cdot s^{-1}}$	estimated volume of missed bubbles 0.291–$0.7488\ \mathrm{mL \cdot s^{-1}}$
velocity	–	$36.135\ \mathrm{cm \cdot s^{-1}}$	–

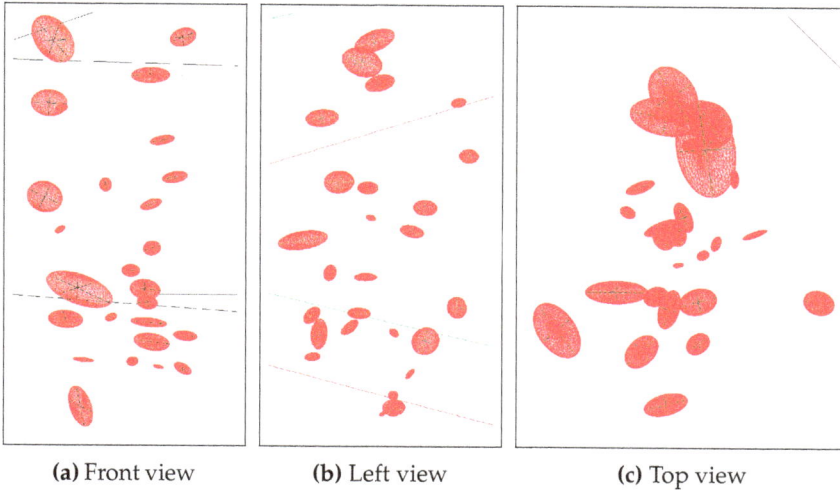

(a) Front view   (b) Left view   (c) Top view

Figure 19. 3D-Reconstruction of a bubble stream.

5. Discussion

The first results presented in the previous section show that the system has the potential for fully automated computation of bubble size distribution, volume, and rise velocity. In case of non-ideal lighting conditions, the images can be improved by the deconvolution method to avoid biased size estimates.

When considering the bubble stream characterization results, it can be seen that the proposed method works well on the synthetic image sequence, while it over-estimates bubble size when using the small-baseline camera setting on the same bubble stream.

On real data however, the proposed method under-estimates the bubble flux, which is due to bubbles not being detected properly mainly due to overlap and stereo matching ambiguities. Overlapping and occluding bubbles in the images are a challenge for all bubble quantification methods, and some examples are shown in Figure 20. In the future, we plan to improve our image processing methods to further disambiguate those cases. For example, we plan to experiment with active contour methods for separating partially overlapping bubbles (refer to [30]). Additionally, the wide baseline camera setup can be utilized to detect bubbles that are completely occluding each other in one image. In those cases, both individual bubbles should be visible in the second image and our ellipsoid intersection method should be able to compute a fairly good estimate of both bubble volumes by intersecting the viewing cones. However, for this, the two bubbles in the second image need to be correctly matched to the overlapping bubble in the first image, *i.e.*, one-to-many and many-to-many matches need to be handled. Additionally, looking at longer bubble trajectories rather than a single stereo pair could allow to further disambiguate overlap or occlusion situations in the future. It should, however, be noted that, compared to a single camera or small baseline system, chances for resolving occlusion scenarios are much better as one of the cameras will see the occluded bubble.

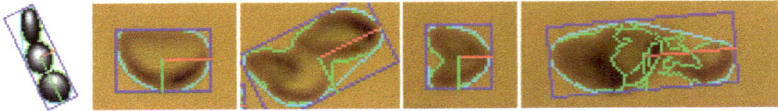

Figure 20. Exemplary bubble detection results. Green shows the detected contours, cyan, the convex hull, and the blue rectangle the final ellipse fitted around the detected contour. From **left** to **right**: erroneous detection on synthetic image; correct detection on real image; erroneous detection of two or three bubbles as one on real image; detection of two bubbles as one on real image; detection of two bubbles as one on real image with additional inner contours.

Where previous 2D methods had to assume a certain symmetric ellipsoid model, it could be shown that it is possible to drop this requirement and rather exploit the two different perspectives onto the ellipsoid, in order to estimate shape and volume. Although the wide baseline stereo setting requires slightly more calibration and synchronization efforts, the approach of using a monocular camera has its own disadvantages: in practical applications, it is very difficult or even impossible to estimate the correct factor for converting pixels into millimeters because this factor depends on the distance between camera and bubbles (20 cm × 20 cm corridor in the bubble box). However, errors in this factor are reflected with the third power in the resulting bubble volume (see Figure 3).

6. Conclusion

In conclusion, the proposed method still has some shortcomings in case of dense and multiple bubble plumes, where a lot of bubbles overlap in the images, but we suggest our measurement and automation will operate with high accuracy and reliability at single bubble seepage sites. Additionally, the hardware setting has the potential to disambiguate overlapping bubbles in the images. Future work will therefore focus on improving bubble detection and stereo matching. Also, the amount of available storage space is one of the limiting factors of the system. A long-term goal is therefore to do as much data processing as possible on site in order to eliminate the need of saving complete images.

Acknowledgments: The design of the bubble box was supported by discussions with Nikolaus Bigalke and Lisa Vielstädte. We highly appreciate the construction and manufacturing skills of Matthias Wieck and the support from Eduard Fabrizius (electronics) and Jan Sticklus (lighting). Adaptions for proper handling with the work class ROV PHOCA were supported by Martin Pieper. We would also like to thank the crew of R/V POSEIDON and the GEOMAR ROV team. For the design, construction, and field work we received funding from the SUGAR II (BMBF and BMWi, grant 3G819), and by the FUTURE OCEAN project (DFG) with grant CP1207. Data processing and algorithm development were developed in the frame of the project QUABBLE (FUTURE OCEAN, grant CP1331).

Author Contributions: All authors contributed to writing the article. Jens Schneider and Kevin Köser were responsible for the hardware design of the bubble box, Anne Jordt, Claudius Zelenka, Reinhard Koch and Kevin Köser for the algorithmic design and setup evaluation.

Conflicts of Interest: The authors declare no conflict of interest.

References

1. Fleischer, P.; Orsi, T.; Richardson, M.; Anderson, A. Distribution of free gas in marine sediments: A global overview. *Geo-Mar. Lett.* **2001**, *21*, 103–122.

2. Judd, A. Natural seabed gas seeps as sources of atmospheric methane. *Environ. Geol.* **2004**, *46*, 988–996.

3. Clift, R.; Grace, J.R.; Weber, M.E. *Bubbles, Drops, and Particles*; Academic Press: Waltham, MA, USA, 1978.

4. Ciais, P.; Sabine, C.; Bala, G.; Bopp, L.; Brovkin, V.; Canadell, J.; Chhabra, A.; DeFries, R.; Galloway, J.; Heimann, M.; *et al.* Carbon and Other Biogeochemical Cycles. In *Climate Change 2013: The Physical Science Basis. Contribution of Working Group I to the Fifth Assessment Report of the Intergovernmental Panel on Climate Change*; Cambridge University Press: New York, NY, USA, 2013.

5. Sommer, S.; Pfannkuche, O.; Linke, P.; Luff, R.; Greinert, J.; Drews, M.; Gubsch, S.; Pieper, M.; Poser, M.; Viergutz, T. Efficiency of the benthic filter: Biological control of the emission of dissolved methane from sediments containing shallow gas hydrates at Hydrate Ridge. *Glob. Biogeochem. Cycles* **2006**, *20*, 650–664.

6. Chadwick, W.W.; Merle, S.G.; Buck, N.J.; Lavelle, J.W.; Resing, J.A.; Ferrini, V. Imaging of CO_2 bubble plumes above an erupting submarine volcano, NW Rota-1, Mariana Arc. *Geochem. Geophys. Geosyst.* **2014**, *15*, 4325–4342.

7. Vielstädte, L.; Karstens, J.; Haeckel, M.; Schmidt, M.; Linke, P.; Reimann, S.; Liebetrau, V.; McGinnis, D.F.; Wallmann, K. Quantification of methane emissions at abandoned gas wells in the Central North Sea. *Mar. Pet. Geol.* **2015**, in press.

8. Schneider von Deimling, J.; Linke, P.; Schmidt, M.; Rehder G. Ongoing methane discharge at well site 22/4b (North Sea) and discovery of a spiral vortex bubble plume motion *Mar. Pet. Geol.* **2015**

9. Leifer, I.; Patro, R.K. The bubble mechanism for methane transport from the shallow sea bed to the surface: A review and sensitivity study. *Cont. Shelf Res.* **2002**, *22*, 2409–2428.

10. McGinnis, D.F.; Greinert, J.; Artemov, Y.; Beaubien, S.E.; Wüest, A. Fate of rising methane bubbles in stratified waters: How much methane reaches the atmosphere? *J. Geophys. Res. Oceans* **2006**, *111*, 141–152.

11. Leifer, I.; Boles, J. Measurement of marine hydrocarbon seep flow through fractured rock and unconsolidated sediment. *Mar. Pet. Geol.* **2005**, *22*, 551–568.

12. Leifer, I.; Culling, D. Formation of seep bubble plumes in the Coal Oil Point seep field. *Geo-Mar. Lett.* **2010**, *30*, 339–353.

13. Rehder, G.; Brewer, P.W.; Peltzer, E.T.; Friederich, G. Enhanced lifetime of methane bubble streams within the deep ocean. *Geophys. Res. Lett.* **2002**, *29*, 21-1–21-4.

14. Merewether, R.; Olsson, M.S.; Lonsdale, P. Acoustically detected hydrocarbon plumes rising from 2-km depths in Guaymas Basin, Gulf of California. *J. Geophys. Res. Solid Earth* **1985**, *90*, 3075–3085.

15. Greinert, J. Monitoring temporal variability of bubble release at seeps: The hydroacoustic swath system GasQuant. *J. Geophys. Res. Oceans* **2008**, *113*, 827–830.

16. Schneider von Deimling, J.; Brockhoff, J.; Greinert, J. Flare imaging with multibeam systems: Data processing for bubble detection at seeps. *Geochem. Geophys. Geosyst.* **2007**, *8*, 57–77.

17. Colbo, K.; Ross, T.; Brown, C.; Weber, T. A review of oceanographic applications of water column data from multibeam echosounders. *Estuar. Coast. Shelf Sci.* **2014**, *145*, 41–56.

18. Schneider von Deimling, J.; Papenberg, C. Technical Note: Detection of gas bubble leakage via correlation of water column multibeam images. *Ocean Sci.* **2012**, *8*, 175–181.

19. Veloso, M.; Greinert, J.; Mienert, J.; de Batist, M. A new methodology for quantifying bubble flow rates in deep water using splitbeam echosounders: Examples from the Arctic offshore NW-Svalbard. *Limnol. Oceanogr. Methods* **2015**, *13*, 267–287.

20. Leifer, I.; de Leeuw, G.; Cohen, L.H. Optical Measurement of Bubbles: System Design and Application. *J. Atmos. Ocean. Technol.* **2003**, *20*, 1317–1332.

21. Thomanek, K.; Zielinski, O.; Sahling, H.; Bohrmann, G. Automated gas bubble imaging at sea floor—A new method of *in situ* gas flux quantification. *Ocean Sci.* **2010**, *6*, 549–562.

22. Wang, B.; Socolofsky, S.A. A deep-sea, high-speed, stereoscopic imaging system for *in situ* measurement of natural seep bubble and droplet characteristics. *Deep Sea Res. I Oceanogr. Res. Pap.* **2015**, *104*, 134–148.

23. Leifer, I. Characteristics and scaling of bubble plumes from marine hydrocarbon seepage in the Coal Oil Point seep field. *J. Geophys. Res. Oceans* **2010**, *115*, 45–54.

24. Sahling, H.; Bohrmann, G.; Artemov, Y.G.; Bahr, A.; Brüning, M.; Klapp, S.A.; Klaucke, I.; Kozlova, E.; Nikolovska, A.; Pape, T.; *et al.* Vodyanitskii mud volcano, Sorokin trough, Black Sea: Geological characterization and quantification of gas bubble streams. *Mar. Pet. Geol.* **2009**, *26*, 1799–1811.

25. Xue, T.; Qu, L.; Wu, B. Matching and 3-D Reconstruction of Multibubbles Based on Virtual Stereo Vision. *IEEE Trans. Instrum. Measur.* **2014**, *63*, 1639–1647.

26. Bian, Y.; Dong, F.; Zhang, W.; Wang, H.; Tan, C.; Zhang, Z. 3D reconstruction of single rising bubble in water using digital image processing and characteristic matrix. *Particuology* **2013**, *11*, 170–183.

27. Kotowski, R. Phototriangulation in multi-media photogrammetry. *Int. Arch. Photogramm. Remote Sens.* **1988**, *27*, B5.

28. Jordt-Sedlazeck, A.; Koch, R. Refractive Calibration of Underwater Cameras. In *Computer Vision–ECCV 2012*; Fitzgibbon, A., Lazebnik, S., Perona, P., Sato, Y., Schmid, C., Eds.; Springer: Berlin, Germany, 2012; Volume 7576, pp. 846–859.

29. Jordt, A. Underwater 3D Reconstruction Based on Physical Models for Refraction and Underwater Light Propagation. Ph.D. Thesis, Kiel University, Kiel, Germany, 2013.

30. Zelenka, C. Gas Bubble Shape Measurement and Analysis. In *Pattern Recognition*, Proceedings of the 36th German Conference on Pattern Recognition, (GCPR 2014), Münster, Germany, 2–5 September 2014; pp. 743–749.

31. Brown, D.C. Close-range camera calibration. *Photogramm. Eng.* **1971**, *37*, 855–866.

32. Schiller, I.; Beder, C.; Koch, R. Calibration of a PMD-camera using a planar calibration pattern together with a multi-camera setup. *Int. Arch. Photogramm. Remote Sens. Spat. Inf. Sci.* **2008**, *21*, 297–302.

33. Treibitz, T.; Schechner, Y.Y.; Kunz, C.; Singh, H. Flat Refractive Geometry. *IEEE Trans. Pattern Anal. Mach. Intell.* **2012**, *34*, 51–65.

34. Agrawal, A.; Ramalingam, S.; Taguchi, Y.; Chari, V. A theory of multi-layer flat refractive geometry. In Proceedings of the IEEE Conference on Computer Vision and Pattern Recognition (CVPR), Providence, RI, USA, 16–21 June 2012; pp. 3346–3353.

35. Harvey, E.S.; Shortis, M.R. Calibration stability of an underwater stereo-video system: Implications for measurement accuracy and precision. *Mar. Technol. Soc. J.* **1998**, *32*, 3–17.

36. Farnebäck, G. Two-frame Motion Estimation Based on Polynomial Expansion. In *Image Analysis*, Proceedings of the 13th Scandinavian Conference, Halmstad, Sweden, 29 June–2 July 2003; Springer-Verlag: Berlin, Germany, 2003; pp. 363–370.

37. Born, M.; Wolf, E. *Principles of Optics: Electromagnetic Theory of Propagation, Interference and Diffraction of Light*, 6th ed.; Pergamon Press: Oxford, UK; New York, NY, USA, 1980.

38. Zelenka, C.; Koch, R. Blind Deconvolution on Underwater Images for Gas Bubble Measurement. *Int. Arch. Photogramm. Remote Sens. Spa. Inf. Sci.* **2015**, *XL-5/W5*, 239–244.

39. Levin, A.; Weiss, Y.; Durand, F.; Freeman, W.T. Understanding Blind Deconvolution Algorithms. *IEEE Trans. Pattern Anal. Mach. Intell.* **2011**, *33*, 2354–2367.

40. Perrone, D.; Favaro, P. Total Variation Blind Deconvolution: The Devil Is in the Details. In Proceedings of the IEEE Conference on Computer Vision and Pattern Recognition (CVPR), Columbus, OH, USA, 23–28 June 2014; pp. 2909–2916.

41. Kotera, J.; Šroubek, F.; Milanfar, P. Blind deconvolution using alternating maximum a posteriori estimation with heavy-tailed priors. In *Computer Analysis of Images and Patterns*; Springer: Berlin, Germany, 2013; pp. 59–66.

42. Fitzgibbon, A.W.; Fisher, R.B. A Buyer's Guide to Conic Fitting. In *Proceedings of the 6th British Conference on Machine Vision*; BMVA Press: Surrey, UK, 1995; Volume 2; pp. 513–522.

43. Hartley, R.; Zisserman, A. *Multiple View Geometry in Computer Vision*, 2nd ed.; Cambridge University Press: New York, NY, USA, 2003.

44. Forbes, K.; Nicolls, F.; de Jager, G.; Voigt, A. Shape-from-silhouette with two mirrors and an uncalibrated camera. In *Computer Vision–ECCV 2006*; Springer: Berlin/Heidelberg, Germany, 2006; pp. 165–178.
45. Kuhn, H.W. The Hungarian method for the assignment problem. *Naval Res. Logist.* **2005**, *52*, 7–21.
46. Angel, E. *Interactive Computer Graphics: A Top-down Approach Using OpenGL*; Pearson/Addison-Wesley: Boston, MA, USA, 2009.

Underwater 3D Surface Measurement Using Fringe Projection Based Scanning Devices

Christian Bräuer-Burchardt, Matthias Heinze, Ingo Schmidt, Peter Kühmstedt and Gunther Notni

Abstract: In this work we show the principle of optical 3D surface measurements based on the fringe projection technique for underwater applications. The challenges of underwater use of this technique are shown and discussed in comparison with the classical application. We describe an extended camera model which takes refraction effects into account as well as a proposal of an effective, low-effort calibration procedure for underwater optical stereo scanners. This calibration technique combines a classical air calibration based on the pinhole model with ray-based modeling and requires only a few underwater recordings of an object of known length and a planar surface. We demonstrate a new underwater 3D scanning device based on the fringe projection technique. It has a weight of about 10 kg and the maximal water depth for application of the scanner is 40 m. It covers an underwater measurement volume of 250 mm × 200 mm × 120 mm. The surface of the measurement objects is captured with a lateral resolution of 150 μm in a third of a second. Calibration evaluation results are presented and examples of first underwater measurements are given.

Reprinted from *Sensors*. Cite as: Bräuer-Burchardt, C.; Heinze, M.; Schmidt, I.; Kühmstedt, P.; Notni, G. Underwater 3D Surface Measurement Using Fringe Projection Based Scanning Devices. *Sensors* **2016**, *16*, 13.

1. Introduction

Recently, the acquisition of 3D surface geometry of underwater objects has attracted increasing interest in various research fields, including archaeology [1–4], biological applications [5–8], or industrial facility inspection tasks [9]. Typical 3D reconstruction methodologies for underwater applications are underwater photogrammetry [5,7,10–13] or laser scanning techniques [9,14–17]. These techniques are time consuming and have limited precision. The increasing requirements of modern application scenarios concerning accuracy and data acquisition speed, however, require new approaches. One such approach is the use of structured light for the illumination of the underwater scene and to capture 3D measurement data using a stereo camera rig. Structured light illumination should be conducted by projection of a certain pattern (for example fringe pattern, stochastic pattern, other patterns) onto the object to be measured.

Recently, structured light (fringe) projection-based techniques have been increasingly used for underwater 3D measurements [18–20]. Some advantages of this technique in contrast to classical photogrammetric measurements are obvious: more measurement points can be generated and a higher 3D point density can be obtained in the measurement result. Additionally, measurement data can be obtained quickly and automatically. Due to the imaging principle, improved measurement accuracy can be expected. The main disadvantage, however, is the restricted measurement field because of the limited illumination conditions under water. Although the small size of the measurement field is a real limitation for certain potential applications we propose a fringe projection based method for underwater 3D data acquisition. Additionally, we introduce a prototype of a handheld underwater 3D scanning device in order to show the power of the structured light projection technique for underwater 3D measurements.

The main motivation for our developments is high precision inspection of industrial facility parts located under water (see e.g., [21]) such as pipeline systems, tanks, and pylons of off-shore windmills. Here a structured light based inspection methodology can help to carry out the maintenance of such facilities with considerably lower effort.

However, underwater 3D measurements using the structured light projection technique are more complicated than measurements in air. Reasons are the light absorbing property of the water, possible water pollution, and light ray fraction at the interfaces of different media (water, glass, air). Other causes are the more sophisticated requirements of the mechanics concerning robustness against impacts, temperature, humidity, and other environmental influences.

Our first goal was the construction of a new, handheld, practicable 3D scanning device for underwater measurements including algorithms for the attainment of accurate measurement results, and the development of a calibration procedure which can be done with low effort, but makes accurate measurement results possible. The second goal was to show the suitability of the structured light projection technique for sophisticated measurement tasks in underwater environments.

In this paper, an overview over the state of the art in optical underwater 3D measurement techniques is given, the challenges of the application of the structured light projection technique are outlined, the principles of accurate underwater 3D measurements are introduced, the development of a new handheld scanning device is described, and the first experimental results are presented and discussed.

2. State of the Art

Photogrammetry has been applied to underwater measurements with increasing importance and frequency for more than 40 years [22,23]. Documentation of archaeological sites [1,2] or sunken objects like boat parts or ship wrecks [10] are some application fields, as well as inspection and surface measurement of industrial facilities (for example pipeline systems [9]) or the measurement of biological objects such as coral reefs [8] or fishes [5–7].

The challenges for photogrammetric underwater applications are mainly the robustness and the accuracy of the captured 3D data. This is due to potentially bad image recording conditions (polluted water, insufficient light sources, difference in absorption behavior of distinct wavelengths, *etc.*) on the one hand and the possibly limited quality of the photogrammetric sensor calibration on the other hand. Water quality and illumination conditions have considerable influence on the measurement result and have been examined by Sedlazek *et al.* [24], Schechner and Karpel [25], and Bianco *et al.* [20].

Many works are concerned with calibration of photogrammetric underwater stereo scanners. An overview is given by Sedlazeck and Koch [12]. There are works considering the significant refraction effects using plane interface glasses of the underwater sensor housing [11,13,26–29]. Other authors neglect the refraction effects and try to approximate the underwater situation using the common pinhole camera model [5,7,30–32]. They propose modifications concerning the principal distance and the lens distortion [33,34].

A third situation appears when spherical dome ports are used in the underwater housing and the projection centers of the cameras are placed exactly into the sphere center points. Then, theoretically, no refraction occurs because of the perpendicular crossing of the interfaces of all vision rays. In this case, the common pinhole model may be applied for underwater applications in the same way as in the normal case. Furthermore, the calibration procedure can be performed under "normal" laboratory conditions in air. However, if the necessary preconditions concerning the exact spherical form of the dome-ports, the homogeneity of the glass material, and the exact placement of the cameras are denied, then measurement errors occur which should be corrected by a suitable method, as for instance additional distortion correction. Authors who have used this model describe a considerable deviation of the expected parameters [20].

The decision concerning the camera model used should depend on the requirements of the measurement accuracy and the accepted calibration effort. Here application of a suitable evaluation strategy seems meaningful which could be the determination of the measurement error using a certain reference measurement object with known geometry.

The use of structured light for photogrammetric underwater measurements has been recently proposed by Bruno *et al.* [18], Zhang *et al.* [19], and Bianco *et al.* [20], respectively. They showed that this technique can be successfully applied to underwater measurements. However, there are still some restrictions (small measurement fields and short measurement distances) of the structured light technique which should be overcome in the near future.

The present work describes the application of the structured light technique using projection of fringe sequences (Gray-code and sinusoidal sequences) in two orthogonal projection directions and the development of the prototype of an underwater measurement device which could be useful for certain applications and should be developed further for more sophisticated applications.

3. Measurement Principles and Underwater Challenges

3.1. Basic Principles

Contactless optical 3D measurement using active or passive photogrammetric methods employing a stereo camera arrangement will be performed as follows: the measurement object will be recorded by two cameras with known intrinsic parameters (principal point, principal distance, and distortion function) and known relative orientation to each other (extrinsic parameters). The intrinsic and extrinsic camera parameters are obtained by calibration according to the selected camera model (see next paragraph).

Measured 3D points are computed by triangulation (see e.g., [35]) which means the intersection of two rays in the 3D space. If the two rays are not parallel, there exists one unique 3D point which is either a proper intersection point of the two rays or that point having the shortest Euclidean distance to both rays. The vision rays can be obtained by the image co-ordinate and the intrinsic and extrinsic camera parameters. In common air applications typically the well-known pinhole camera model (PM) will be used [35]. However, also other, more general camera models can be applied such as the "raxel" model [36] or the ray-based camera model (RM) as described by Bothe *et al.* [37]. Similarly, whether the ray-based model is used or the pinhole model, calculation of resulting 3D points is obtained by triangulation.

3.2. Correspondence Finding and 3D Point Calculation

The determination of point correspondences may be achieved with the help of markers on the measurement object or by finding some characteristic structure(s) in the images. By using structured illumination one projects such characteristic structures onto the measurement object and increases significantly the number of correctly identifiable corresponding points. It additionally provides the possibility of better sub-pixel exact determination of image co-ordinates of the corresponding points. The projected patterns may be sequences of Gray-codes, sinusoidal fringe patterns, or series of stochastic patterns (see e.g., [38–42]).

The principle of fringe projection based sensors for the 3D surface acquisition is called phasogrammetry [39] or fringe projection profilometry [40] and means the mathematical conjunction of fringe projection and photogrammetry. A projection unit produces sequences of fringe pattern which will be observed by one or more cameras and transformed into so called phase images. Using stereo camera equipment as in our case, the phase images may serve purely for the identification of corresponding points.

From the sequence of sinusoidal fringe images the rough phase will be calculated which is periodically distributed according to the period length over the image. The Gray-code enables the unique identification of the period number and leads together with the rough phase to the fine or unwrapped phase value. Determination of the phase values in two orthogonal projection directions [39] or the use of epipolar geometry information [35] make the phase values unique. Because the epipolar geometry can be applied only approximatively in case of refracted rays in underwater application, two perpendicular sequences of fringes should be used. In the case of underwater application the triangulation is performed between the vision rays in the water together with the calibration parameters of a virtual camera (see Section 3.3).

3.3. Extended Camera Model

In contrast to classical photogrammetric modeling the pinhole camera model is not valid for underwater use because of the fact that there is no single viewpoint (the projection center) where all the image points representing rays pass through. This happens because of the refraction of the vision rays at the interfaces between air and housing glass (viewport) inside and between viewport and water outside the scanner housing (see Figure 1) following Snell's law. An exception would be the usage of spherical dome ports and an exact positioning of the camera's projection center into the center-point of the sphere (see Figure 1). Here (theoretically) no refraction occurs, and the pinhole model can be used as well as the same camera parameters as in air. However, in practice the exact positioning of the cameras is difficult to achieve. Experiments of other researchers [10,20] have showed considerable differences in the calibration parameters when performed both in air and under water. These differences lead to additional errors and distortion effects which must be corrected correspondingly.

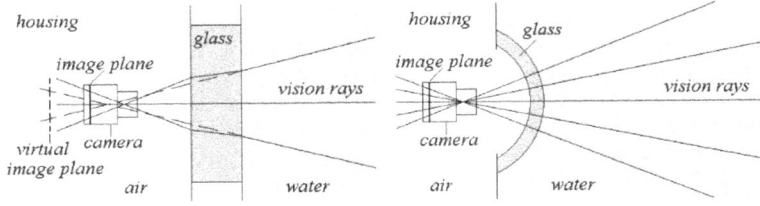

Figure 1. Geometry of rays in case of plane glasses (**left**) and spherical dome ports (**right**).

Let us consider the case of plane interface ports again. The refraction at the two interfaces leads to the effect that the vision rays corresponding to the different image points do not intersect in one single point (the projection center according to PM). Additionally, the principal distance c becomes image point dependent. For illustration see Figure 2. The refraction of the rays is shown for the case of perpendicular orientation of the camera according to the glass interface.

In this case the pinhole camera model can be applied only as an erroneous approximation. In order to avoid these errors the refraction should be taken into account. The description of the camera model can be given accordingly to the ray-based model given by Bothe *et al.* [37] or corresponding to other proposals (see e.g., [11,13,26,28,33,43]).

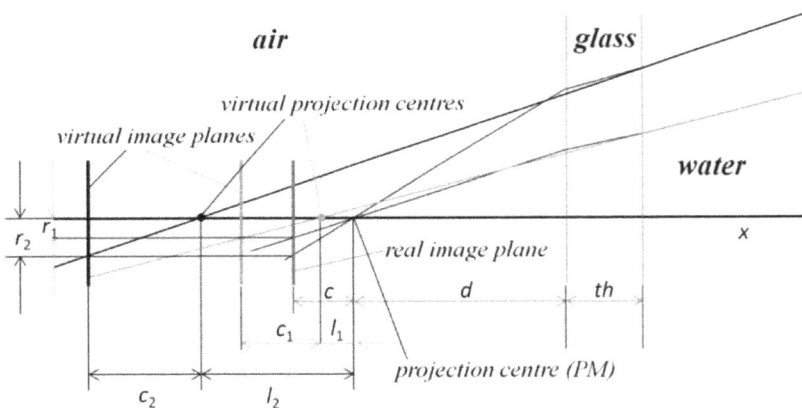

Figure 2. Ray geometry for two image points with radial distances r_1 and r_2 from the principal point using plane glasses (according to [44]).

Our approach is to use the parameters of the air calibration and describe the direction of the vision rays according to refraction and the additional parameters glass thickness th and interface distance d. Hence, one task of the underwater calibration is to find the values for the additional parameters. The other one is the formulation of the correct calculation of 3D measurement points based on found corresponding image points in the two cameras.

3.4. Approach for Underwater 3D Measurement

Our new approach to obtain accurate underwater 3D measurements using structured light illumination is the following. The exact modeling of the geometric situation of the cameras should be performed by consideration of the refraction effects and use of an accurate air calibration of the stereo scanner. An additional requirement is a relative low effort of the calibration procedure. This means that only few recordings of certain calibration objects should be acquired under water as input of the calibration procedure. The calibration for the underwater stereo scanner will be performed by the following steps:

1. Air calibration using the pinhole model without underwater housing
2. Determination of additional parameters (refraction indices, glass thickness, interface distance)
3. Ray-based model description of the underwater scanner inside housing
4. Determination of additional distortion matrices

When calibration is complete, 3D measurements can be performed under water. The 3D surface data of the measurement objects are obtained by finding corresponding points in the two camera images and performing a triangulation using the extended camera model and the parameters obtained by calibration. In the next section we will describe the calibration process in detail.

3.5. Calibration Procedure

The calibration of the underwater scanner can be performed according to the steps described in the previous section. For the first step any suitable calibration procedure for stereo camera scanners using the PM including distortion description can be used. For the second step we suggest the following procedure: our approach is to use the parameters of the air calibration and describe the direction of the vision rays according to refraction and the additional parameters interface distance d and glass thickness th. The indices of refraction for air n_a, water n_w, and glass (acrylic) n_g are assumed to be sufficiently exact known. The camera orientations concerning the glass interfaces are supposed to be perpendicular. Possible deviations should be compensated by final application of the additional distortion matrix.

Hence, one task of the underwater calibration is to find the values for the additional parameters. The other one is the formulation of the correct calculation of 3D measurement points based on found corresponding image points in the two cameras.

We assume for simplification an orthogonal normal angle regarding the glass surface. By consideration of the geometry as depicted in Figure 2 we get a shift l of the projection center in direction of the optical axis and a modified principal distance c' depending on the Euclidean distance r of the image points to the principal point:

$$l\left(r\right) = \frac{d \cdot r}{c \cdot \tan\left(\arcsin\left(\frac{\sin\left(\arctan\left(r/c\right)\right)}{n_w}\right)\right)} - d - th \cdot \frac{r}{c} \cdot \left(\frac{r}{c} - \tan\left(\arcsin\left(\frac{\sin\left(\arctan\left(r/c\right)\right)}{n_g}\right)\right)\right) \quad (1)$$

$$c'\left(r\right) = \frac{r}{\tan\left(\arcsin\left(\frac{\sin\left(\arctan\left(r/c\right)\right)}{n_w}\right)\right)} \quad (2)$$

Equations (1) and (2) are applied at calculation of the 3D points by triangulation complementing the known formulas of the pinhole camera case. By this extension we, in fact, obtain a ray-based camera model (RM).

With Equations (1) and (2) we can uniquely describe the rays corresponding to the image points $q = (x, y)$ according to the RM by the two 3D points $O_i(x, y)$ and $Q_i(x, y)$.

$$O_i\left(x, y\right) = O_i - l\left(r\right) \cdot R \cdot e \quad (3)$$

$$Q_i\left(x, y\right) = O_i\left(x, y\right) - c'\left(r\right) \cdot R \cdot \begin{pmatrix} x \\ y \\ -c'\left(r\right) \end{pmatrix} \quad (4)$$

where O_i is the initially determined projection center of the camera outside housing according to the pinhole model, e is the unit vector, and R is the rotation matrix of the air calibration (see also [45]). For simplification we assume without loss of generality no distortion here.

The next step is the determination of the glass thickness th. In our case we measured it tactile before mounting the underwater housing. If this would not be possible, e.g., the method proposed by Chen and Yang [46] can be applied.

Finally, the interface distances d_1 and d_2 for both cameras had to be determined. In order to obtain this, the following algorithm for underwater interface distance determination (UIDD) using four measurements (M_1, M_2, M_3, and M_4) is applied:

- Underwater recording (M_1, M_2) of specimen with known (calibrated) length information L_r (ball bar—see Figure 3) in two different positions
- Underwater recording (M_3, M_4) of a plane surface in two different distances (minimum and maximum distance in the measurement volume) according to the scanner

- Definition of the error function for the test statistics T utilizing length measurement error and flatness error of the plane according to Equations (6) and (7)
- Determination of the searched parameters by minimization of T

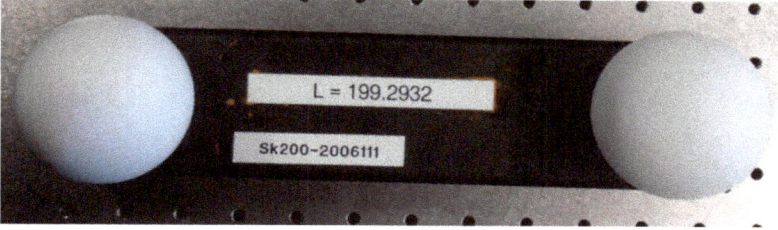

Figure 3. Specimen for calibration: calibrated ball bar.

Minimization of T can be achieved by a systematic search in the parameter space of d_1 and d_2 with meaningful search interval limits and step-widths. Having only two parameters, systematic search may be considerable more effective than trying to formulate an optimization function, because of the trigonometric functions in the Equations (1) and (2). The test quantity T is defined as:

$$T(d_1, d_2) = (E_{L,M1})^2 + (E_{L,M2})^2 + (E_{F,M3})^2 + (E_{F,M4})^2 \tag{5}$$

with relative length error E_L and relative flatness error E_F defined as:

$$E \left| \frac{L_a - L_r}{L_r} \right| \tag{6}$$

and:

$$E_F = \frac{|d_{\max}| + |d_{\min}|}{d_r} \tag{7}$$

Here, L_r is the calibrated reference distance between the two spheres of the ball bar and L_a is the measured distance. Quantity d_r represents the maximal possible distance in the observed region (e.g., diameter of a circle area or diagonal of a rectangular region) at the plane measurement. It is a constant which is fixed according to the measurement field size and which should have a length of at least half the diagonal of the measurement volume. The term $|d_{max}| + |d_{min}|$ denotes the flatness measurement error, which is defined as the range of the signed distances of the measured points from the best fit plane as defined in [47]. Application of this algorithm leads to the searched parameters d_1 and d_2. A more detailed description of the calibration procedure can be found in [44]. For all measurements it is allowed to remove a maximum of 0.3% outliers according to [46].

165

Application of this calibration procedure yielded good and meaningful results (see Section 5). However, the E_F error seemed to be too high, and the shape deviation of the plane was quite systematic (bowl-shape). This led to the introduction of an additional or alternative, respectively, distortion function and a corresponding matrix operator D for distortion correction. This operator will be obtained as follows. The two plane measurements (M_3 and M_4) are used for the determination of D. For both a fitting of a plane to the 3D points is performed leading to the plane parameters $E_1 = \{A_1, B_1, C_1, D_1\}$ and $E_2 = \{A_2, B_2, C_2, D_2\}$, respectively. Now, every calculated 3D point P is replaced by the nearest point P' in the fitted plane. Using P', new residuals for the corresponding image points are determined. Finally, the new distortion function is obtained by approximation of the residuals (from both planes) by a polynomial function of third degree leading to ten parameters (see [47] for more details).

In order to compensate new arising scaling errors after application of the new distortion compensation operator, last step of UIDD algorithm is performed again. If necessary, this procedure can be performed iteratively. The improvement in the resulting calibration is described in Section 5. Additionally, first results of the underwater measurements are presented.

4. New Underwater 3D Scanner

The paragon for the inner parts of the scanner was the handheld "kolibri Cordless" mobile scanning device [48]. For the underwater application, however, certain functional units such as the PC technique, the included display, and considerably more robust mechanics had to be developed completely anew. Additionally, an underwater resistant housing was necessary. The desired technical system parameters were similar to those of the "kolibri Cordless" and are listed in Table 1.

Table 1. Desired system parameters.

Property	Desired Parameter
Measurement volume (MV)	$250 \times 200 \times 150 \text{ mm}^3$
Working distance	500 mm
Camera resolution	1600×1200 pixel
Lateral resolution in the MV	150 μm
Noise of the 3D points	10 μm ... 50 μm
Frame rate	60 Hz
Recording time per scan	350 ms
Maximal water depth	40 m
Sensor weight (without housing)	2 kg
Sensor weight with housing	10 kg

166

A new principle is the direct connection and insertion of the control and analysis PC and the display into the underwater housing. This allows the omission of a connection between the scanner and an external PC, which would be otherwise carried in a backpack or external case. Additionally, underwater handling of the scanner becomes easier. The consequently higher weight of the device is no disadvantage in underwater use because it is only important that neither sinking nor upwelling of the scanner occur. The following criteria were essential for the scanner construction:

- Desired technical system parameters (see Table 1)
- Easy handling at use and also at the process of mounting into the housing
- Compactness of the scanner including housing
- Low weight
- Suitable heat exhausting at underwater use
- Easy navigation of the scanner under water

The main components of the scanner, which should be connected compactly for mounting into the housing, are the projection unit, two cameras, the PC, the display, cooling elements, and mechanical elements. The projection unit including the lens causes the structured illumination of the scene which is observed by the two cameras. The PC controls the projection and observation, stores the recorded image data, and performs data processing. A rough presentation of the resulting 3D point cloud on the display allows the user to evaluate the quality of the measurement. Cooling elements such as heatsinks, heatpipes, and cooling ribs are responsible for heat dissipation from the housing. Mechanical elements are necessary for the connection of the principal parts.

The projection unit was built up using a commercially available beamer (HX301G, LG, Seoul, Korea) with a frame rate of 60 Hz. The power consumption of the beamer is less than 100 W and the luminous flux is 300 Lm. The maximal pixel resolution is 1680×1050, but for fringe pattern projection the native resolution of 1024×768 pixels was used.

A sequence of fringes (Gray-code and sinusoidal fringes) in two orthogonal directions is projected onto the scene by the projection unit (see Figure 4). Typically, a fringe period length of 16, 32, or 64 projector pixels is used. This leads to 64, 32, or 16 vertical and 48, 24, or 12 horizontal projected fringes and seven, six, or five Gray-code images per projection direction. Using a 90° phase shift four sinusoidal fringe images per projection direction are necessary. Hence, one complete projection sequence consists of 24, 22, or 20 single fringe images taking 400, 367, or 333 ms time, respectively.

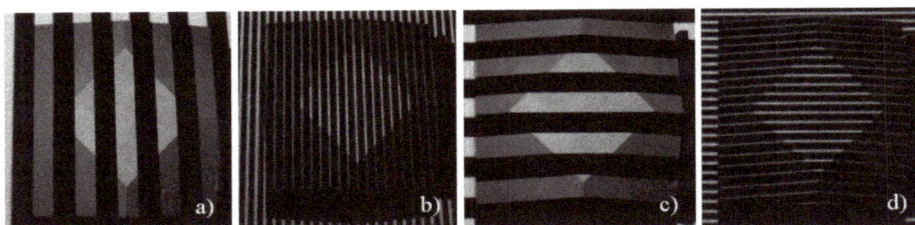

Figure 4. Selected images from fringe sequence: parts of Gray-code sequence (**a,c**) and sinusoidal fringes (**b,d**) taken from air measurements of a frustum of a pyramid.

The fringe image sequence is synchronously recorded by both cameras. From these image sequences the so called phase images are calculated, which are used for the determination of the 3D data (see e.g., [39]). The 3D calculation is performed on the PC, which also has the task to control the image recording. The measurement results are also indicated on the display (see Figure 5). Additional components are two laser pointers for checking the current measurement distance. Figure 5 shows the main components of the scanner in two construction drawings.

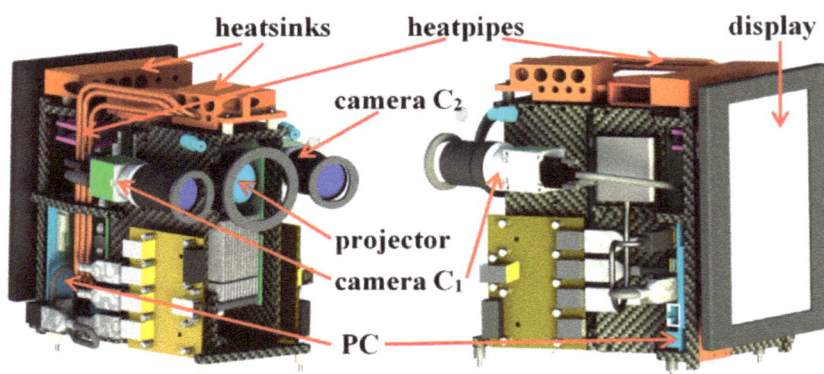

Figure 5. Principal components in two views of a construction drawing.

The underwater housing was produced using the synthetic material PA 2200 (see [49]) and developed by the 4h Jena Engineering GmbH (see [50]). It can be used in both fresh and salt water. It can distinctly withstand the water pressure at a diving depth of 40 m. The interfaces for the projector, the cameras, and the laser beams were made from sapphire glass, whereas the window for the display is made from polycarbonate. The planar windows for the cameras and the projector do not lie in a common plane but are tilted according to the directions of the optical axis. This should simplify the calibration procedure (see Section 3.5).

168

One major problem of the scanner was the adequate heat dissipation from the housing. A base plate with cooling ribs and appropriate heat sinks were constructed. For power supply and signal lines a separable under water plug-in connector cable was selected and employed. Separable inductive switching boards including control keys and interfaces to the scanner were developed in order to provide a correct handling also with diver gloves (see Figure 6). A construction drawing of the scanner is shown in Figure 7 as well as a photograph of the scanner in the housing without back panel with the display.

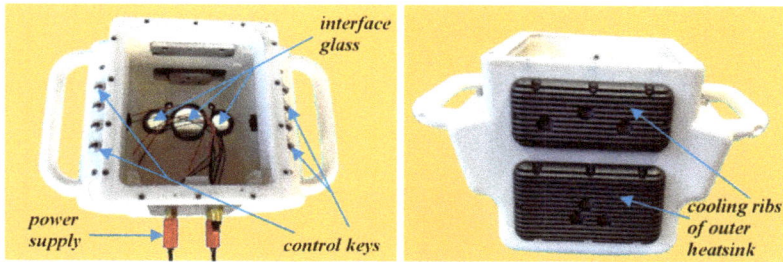

Figure 6. Two housing views showing certain sensor elements.

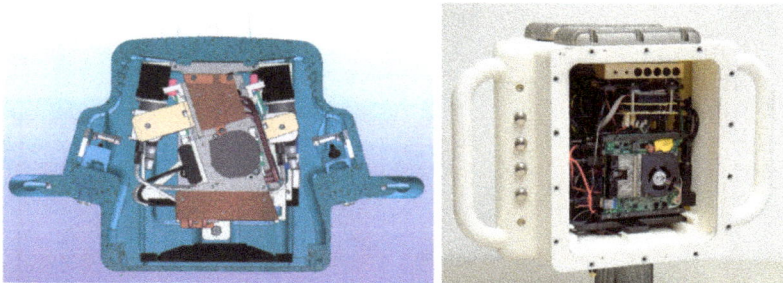

Figure 7. Housing views: construction draw, view from above (**left**), sensor inside housing without back-panel (**right**).

5. Measurements

5.1. Calibration Evaluation

After the four measurements which were used for the calibration of the underwater parameters we made several underwater measurements of certain specimen in order to evaluate the quality of the calibration. According to the suggestions of the VDI/VDE guidelines [46] the measurement objects ball bar and plane were placed in different positions in the measurement volume. Additionally, a frustum of a pyramid (see Figure 8) was measured. All measurements were performed in clear water using a rain barrel (see Figure 8). Figure 9 shows an example for the projected fringe patterns under water. Compared to Figure 4 differences in the image quality can be hardly detected.

For evaluation the quantities relative length deviation and relative flatness deviation defined by Equations (5) and (6) were used. For comparison, all measurements and calculations were performed

- With the sensor outside the housing in the laboratory (Air)
- Underwater with the pinhole camera model with modified parameters (PM)
- Underwater with the extended model, but without additional distortion matrices (RM)
- Underwater with the extended model including the additional distortion matrices (RMD)

Figure 8. Frustum of pyramid in a rain barrel (**left**) and scanner in underwater use (**right**).

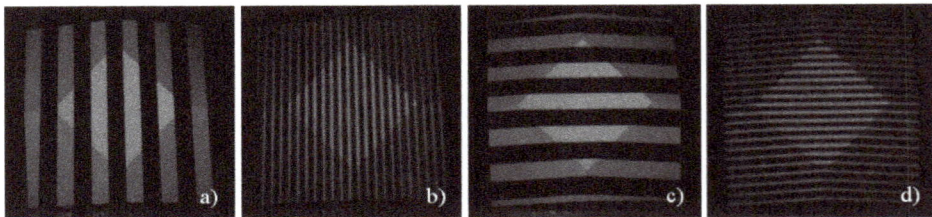

Figure 9. Selected images from fringe sequence: parts of Gray-code sequence (**a,c**) and sinusoidal fringes (**b,d**) taken from underwater measurements of a frustum of a pyramid.

In all cases at least seven measurements were performed. Noise determination was performed additionally. It was defined as standard deviation of the measured 3D points from a small locally fitted plane (or sphere, respectively). The obtained maximum error results are documented by Table 2. Note that for E_L and E_F the percentage values are given in contrast to Equations (6) and (7), and *noise* is an averaged value over all noise measurements.

Table 2. Calibration evaluation results (given in %) using specimen ball bar (E_{Lb}), frustum of pyramid (E_{Lp}), and plane (E_F), and averaged 3D noise values.

Location/Quantity	E_{Lb} (%)	E_{Lp} (%)	E_F (%)	*noise* (mm)
Air, PM	0.1	0.4	0.15	0.02
Water, PM	1.0	1.45	1.7	0.05
Water, RM	0.4	0.6	0.65	0.04
Water, RMD	0.35	0.35	0.4	0.03

The results show that pinhole approximation in water is not sufficient to obtain acceptable measurement accuracy. Although the modeling without additional distortion correction should completely describe the ray geometry, the results are not yet as precise as desired. RMD results are acceptable, but should also be improved at future measurements after certain enhancements which can be obtained, for instance, by using a free ray-based camera model. However, for improved ray-based model representation an additional calibration step must be performed. An approach of this idea will be given in the outlook section.

5.2. Examples of Underwater Measurements

Underwater measurements were performed next by application of the scanner in a water basin. It was handled by a diver (Figure 8). The first measurement objects were a pipe (Figures 10 and 11), a fossil sea shell (Figure 12), and stones (Figure 13).

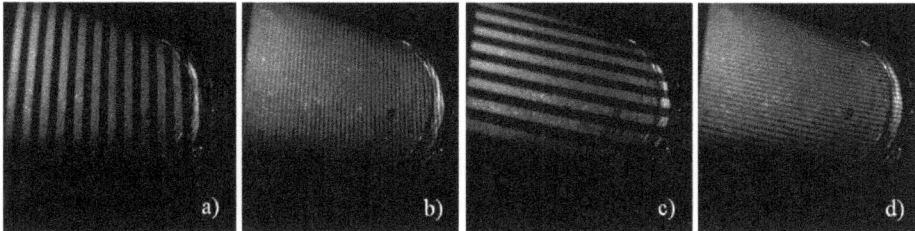

Figure 10. Selected images from fringe sequence: parts of Gray-code sequence (**a,c**) and sinusoidal fringes (**b,d**) taken from underwater measurements of a pipe.

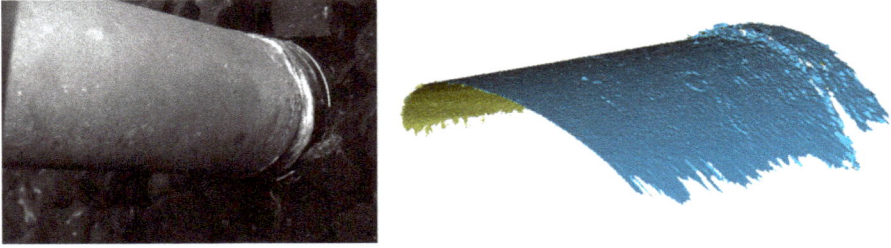

Figure 11. Underwater measurement of a pipe (**left**) and 3D result representation (**right**).

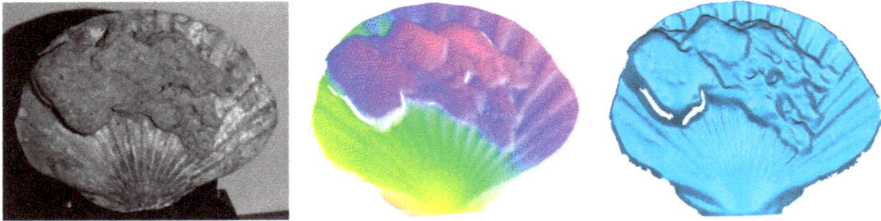

Figure 12. Fossil seashell: photograph (**left**), color-coded 3D- (**middle**), and 3D-surface presentation (**right**).

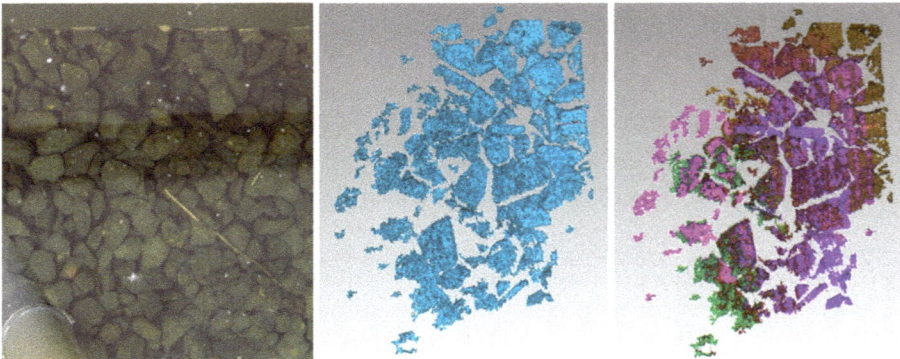

Figure 13. Example underwater measurement stones: photograph (**left**), surface 3D representation (**middle**), and color coding of six different scans (**right**).

The examples can give a first impression of the capabilities of the underwater scanner. The scan results are obtained few seconds after the recording. It can be assumed that the measurement precision is sufficient for a number of application scenarios. However, for bigger objects the 3D point clouds must be merged. Here, additional software is necessary.

6. Summary, Discussion and Outlook

The principle of optical 3D surface measurement based on the fringe projection technique for underwater applications has been described in this work and the challenges of the underwater use of this technique are shown and discussed in contrast to the classical application.

A new fringe projection based 3D underwater scanner which covers a measurement volume of 250 mm × 200 mm × 120 mm for diving depths up to 40 m was introduced. It can record one 3D scan in approximately 350 ms with a lateral resolution of 150 μm in the object space. Measurement objects larger than the measurement volume should be scanned in several views and merged subsequently.

The presented scanning device for 3D surface measurements is one of the first underwater scanners based on fringe projection technique. It is quite complicate to perform a comparison of technical details and measurement capability to other systems based on this technique [18–20]. These systems are laboratory setups and the projector is not included in the camera housing. Certainly, what can be assessed is that our scanner has a smaller measurement field. Measurement accuracy, although also difficult to compare, seems to be similar. Compared to photogrammetric underwater systems (e.g., [13,26]) the measurement precision has the same magnitude. The main differences with photogrammetric systems are the smaller measurement volume and the higher measurement point density.

During the image recording time of 0.33 or 0.4 s operator movements may disturb the image recording. Effective movement correction algorithms could manage this, but, unfortunately such methods are not available for this scanner. However, we have developed an algorithm which detects vibrations and movements during the image recording and suggests the rejection of the affected current measurement and its repetition.

If moving dust clouds or dirt particles disturb the image acquisition, it must be decided whether to wait for better recording conditions or to take the images and to accept decreased image quality. In order to make a more exact statement possible, appropriate experiments should be performed in the future.

The first quantitative measurements showed acceptable measurement accuracy. However, best results could be achieved only by application of additional (heuristically found) distortion corrections which cannot be explained by the model. They perhaps compensate partly the effects of the deviations of the current modelling from the real situation. The search of the origins of these effects may be a part of our future work.

First measurements and experiments were performed in clear freshwater in the laboratory and in an outdoor water basin with slightly more pollution. This, however, is not a practice-oriented situation. It is expected that in a sea environment with certain turbidity the robustness of the object surface acquisition will decrease and the noise of the measurement values will increase. Measurement accuracy, however, will be affected only by a slightly higher random error. The corresponding experiments should confirm and quantify these expectations.

The main part of our future work should be occupied by experiments concerning the dependence of the measurement accuracy on the water quality. Here the influence of difference levels of turbidity to the robustness and accuracy of the measurements must be analyzed. Additionally, comparisons between measurements in fresh and salt water will be made.

In order to avoid the necessity of additional distortion compensation by special operator, the complete ray-based representation of the observation system should be used in the future. Here, an additional calibration step with underwater recordings of a plane must be applied similarly to the method described by Bothe *et al.* [37]. With this final transition to the ray-based camera model a further reduction of the remaining length error and flatness deviation should be achieved.

As it was already mentioned before, the size of the measurement volume is not very large and seems to be too small for certain applications such as pipeline system inspection or survey of archaeological sites. The possible stitching of resulting 3D datasets is sophisticated, time consuming, and prone to errors. Hence, scanning devices with considerable larger measurement volume are desired. This requires stronger illumination power of the projection unit with coincident more efficient heat dissipation. The design of such a more powerful underwater 3D scanner will be also a part of our future work. Perhaps, the underwater housing for the projector should be separated from the housing for the cameras because of the heat dissipation.

Acknowledgments: This work was supported by the state Thuringia and the European Union (EFRE) under grant label TNA VIII-1/2012.

Conflicts of Interest: The authors declare no conflict of interest.

References

1. Roman, C.; Inglis, G.; Rutter, J. Application of structured light imaging for high resolution mapping of underwater archaeological sites. In Proceedings of the 2010 IEEE OCEANS, Sydney, Australia, 24–27 May 2010; pp. 1–9.
2. Drap, P. Underwater Photogrammetry for Archaeology. In *Special Applications of Photogrammetry*; Carneiro Da Silva, D., Ed.; InTech: Rijeka, Croatia, 2012; pp. 111–136.
3. Eric, M.; Kovacic, R.; Berginc, G.; Pugelj, M.; Stopinsek, Z.; Solina, F. The Impact of the Latest 3D Technologies on the Documentation of Underwater Heritage Sites. In Proceedings of the IEEE Digital Heritage International Congress 2013, Marseille, France, 28 October–1 November 2013; Volume 2, pp. 281–288.

4. Canciani, M.; Gambogi, P.; Romano, F.G.; Cannata, G.; Drap, P. Low cost digital photogrammetry for underwater archaeological site survey and artifact insertion. The case study of the Dolia wreck in secche della Meloria-Livorno-Italia. *Int. Arch. Photogramm. Remote Sens. Spat. Inf. Sci.* **2003**, *34 Pt 5*, 95–100.

5. Harvey, E.; Cappo, M.; Shortis, M.; Robson, S.; Buchanan, J.; Speare, P. The accuracy and precision of underwater measurements of length and maximum body depth of southern bluefin tuna (Thunnus maccoyii) with a stereo–video camera system. *Fish. Res.* **2003**, *63*, 315–326.

6. Dunbrack, R.L. *In situ* measurement of fish body length using perspective-based remote stereo-video. *Fish. Res.* **2006**, *82*, 327–331.

7. Costa, C.; Loy, A.; Cataudella, S.; Davis, D.; Scardi, M. Extracting fish size using dual underwater cameras. *Aquac. Eng.* **2006**, *35*, 218–227.

8. Bythell, J.C.; Pan, P.; Lee, J. Three-dimensional morphometric measurements of reef corals using underwater photogrammetry techniques. *Coral Reefs* **2001**, *20*, 193–199.

9. Tetlow, S.; Allwood, R.L. The use of a laser stripe illuminator for enhanced underwater viewing. *Proc. SPIE* **1994**, *2258*, 547–555.

10. Korduan, P.; Förster, T.; Obst, R. Unterwasser-Photogrammetrie zur 3D-Rekonstruktion des Schiffswracks "Darßer Kogge". *Photogramm. Fernerkund. Geoinf.* **2003**, *5*, 373–381.

11. Kwon, Y.H.; Casebolt, J. Effects of light refraction on the accuracy of camera calibration and reconstruction in underwater motion analysis. *Sports Biomech.* **2006**, *5*, 315–340.

12. Sedlazeck, A.; Koch, R. Perspective and non-perspective camera models in underwater imaging—Overview and error analysis. In *Theoretical Foundations of Computer Vision*; Springer: Berlin, Germany, 2011; Volume 7474, pp. 212–242.

13. Telem, G.; Filin, S. Photogrammetric modeling of underwater environments. *ISPRS J. Photogramm. Remote Sens.* **2010**, *65*, 433–444.

14. Moore, K.D. Intercalibration method for underwater three-dimensional mapping laser line scan systems. *Appl. Opt.* **2001**, *40*, 5991–6004.

15. Narasimhan, S.G.; Nayar, S.K. Structured Light Methods for Underwater Imaging: Light Stripe Scanning and Photometric Stereo. In Proceedings of the 2005 MTS/IEEE OCEANS, Washington, DC, USA, 17–23 September 2005; Volume 3, pp. 2610–2617.

16. Tan, C.S.; Seet, G.; Sluzek, A.; He, D.M. A novel application of range-gated underwater laser imaging system (ULIS) in near-target turbid medium. *Opt. Lasers Eng.* **2005**, *43*, 995–1009.

17. Massot-Campos, M.; Oliver-Codina, G. Underwater laser-based structured light system for one-shot 3D reconstruction. In Proceedings of the 5th Martech International Workshop on Marine Technology, Girona, Spain, 2–5 November 2014.

18. Bruno, F.; Bianco, G.; Muzzupappa, M.; Barone, S.; Razionale, A.V. Experimentation of structured light and stereo vision for underwater 3D reconstruction. *ISPRS J. Photogramm. Remote Sens.* **2011**, *66*, 508–518.

19. Zhang, Q.; Wang, Q.; Hou, Z.; Liu, Y.; Su, X. Three-dimensional shape measurement for an underwater object based on two-dimensional grating pattern projection. *Opt. Laser Technol.* **2011**, *43*, 801–805.

20. Bianco, G.; Gallo, A.; Bruno, F.; Muzzupappa, M. A comparative analysis between active and passive techniques for underwater 3D reconstruction of close-range objects. *Sensors* **2013**, *13*, 11007–11031.

21. McLeod, D.; Jacobson, J.; Hardy, M.; Embry, C. Autonomous inspection using an underwater 3D LiDAR. In Proceedings of the Ocean in Common, San Diego, CA, USA, 23–27 September 2013.

22. Höhle, J. Zur Theorie und Praxis der Unterwasser-Photogrammetrie. Ph.D. Thesis, Bayerische Akademie der Wissenschaften, München, Germany, 1971.

23. Moore, E. J. Underwater photogrammetry. *Photogramm. Rec.* **1976**, *8*, 748–163.

24. Sedlazeck, A.; Koser, K.; Koch, R. 3D reconstruction based on underwater video from rov kiel 6000 considering underwater imaging conditions. In Proceedings of the 2009 OCEANS—Europe, Bremen, Germany, 11–14 May 2009; pp. 1–10.

25. Schechner, Y.Y.; Karpel, N. Clear underwater vision. In Proceedings of the 2004 IEEE Computer Society Conference on Computer Vision and Pattern Recognition, Washington, DC, USA, 27 June–2 July 2004; Volume 1, pp. 536–543.

26. Li, R.; Tao, C.; Curran, T.; Smith, R. Digital underwater photogrammetric system for large scale underwater spatial information acquisition. *Mar. Geod.* **1996**, *20*, 163–173.

27. Maas, H.G. New developments in multimedia photogrammetry. In *Optical 3-D Measurement Techniques III*; Grün, A., Kahmen, H., Eds.; Wichmann Verlag: Karlsruhe, Germany, 1995.

28. Sedlazeck, A.; Koch, R. Calibration of housing parameters for underwater stereo-camera rigs. In Proceedings of the 22nd British Machine Vision Conference, Dundee, UK, 29 August–2 September 2011.

29. Kawahara, R.; Nobuhara, S.; Matsuyama, T. A Pixel-wise Varifocal Camera Model for Efficient Forward Projection and Linear Extrinsic Calibration of Underwater Cameras with Flat Housings. In Proceedings of the 2013 IEEE International Conference on Computer Vision Workshops, Sydney, Australia, 2–8 December 2013; pp. 819–824.

30. Shortis, M.R.; Harvey, E.S. Design and calibration of an underwater stereo-video system for the monitoring of marine fauna populations. *International Archives Photogramm. Remote Sens.* **1998**, *32*, 792–799.

31. Fryer, J.G.; Fraser, C.S. On the calibration of underwater cameras. *Photogramm. Rec.* **1986**, *12*, 73–85.

32. Bryant, M.; Wettergreen, D.; Abdallah, S.; Zelinsky, A. Robust camera calibration for an autonomous underwater vehicle. In Proceedings of the Australian Conference on Robotics and Automation (ACRA 2000), Melbourne, Australia, 30 August–1 September 2000.

33. Lavest, J.M.; Rives, G.; Lapreste, J.T. Underwater camera calibration. In Proceedings of the 6th European Conference on Computer Vision (ECCV 2000), Dublin, Ireland, 26 June–1 July 2000; pp. 654–668.

34. Lavest, J.M.; Rives, G.; Lapreste, J.T. Dry camera calibration for underwater applications. *Mach. Vis. Appl.* **2003**, *13*, 245–253.

35. Luhmann, T.; Robson, S.; Kyle, S.; Harley, I. *Close Range Photogrammetry*; Wiley Whittles Publishing: Caithness, UK, 2006.

36. Grossberg, M.D.; Nayar, S.K. The raxel imaging model and ray-based calibration. *Int. J. Comput. Vis.* **2005**, *61*, 119–137.

37. Bothe, T.; Li, W.; Schulte, M.; von Kopylow, C.; Bergmann, R.B.; Jüptner, W. Vision ray calibration for the quantitative geometric description of general imaging and projection optics in metrology. *Appl. Opt.* **2010**, *49*, 5851–5860.

38. Sansoni, G.; Carocci, M.; Rodella, R. Three-Dimensional Vision Based on a Combination of Gray-Code and Phase-Shift Light Projection: Analysis and Compensation of the Systematic Errors. *Appl. Opt.* **1999**, *38*, 6565–6573.

39. Schreiber, W.; Notni, G. Theory and arrangements of self-calibrating whole-body three-dimensional measurement systems using fringe projection techniques. *Opt. Eng.* **2000**, *39*, 159–169.

40. Zhang, S. Recent progresses on real-time 3D shape measurement using digital fringe projection techniques. *Opt. Lasers Eng.* **2010**, *48*, 149–158.

41. Schaffer, M.; Große, M.; Harendt, B.; Kowarschik, R. Statistical patterns: An approach for high-speed and high-accuracy shape measurements. *Opt. Eng.* **2014**, *53*.

42. Salvi, J.; Fernandez, S.; Pribanic, T.; Llado, X. A state of the art in structured light patterns for surface profilometry. *Pattern Recognit.* **2010**, *43*, 2666–2680.

43. Yau, T.; Gong, M.; Yang, Y.H. Underwater Camera Calibration Using Wavelength Triangulation. In Proceedings of the 2013 IEEE Conference on Computer Vision and Pattern Recognition (CVPR), Portland, OR, USA, 23–28 June 2013; pp. 2499–2506.

44. Bräuer-Burchardt, C.; Kühmstedt, P.; Notni, G. Combination of air- and water-calibration for a fringe projection based underwater 3D-Scanner. In Proceedings of the 16th International Conference (CAIP 2015), Valletta, Malta, 2–4 September 2015; pp. 49–60.

45. Chen, X.; Yang, Y.H. Two view camera housing parameters calibration for multi-layer flat refractive interface. In Proceedings of the 2014 IEEE Conference on Computer Vision and Pattern Recognition (CVPR), Columbus, OH, USA, 23–28 June 2014; pp. 524–531.

46. The Association of German Engineers (VDI). Optical 3D-Measuring Systems. In *VDI/VDE Guidelines, Parts 1–3*; VDI/VDE 2634; VDI: Duesseldorf, Germany, 2008.

47. Bräuer-Burchardt, C.; Kühmstedt, P.; Notni, G. Ultra-Precise Hybrid Lens Distortion Correction. In Proceedings of the International Conference on Image and Vision Computing (ICIVC 2012), Venice, Italy, 14–16 November 2012; Volume 71, pp. 179–184.

48. Munkelt, C.; Bräuer-Burchardt, C.; Kühmstedt, P.; Schmidt, I.; Notni, G. Cordless hand-held optical 3D sensor. In Proceedings of the SPIE Optical Metrology 2007, Munich, Germany, 18–22 June 2007; Volume 6618.

49. Eos e-Manufacturing Solutions. Available online: http://www.eos.info/material-p (accessed on 3 December 2015).

50. 4h Jena. 2015. Available online: http://www.4h-jena.de/ (accessed on 3 December 2015).

Sinusoidal Wave Estimation Using Photogrammetry and Short Video Sequences

Ewelina Rupnik, Josef Jansa and Norbert Pfeifer

Abstract: The objective of the work is to model the shape of the sinusoidal shape of regular water waves generated in a laboratory flume. The waves are traveling in time and render a smooth surface, with no white caps or foam. Two methods are proposed, treating the water as a diffuse and specular surface, respectively. In either case, the water is presumed to take the shape of a traveling sine wave, reducing the task of the 3D reconstruction to resolve the wave parameters. The first conceived method performs the modeling part purely in 3D space. Having triangulated the points in a separate phase via bundle adjustment, a sine wave is fitted into the data in a least squares manner. The second method presents a more complete approach for the entire calculation workflow beginning in the image space. The water is perceived as a specular surface, and the traveling specularities are the only observations visible to the cameras, observations that are notably single image. The depth ambiguity is removed given additional constraints encoded within the law of reflection and the modeled parametric surface. The observation and constraint equations compose a single system of equations that is solved with the method of least squares adjustment. The devised approaches are validated against the data coming from a capacitive level sensor and on physical targets floating on the surface. The outcomes agree to a high degree.

Reprinted from *Sensors*. Cite as: Rupnik, E.; Jansa, J.; Pfeifer, N. Sinusoidal Wave Estimation Using Photogrammetry and Short Video Sequences. *Sensors* **2015**, *15*, 30784–30809.

1. Introduction

Attempts to characterize the water surface with optical methods date back to the beginning of the 20th century [1,2]. The interest in a quantitative description of the surface with light came from the field of oceanography and the use of photography to map the coastlines. This prompted further applications, namely the use of photography to quantify ocean waves and to exploit these parameters in, e.g., shipbuilding, to engineer structures of the appropriate strength [3–5].

The same drivers disseminated optical methods among other applications, in river engineering and the oceanographic domain. Understanding river flow allows for a better riverbed management and mitigation of floods through combined fluid dynamics modeling and experimental testing. Additionally, the knowledge of the dispersive processes gives an insight into the way pollution and sediments are transported [6–9].

In the coastal zones, optical methods became a good alternative to *in situ* measurements, which require substantial logistical commitments and offer low spatial, as well as temporal resolution. The dynamics of the water, hence the energy it carries, influences the nearshore morphology, which is of significance for both coastal communities and marine infrastructure, e.g., wharfs and mooring systems [10–13].

Last, but not least, the roughness of the ocean's surface is meaningful from the viewpoint of exchange processes with the atmosphere. The oceans, accounting for two-thirds of the Earth's surface, contain high concentrations of carbon dioxide and thereby influence the global climate system through the evaporation and absorption phenomena. The optically-measured wave slope can be used to parametrize the relationship between the surface roughness and gas transfers, as well as the wind speed and direction [14,15].

Other recent and operational approaches to ocean surface observations are radar altimetry [16] and GNSS reflectometry [17].

1.1. Contributions

The objective of this work was to observe the motion of a floating platform, simultaneously providing information on the forces generating the motion, *i.e.*, the shape of the water waves (*cf.* Figure 1). The waves were generated with a linear mechanical arm, occupying the entire width of the basin. The wave propagation followed a single direction along the basin's longer axis. As such, the arm moved at equal time intervals producing sinusoidal waves at various, but constant frequencies (f), periods (ω) and wavelengths (λ); *cf.* Figure 1. The water was rendered a mirror-like surface, while the testing facility was in large part occupied by a model basin and offered little space around the measurement volume. Hence, the challenges were split between the uncommon surface characteristics, the workplace constraints, as well as the non-professional, budget-conscious imaging equipment employed for the timely observations. The contributions of this article include: (i) a concept for exploiting specular reflections of light sources to model the water surface, (ii) a method for measuring and modeling specular reflections of light sources in an image and 3D reconstruction of the dynamic wave shape from it, plus; (iii) an evaluation of the suggested method using the capacitive level sensor and physical targets freely floating on the surface's top.

Two methods are proposed, treating the water as a diffuse and as a specular surface, respectively. In either case, the water is presumed to take the shape of a traveling sine wave, reducing the task of the 3D reconstruction to resolve the wave parameters. This was accomplished in a two-fold manner, by observing: (i) a few

179

physical targets floating on the top of the surface (Method 1 in Section 3.1), and (ii) the apparent motion of specular reflections coming from points of known coordinates (Method 2 in Section 3.2). In either case, the transfer from point-based measurements to a surface was possible thanks to the prior knowledge of the wave excitation.

The determination of the platform's motion is out of scope of this publication and was presented in [18].

1.2. Related Works

One can distinguish between three groups of measurement methods, each exploiting different characteristics of the water medium. The water surface can be regarded as (1) diffuse-like (quasi-Lambertian), (2) mirror-like (specular reflection) or else (3) as a transparent medium (refraction). When the wavelength of the incident radiation is smaller or equal to that of the surface roughness, the surface will appear as diffuse. On the other hand, smooth surfaces will produce specular reflections, and their appearance will be highly dependent on the position of the observer, as well as the radiation source [19]. The majority of approaches utilize passive sensors in reconstructing the surface shape. Active illumination techniques applied to highly reflecting surfaces produce systematic errors in depth estimation. There are few recorded attempts to adapt the scanning techniques to deal with the specularity effects, in particular by space-time analyses of the signal and filtering procedures [20].

1.2.1. Water as a Diffuse Surface

In practice, there are two conditions when the water is rendered diffuse-like: first, when the surface is disturbed with small waves (e.g., rough sea in the coastal zone); second, when artificial targeting is employed. The most common targeting techniques use physical material, like powder, Styropor, oil or optical projections in the forms of a laser sheet, a grid of points or sinusoidal patterns [9,10,21–23]. Specular reflections are inevitable and are often the source of errors in the estimated depths. To avoid the corrupted measurements, specular highlights can be: (1) removed in the image preprocessing step, (2) eliminated in multiview setups during the processing (the appearance of glints in images is view dependent; with the third or n-th view, every identified feature can be verified, and glints can be eliminated; the method is apt for scenes with single or few glints) [24], (3) filtered with the help of either polarized or chromatic filters (the filters are mounted in front of the camera lens; hence, there is no restriction on the number of present glints) [25] and (4) in industrial photogrammetry of rigid objects, attenuated through the use of special targets, e.g., doped with fluorescing dye, that respond to a wavelength other than the wavelength of the specular reflections [26].

Alternatively, the water being itself a source of infrared radiation can be observed with thermal cameras. Because the heat distribution is heterogeneous across the surface, it provides a good base for the correspondence search. A surface treated in this way can be measured with classical stereo- or multi-view photogrammetric approaches. If qualitative results are expected, it is sufficient to acquire single images and to proceed with data evaluation in the image space only [27].

1.2.2. Water as a Specular Surface

Sometimes, it is advantageous to exploit the inherent optical characteristics of water, *i.e.*, the total reflection and refraction, for measurement purposes. Contaminating the liquid with physical material is cumbersome, because it becomes (1) unfeasible for large areas and field surveys, (2) difficult to keep a homogeneous point distribution and (3) it may influence the response of the water by interacting with it. In such situations, and depending on the working environment, whether in a lab or out in the field, it is possible to derive the surface shape by mere observation of a reflection of a source light or a pattern whose distortion corresponds to the surface slope and height

Reference [28] pioneered the characterization of water surface slopes with a reflection-based method. Their motive was to analyze slope distributions under different wind speeds through observing the Sun's glitter on the sea surface from an aerial platform. Variations of the Cox and Munk method include: replacing the natural illumination with one or more artificial light sources, also known as reflective (stereo) slope gauge (RSSG) [3,29], and using entire clear or overcast sky to derive surface slope information for every point in the image, also known as Stilwell photography [30].

A combination of stereo- and reflection-based techniques (RSSG) was proven to be a sound way to characterize not only the surface slopes, but also the water heights. In a typical stereo setting, one is faced with a bias in the corresponding features seen by the left and the right cameras; the reason being that the cameras will record a change from the spots on the water whose normals' are in the line of sight of the given cameras. Naturally, the steeper the observed wave, the less the systematic error. Using the Helmholtz reciprocity principle, *i.e.*, upgrading the method to employ light sources placed right next to the cameras, eliminates the correspondence ambiguity. The identified features in image spaces are then bound to be a unique feature in the object space [3,4].

In the field of computer vision, two principal classes of algorithms are shape from distortion and shape from specularity, which, e.g., inspect single or multiple highlights with a static or a moving observer and emitter [31–33], observe known or unknown intensity patterns reflected from mirror-like surfaces [34–39], directly measure the incident rays [40], exploit light polarization [41,42] and make assumptions on the surface's bidirectional reflectance distribution [43–45]. Further interesting approaches that fall outside the scope of the adopted categories exploit other (than the visible) parts of the electromagnetic spectrum, such as infra-red [46], thermal [47] or UV. The principle resembles that of [48,49], *i.e.*, one searches for a wave spectrum, in which the information/signal received from the problematic surfaces, either doped or hit by the energy portion, is maximized, while minimizing the close-by, disturbing signals. For a good overview of the techniques, refer to [50,51].

1.2.3. Water as a Refractive Medium

Refraction-based techniques are more complex due to the fact that the light path is dependent not only on the surface normal, but also the medium's refraction index. The development of shape from refraction goes hand in hand with the shape from reflection methods and, therefore, has an equally long history. The work in [52], again, first experimented with light refraction to derive surface slopes from an intensity gradient emerging from beneath the water surface. Today, a successor of this technique, called imaging slope gauge, alternatively shape from refractive irradiance, is considered a highly reliable instrument for measuring wind-induced waves in laboratory environments. In contrast to the reflection-based techniques, refraction of a pattern through the water maintains a quasi-linear relationship between the slope and the image irradiance [14,53–55].

White light can also be replaced with active laser lighting. The laser offers a greater temporal resolution at the expense of a lower spatial resolution. The first examples of laser imaging gauges employed a single beam, thereby delivering information about a single slope [56,57]. Over time, they evolved to scanning systems that could capture spatial information over an area [58]. A laser slope gauge is used in the laboratory and field environment, however being most apt for the characterization of longer waves, as the literature reports.

In the field of computer vision, dynamic transparent surfaces have been analyzed with single- and multi-image approaches. The work in [59] first introduced a single view shape from the motion of a refractive body. By assuming that the water's average slope is equal to zero, *i.e.*, it regularly oscillates around a flat plane, the author proposed a method for reconstructing the shape of an undulating water surface by inverse ray tracing. The work in [60] developed a method that recovers complete information of the 3D position and orientation of dynamic, transparent surfaces using stereo observations. The authors formulate an optimization procedure that minimizes a so-called refractive disparity (RD). In short, for every pixel in an image that refracts from an underwater pattern, the algorithm finds a 3D point on the water surface that minimizes the conceived RD measure. This measure expresses Snell's law of refraction by enforcing that the normal vector of the computed 3D position must be equal when computed from the left and the right images of the stereo image. The two flag examples of computer vision approaches are limited to use in laboratory conditions due to (1) the need to locate a reference pattern under the water and (2) the demand of clear water for a reliable pattern to image correspondence retrieval. For more information, refer to [50,51].

2. Preliminaries

The traveling 3D sine wave is deemed (note that the shift term H_0 in Equation (2) is omitted; this is possible if the translation from $GCS \rightarrow LCS$ already compensates for that shift):

$$\mathbf{y}(t_i) = \mathbf{y}_i \qquad (1)$$

$$\mathbf{z}(t_i) = A \, \sin \left[2\pi \left(\frac{t_i}{T} - \frac{\mathbf{y}_i}{\lambda} \right) + \phi \right], \quad i = 1 \dots N \qquad (2)$$

where ϕ is the phase of the wave-front and N is the duration of the measurement in seconds. The traveling sine wave is a special surface, and as such, it is advantageously modeled in a *local coordinate system* (*LCS*), that is parallel to the wave propagation direction and shifted to the mean level of oscillations (*cf.* Figures 1 and 6). In order to link the *LCS* with the *global coordinate system* (*GCS*) where the cameras are defined, a 3D spatial similarity transformation (the scale factor is unity; thus, the transformation reduces to a 3D rigid transformation) is formulated. The parameters of this transformation (*i.e.*, three components of the translation vector and three rotation angles) could likewise be included within the adjustment.

The methods invented devise the above-defined model in a least square approach, that is trying to minimize the residuals between the nominal observations and the observations predicted by that model. To solve the non-linear least squares problem, one must know the starting values of all parameters involved. Presented below is the strategy for retrieving the wave amplitude A, the wavelength λ, period T, phase-shift ϕ and the rigid transformation of the wave *LCS* to the camera *GCS*. Experience has shown that unless the amplitude is infinitesimally small and the wave infinitely long, even very rough estimates of the unknown wave parameters assure convergence of the system of equations.

Throughout the text, the notion of real and virtual points appears. The points are real if their projections in images come directly from the points (e.g., physical targets) or virtual (*i.e.*, specular highlights) if the camera sees merely the reflection of a real point. Virtual points are always single-image observations, as their positions in space depend on the shape of the surface from which it is reflected and the view angle of the camera (*vide* the law of reflection).

183

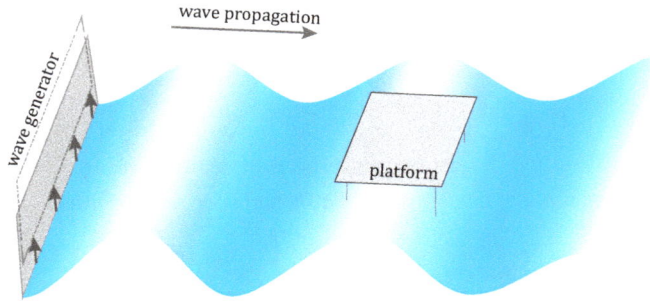

Figure 1. A simplistic view of the model basin. The platform is placed at a distance from the wave generator (on the left) and parallel to the wave propagation direction.

2.1. Derivation of Approximate Wave Parameters

The wave amplitude can be: (i) estimated visually, on-site, during the measurement taking place, (ii) recovered from the wave probes, as these are a common place in any ship model basin, or (iii) derived from image-based measurement, provided there are observed real points found on the water surface (adopted by the authors). When the image-based approach is undertaken, the triangulation step must be followed to obtain the 3D coordinates of the real points (*cf.* Section 3.1.2). The amplitude can be recovered with the complex amplitude demodulation method (AMD) or merely by removing the trend from a point's response and taking the halved maximum bound to be the starting A value. The collateral benefit of the AMD is that, were there is a varying amplitude signal, a slope instead of a horizontal line outcome would be observable. Accordingly, the A shall be replaced with a (linear) time-varying function, the parameters of which may be engaged in the total adjustment, as well.

Similarly, the value of the period T can be approximated either from on-site visual impressions, using the image data in post-processing (adopted by the authors), or taken directly from the wave probe data. The dominant period is then restored with the help of the spectral analyses, *i.e.*, the periodogram power spectral density estimate.

The wavelength might be inspected visually or with the help of the method presented in Section 3.1 (adopted by the authors). The wave probe devices, being single-point-based measurements, do not deliver enough data to indicate the length of the traveling wave.

The remaining wave parameter is the phase shift. Its computation requires a real point floating on the water surface, or otherwise, the wave probe data can become useful if its position is known in the reference coordinate system (CS) (adopted by the authors). If one moves the CS to that point, the term $\frac{y_j}{\lambda}$ in Equation (2) cancels out. If one further takes the starting frame $t_1 = 0$, also the term $\frac{t_1}{T}$ is gone, and the phase shift can be computed from $\phi = \arcsin \frac{z_j}{A}$, where j denotes the point at the origin of the translated CS. A slightly more elegant way to solve the equation for the initial phase shift is to use, again, the demodulation technique.

As for the wave transformation parameters $(\omega, \Phi, \kappa; \mathbf{T})^{GCS \to LCS}$, one usually tries to define the global system that is quasi-parallel to the wave propagation direction or in best case that aligns with it. If so, the rigid transformation parameters can be set to zero values at the beginning of the adjustment and receive their corrections, which compensate for the inaccurate alignment. The image data and the scene context must be exploited to find the transformation relating the two coordinate system by identifying, e.g., the minimum of three common points or a 3D line and a point.

2.2. Optimization Technique

In the computational part, the Gauss–Markov least squares adjustment with conditions and constraints was adopted. The adjustment workflow proceeds in repetitive cycles of five steps, *i.e.*, (i) the generation of current approximate values of all unknown parameters, (ii) the calculation of the reduced observation vector, (iii) the calculation of the partial derivatives of the functional model with respect to all current parameters, (iv) the construction of the normal equations and (v) the solution of the system of equations.

The partial derivatives that make up the Jacobian matrix are always evaluated at the current values of the parameters. If the system of equations includes both condition and constraint equations, it does not fulfil the positive-definite requirement put by, e.g., the Cholesky decomposition. Indeed, on the diagonal of the equation matrix in Method 2, zero entries are present. The vector of solutions is then retrieved with the Gauss elimination algorithm.

3. Methods

3.1. Water as a Diffuse Surface

The following method sees the water as a diffuse surface. It is converted to such, owing to artificial targeting. A set of retro-reflective targets floated on the water surface were tracked during the measurements (*cf.* Figure 2). The targets were in-house produced using a diamond grade reflective sheeting ($3M^{TM}$, Minneapolis, Minnesota, USA). The sheet, thanks to the cube corner reflectors' composition, allowed for an efficient light return for very wide entrance angles, thereby assuring good visibility in the oblique looking camera views. The targets were interconnected with a string, so as to avoid their collision and dispersion; their spacing equaled *ca*. 30 cm.

Reconstruction of a singular point provides for all but one wave parameter: the wavelength λ. Combining the responses of the minimum of two such points allows for the possibility of recovering also the remaining λ. It is presumed that the 3D data have already been transformed to the *LCS*, and to shorten the discussion, this is excluded from the mathematical model. A complete description on how to include this information in the model is given in Section 3.2.

Since the parameters are found in a least squares approach, the discussion commences with the initial parameter retrieval. In the next step, the adjustment mathematical model is outlined. Evaluation of the results continues in Section 5.

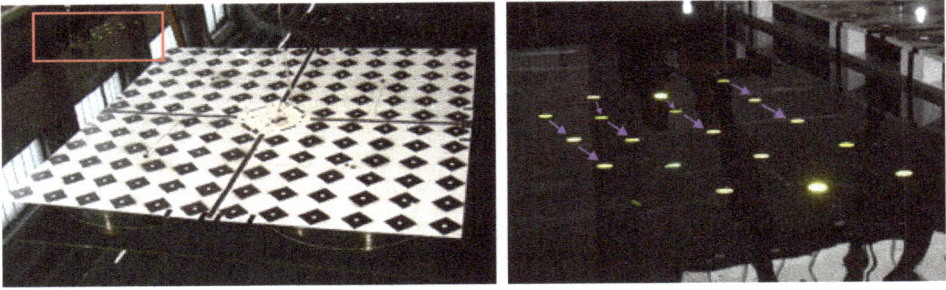

Figure 2. Left: physical, retro-reflective targets floating in the vicinity of the platform are marked with the red box. Right: a close-up of the targets; the arrows indicate established cluster pairs.

3.1.1. Mathematical Model

The functional model describes analytically the prior knowledge on the wave shape expressed in Equation (2). The modeling part, unlike in Method 2 in Section 3.2, is formulated purely in 3D space. The point triangulation is treated as a separate and unrelated phase, even though the image measurements indirectly contribute to the eventual outcome, and a joint treatment might be suggested. As a matter of fact, the 3D space is reduced to a 2D space, building on the fact that the transformation from *GCS* to *LCS* has taken place in advance, and the *x*-coordinate can take arbitrary values. All parameters are present in the adjustment as observations and unknowns; see the adjustment workflow in Figure 3. The stochastic model is formalized within the weight matrix and conveys the observation uncertainty. The matrix holds non-zero elements on its diagonal, and they may take the following form: $w_i = \frac{\sigma_0^2}{\sigma_i^2}$, where σ_i signifies the *a priori* standard deviation of an observation and σ_0 is the *a priori* standard deviation of a unit weight. Within the experiments, the σ_0 was set to unity, whereas $\sigma_A = 2$ mm, $\sigma_T = 0.05$ s, $\sigma_\lambda = 100$ mm, $\sigma_\phi = 0.25$ rad. These values ɛre rough and rather pessimistic estimates of the uncertainty of the approximate parameters.

The condition equations are formed by all parameters that are regarded as observed, i.e., y_i, z_i, A, T, λ, ϕ, and follow Equation (6). Every observed point i provides three condition equations: $\hat{y}_i = y_i$, $\hat{z}_i = z_i$ and Equation (2).

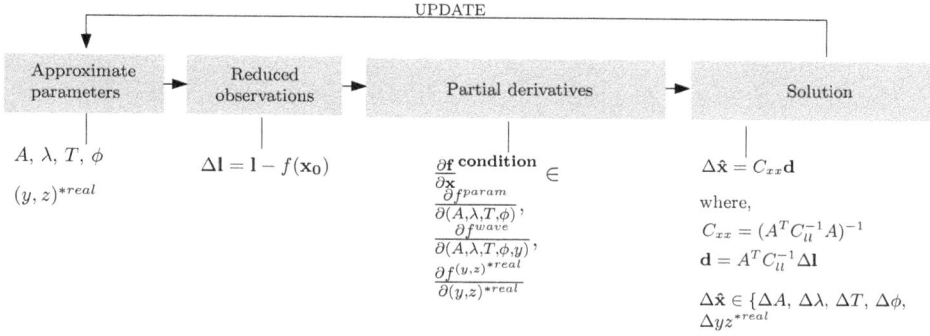

Figure 3. The least squares adjustment workflow of Method 1. The cycle repeats until the stop criterion is reached. $\partial \mathbf{x}$ denotes the entire vector of parameters; A is the Jacobian in observation equations; and C_{ll} is the covariance matrix. See f^{wave}, f^{param} in Equations (2) and (6).

3.1.2. Image-Based Approximate Wave Retrieval

The starting point is to measure the targets in images, for instance with centroiding methods, ellipse fitting or cross-correlation [61–63]. The authors adopted the intensity centroiding method to detect points in the initial frame and the sub-pixel cross-correlation to track them in time [64]. The 2D image measurements are then transferred to 3D space in a regular bundle adjustment and exploited to recover the initial wave parameters. The developed pipeline is fully automatic and summarized in the following order:

1. clustering of 3D points;
2. coupling of neighboring clusters;
3. calculation of mean A, T and ϕ from the clusters;
4. calculation of mean λ from the couples of clusters.

The reconstructed targets in time are considered an unorganized point cloud; thus, their clustering is carried out up front (*cf.* Figure 2). The retrieved clusters are equivalent to the responses of a single target floating on the water surface. The clustering *per se* is not required, as this piece of information is already carried within the point naming convention. Nonetheless, because this may not always be the case, the clustering is a default operation. It creates boundaries in 3D space based on some measure of proximity; in our case, the Euclidean measure was chosen. The algorithm was invented by [65] and implemented in [66].

Now, the coupling establishes a neighborhood relationship between closest clusters (*cf.* Figure 2). Given a *kd*-tree representation of all clusters' centroids, the algorithm searches for the neighbors within a desired radius and ascertains that the selected pair has an offset along the wave propagation direction. This later permits for the computation of the wavelength λ. The selected radius should not be too small, but cannot be greater than the wavelength, to be able to resolve its length.

The mean A, T and ϕ are found within each single cluster, as pointed out in Section 2.1. To find the mean λ, the requirement is to know: (i) the period T, (ii) the direction of the wave propagation, and (iii) the relation between the *GCS* and the *LCS*, where the wave is defined. For every couple of clusters, one counts how much time Δt a wave crest takes to travel between the clusters (*cf.* Figure 4). The distance Δd (in *LCS*) between them is known and so is the period T, hence the wavelength estimate results from the trivial proportion:

$$\frac{\Delta d}{\Delta t} = \frac{\lambda}{T} \tag{3}$$

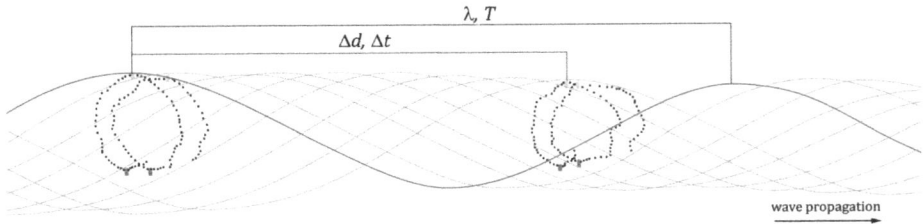

Figure 4. Image-based approximate wave retrieval. For this cluster pair, the wave crest takes eight frames (Δt) to travel between the neighbors. If the distance between the clusters is $\Delta y = 10$ cm and the period $T = 12$, then the wavelength $\lambda = 15$ cm.

3.2. Water as a Specular Surface

The second developed method exploits the fact that water under some conditions can be perceived as specularly reflecting, *i.e.*, manifesting no scattering on its surface. Two parallel arrays of linear lamps hang above the measurement space (*cf.* Figure 8). Their reflections on top of the water surface could be observed as static, in the steady wave condition, and dynamic, when the water was excited. In the latter case, the reflections project to distorted shapes (*cf.* Figure 5). Such deformations implicitly carry information about the instantaneous shape of the water surface and are investigated in the successive paragraphs.

188

Because the specularities (also known as highlights) travel with the observer and depend on the surface geometry, no corresponding features in different cameras are present [33,51,67]. As a result, no stereo or multi-view measurements are made possible. Unless one is able to directly identify 3D rays that would intersect at the interface of a surface [40], the alternative solution to the depth-normal ambiguity is to add control information and/or impose appropriate constraints in 3D object space.

In the developed approach, the images of specular highlights and a number of parameter constraints are combined together to recover the water's instantaneous state. This method solves a least squares problem, simultaneously determining all parameters of interest. The discussion opens with the condition equations and imposed constraints, which constitute the functional model of the LS problem. Next, the adjustment procedure is explicitly given, including: (i) the derivation of approximate values for all unknowns, (ii) the stochastic model, (iii) the system equation forming, as well as (iv) the collection of control information. Lastly, the experimental section presents the results followed by a compact conclusive paragraph.

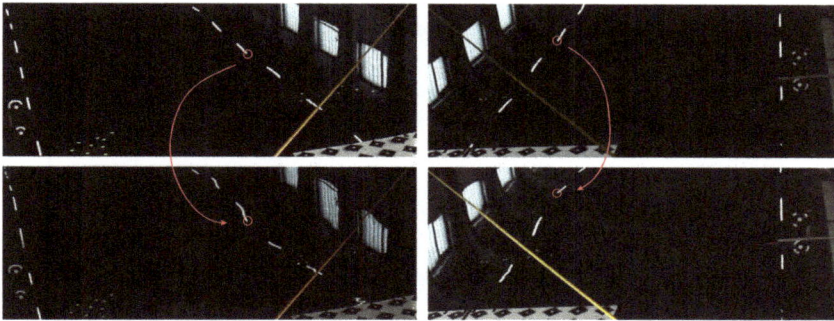

Figure 5. Specular reflections of the control information seen by two cameras. **Top**: calm water condition; the water can be considered a plane, and linear features map to linear reflections. **Bottom**: water in the excited condition; linear features map to deformed reflections.

3.2.1. Mathematical Model

The functional model comprises the mathematical description of two observed phenomena, that is the perspective imaging associated with the camera system, as well as the shape of the induced waves, which in turn associates with a wave maker.

The camera to object points relation was modeled with the collinearity equations. The shape of the induced waves was modeled with Equation (2). The defined wave model is accompanied by three constraint equations. They impose that: (i) virtual points lie on the wave surface (f_{surf}) (their distance from the surface = 0), (ii) for all virtual points, the incident and reflected rays make equal angles with respect to the tangent/normal at that point (f_{refl}) (compliance with the law of reflection), and (iii) the vector from the camera to a virtual point, the normal at the point and the vector from that point towards its 3D real position are coplanar (f_{copl}) (compliance with the law of reflection). The real points are always considered as ground control information; therefore, the developed method belongs to the class of calibrated environment methods. See the adjustment workflow in Figure 7.

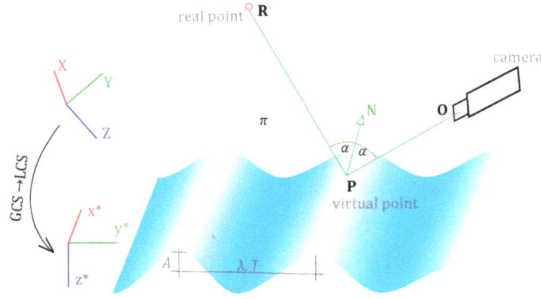

Figure 6. Graphical representation of the functional model. The camera captures the scene following the laws of perspective projection. The wave model, displayed at a time instance, is defined through parameters A, λ, T and ϕ in LCS. The camera (O), the virtual point **P** and real point **R** lie on a common plane π, whereas the incident and reflecting rays are symmetric *w.r.t.* the normal **N**.

Apart from what has been so far discussed in Section 3.1.1, the stochastic model avoids having the solution driven by the observations that are most abundant. It limits the influence of a particular group of observations with the help of a second weighting matrix N_{max}. Here, every group of observations was assigned a value n_{max} that limits its participation in the adjustment to below n_{max} observations. The diminishing effect is realized by the expression in Equation (4) and found on the diagonal of the matrix. The n_{obs} is the cardinality of the observations within a group. The ultimate weight matrix W' is a multiplication, $W' = W \cdot N_{max}$. The bespoke weighting strategy is implemented within MicMac, an open source bundle adjustment software [68,69].

$$n_{max}^{obs,i} = \frac{\frac{n_{obs}\, n_{max}}{n_{obs}+n_{max}}}{N_{obs}}, \quad i = 1 \ldots N \tag{4}$$

Within the experiments, the σ_0 value was always set to unity, whereas $\sigma_A = 1$ mm, $\sigma_\lambda = 10$ mm, $\sigma_T = 0.05$ s, $\sigma_\phi = 0.25$ rad, $\sigma_{xy} = 0.5$ pix, $\sigma_{XYZ}^{real} = 5$ mm, $\sigma_{XYZ}^{virtual} = 10$ mm, $\sigma_T^{GCS \rightarrow LCS} = 25$ mm and $\sigma_{\omega,\phi,\kappa}^{GCS \rightarrow LCS} = 0.01$ rad. In analogy to Method 1, these values are rough estimates of the estimated approximate parameters' values. If the parameter setting shall be unclear, a means of assessing the correctness of the *a priori* values must be employed, e.g., the variance component analysis.

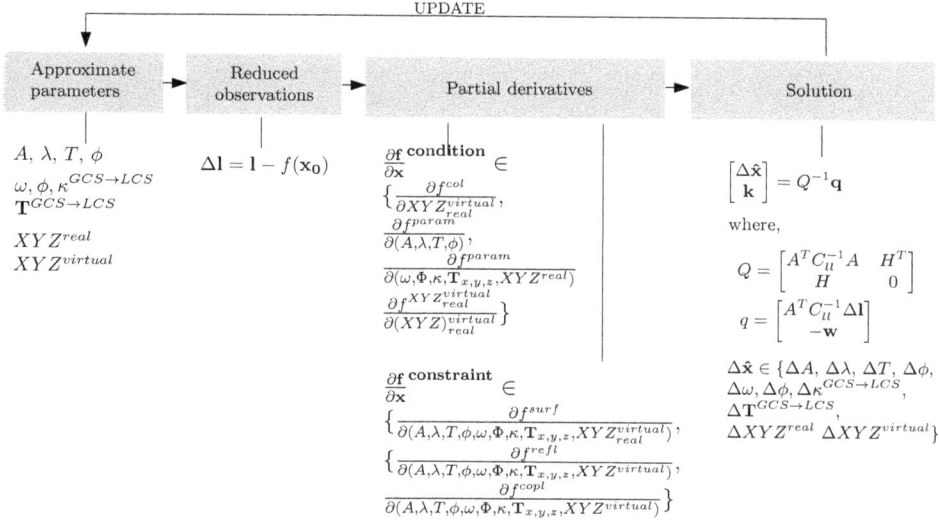

Figure 7. The least squares adjustment workflow of Method 2. The cycle repeats until the stop criterion is reached. ∂x denotes the entire vector of parameters; A is the Jacobian in observations equations; H is the Jacobian in constraint equations; k is the Lagrange multiplier; and C_{ll} is the covariance matrix. See f^{param}, f^{surf}, f^{refl}, f^{copl} in Equations (6)–(9), respectively.

Condition equations are functions of observations and parameters. Collinearity equations are self-evidently observations as a function of parameters: the 3D coordinates of the real or virtual point (*IOR* and *EOR* in the developed implementation were treated as constants). Equation (5) renders the collinearity expanded into Taylor series around the N initial estimates XYZ_i^0:

$$\mathbf{f}^{col}_{(XYZ)} = f^{col}(XYZ_i^0) + \frac{\partial f^{col}}{\partial (XYZ)_i^0}\, d(XYZ)_i, \quad i = 1 \dots N \qquad (5)$$

Optionally, one may define originally free parameters as observed unknowns. This trick helps to include any available knowledge of the unknowns into the pipeline, as well as to avoid surplus parameter updates, *i.e.*, steer the rate of parameter change along the iterations. The parameters to control the rate are the entries of the weight matrix W. Our implementation allows all parameters to be regarded as observed; therefore, any parameter in Figure 7 can be replaced with the *param* in Equation (6). For instance, if an X-coordinate is observed, the condition equation $\hat{X} = X^{obs} + v_x$

and the correction equation $X = X^{obs} + 1 \cdot dX$ are written down, where v_x and dX are the observation and the parameter corrections.

$$\mathbf{f}^{param} = f_{0,i}^{param} + \frac{\partial f^{param}}{\partial (param)_i^0} \, d(param)_i, \quad i = 1 \ldots N_{param} \tag{6}$$

Constraint equations do not involve observations, but pure parameters. The conceived wave model renders three constraint equations:

(i) $\mathbf{f}^{surf}$,

$$\mathbf{f}^{surf} = -\mathbf{z}^*(t_i) + A \, \sin\left[2\pi \left(\frac{t_i}{T} - \frac{y_i^*}{\lambda} \right) + \phi \right] = 0, \tag{7}$$

(ii) $\mathbf{f}^{refl}$,

$$\mathbf{f}^{refl} = \mathbf{ff}^{OP} - \mathbf{ff}^{PR} = \arccos\left(\frac{\overrightarrow{OP} \, \overrightarrow{N}}{|\overrightarrow{OP}| \, |\overrightarrow{N}|} \right) - \arccos\left(\frac{\overrightarrow{PR} \, \overrightarrow{N}}{|\overrightarrow{PR}| \, |\overrightarrow{N}|} \right) = 0 \tag{8}$$

(iii) $\mathbf{f}^{copl}$,

$$\mathbf{f}^{copl} = \overrightarrow{OP} \, (\overrightarrow{PR} \times \overrightarrow{N}) = \begin{vmatrix} OP_{x*} & OP_{y*} & OP_{z*} \\ RP_{x*} & RP_{y*} & RP_{z*} \\ N_{x*} & N_{x*} & N_{x*} \end{vmatrix} = 0 \tag{9}$$

where $\overrightarrow{OP} = [x_O^* - x_P^* \quad y_O^* - y_P^* \quad z_O^* - z_P^*]$, $\overrightarrow{PR} = [x_R^* - x_P^* \quad y_R^* - y_P^* \quad z_R^* - z_P^*]$, $y^* = y^*(X, Y, Z, \omega, \Phi, \kappa; \mathbf{T})$ and $z^* = z^*(X, Y, Z, \omega, \Phi, \kappa; \mathbf{T})$. The constraints are defined locally; thus, coordinate quantities are annexed with the symbol *. The values determined in LCS are not considered in the adjustment, but are obtained after a 3D rigid transformation with the parameters $(\omega, \Phi, \kappa; \mathbf{T})^{GCS \rightarrow LCS}$.

The linearized forms of the above equations, expanded into Taylor series, are presented in Appendix A. Note that the local coordinate quantities x^*, y^*, z^* are functions of their positions X, Y, Z in the GCS, as well as the parameters of the 3D rigid transformation. As a result, the derivatives are calculated for a composition of functions and must obey the chain rule.

3.2.2. Derivation of Control Information

The control information was not acquired physically prior to nor during the measurements. Not even posterior efforts were undertaken to collect the ground truth. The position of the linear lamps (*cf.* Figure 8), which served as the ground truth information, was recovered solely using the image data, under the condition that the reflecting water is globally a plane. As each measurement started with the calm water condition, the planarity condition was valid at numerous times. The imaging situation is depicted in Figure 9.

Figure 8. Two arrays of lamps are marked with red bounding boxes. The end points of the single lamps served as the control information.

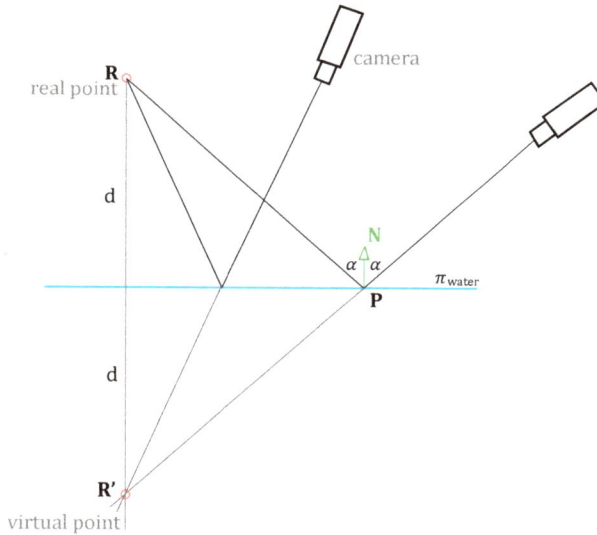

Figure 9. The imaging configuration taking place when deriving the control information. $\mathbf{R}$ represents the end-point of a linear lamp; $\mathbf{R'}$ is its virtual location, at d distance behind the water plane π. The camera to real-point rays follow the law of reflection, *i.e.*, the incident and reflecting rays form equal angles with the plane normal $\mathbf{N}$.

The calculation of the XYZ coordinates of the control information in their real locations divides into: (i) the water plane derivation (from real points), (ii) the identification of homologous points across views and triangulation (virtual points), and lastly (iii) the flipping of the virtual points to their real positions.

193

The plane π of the water was recovered thanks to well-distributed dust particles present on its surface. Their appearance was sufficiently discriminative for their identification across views. Alternatively, one could place artificial targets on top of the water to avoid potential identification problems. Given a few (≥ 3) pairs or triples of 2D image points corresponding to real features, their 3D position is found by intersection. The searched plane defined analytically as $Ax + By + Cz + D = 0$ is then established by singular value decomposition (SVD).

The end points of the linear lamp reflections were identified and measured in images manually only in the initial frame. The subsequent tracking in time was realized by the flagship cross-correlation technique. Having found and measured the reflections, their 3D locations are triangulated ($\mathbf{R}'$ in Figure 9), ignoring the fact that the observed features are not real. The 3D points emerge on the wrong side of the water plane; thus, they must be flipped to their real positions. The flipping is done with respect to an arbitrary plane, being the water plane and determined by the coefficients A, B, C and D. The transformation performing that operation works by:

(i) roto-translating the global coordinate to a local coordinate system that aligns with the flipping plane,

$$T \cdot R_1 = \begin{bmatrix} 1 & 0 & 0 & 0 \\ 0 & 1 & 0 & 0 \\ 0 & 0 & 1 & 0 \\ -X_i & -Y_i & -Z_i & 1 \end{bmatrix} \cdot \begin{bmatrix} \frac{\lambda}{|\mathbf{N}|} & 0 & \frac{n_x}{|\mathbf{N}|} & 0 \\ \frac{-n_x n_y}{\lambda |\mathbf{N}|} & \frac{n_z}{\lambda} & \frac{n_y}{|\mathbf{N}|} & 0 \\ \frac{-n_x n_z}{\lambda |\mathbf{N}|} & -\frac{n_y}{\lambda} & \frac{n_z}{|\mathbf{N}|} & 0 \\ 0 & 0 & 0 & 1 \end{bmatrix} \tag{10}$$

where $\mathbf{N} = [n_x\ n_y\ n_z]$, $\lambda = \sqrt{n_y^2 + n_z^2}$, and $[X_i\ Y_i\ Z_i]$ are 3D coordinates of any points lying within the flipping plane.

(ii) performing the actual flipping over the local $XY-$plane:

$$R_2 = \begin{bmatrix} 1 & 0 & 0 & 0 \\ 0 & 1 & 0 & 0 \\ 0 & 0 & -1 & 0 \\ 0 & 0 & 0 & 1 \end{bmatrix} \tag{11}$$

(iii) and bringing the point back to the global coordinate system with R_1^{-1} and T^{-1}.

The entire procedure committed in a single formula renders:

$$M = T \cdot R_1 \cdot R_2 \cdot R_1^{-1} \cdot T^{-1} \tag{12}$$

3.2.3. Derivation of the Approximate Highlight Position

The highlights' coordinates (**P** in Figure 6) result from the intersection of the approximate wave model with the vectors anchored in the image measurements, passing through the camera perspective center and extending into the object space. The highlights are single-image observations, so there exist no correspondences across images, as in the case of real points. The intersection points are found first by intersecting with the mean water plane and then by iteratively improving the results with Newton's method. The points are first found in the *LCS* and subsequently transferred to the *GCS* given the approximate parameters of the rigid transformation. The algorithm is presented below.

Given the 3D vector defined by points (x_1^*, y_1^*, z_1^*) and (x_2^*, y_2^*, z_2^*) at the camera center and observed in image space, respectively, the 3D line parametric equation takes the form:

$$
\begin{bmatrix} x^* \\ y^* \\ z^* \end{bmatrix} = \begin{bmatrix} x_1^* \\ y_1^* \\ z_1^* \end{bmatrix} + t \cdot \begin{bmatrix} x_2^* - x_1^* \\ y_2^* - y_1^* \\ z_2^* - z_1^* \end{bmatrix}
\tag{13}
$$

The sought y^*-coordinate of the intersection is then equal to $y^* = y_1^* + t \cdot (y_2^* - y_1^*)$ where $t = (z^* - z_1^*)/(z_2^* - z_1^*)$. The unique solution can be obtained when the z in the preceding equation is replaced with the mean water level, e.g., $H_0 = 0$ in the *LCS*. A better approximation can be accomplished if the intersection is performed with a more realistic model than the plane: the observed sine wave. Combining Equation (2) with the last row of Equation (13), such that the z^* terms are equal, brings about the following relationship:

$$
g(y^*) = y^* - y_1^* - \frac{(y_2^* - y_1^*)}{(z_2^* - z_1^*)} \cdot (H_0 + A \sin[2\pi\,(\frac{t}{T} - \frac{y^*}{\lambda}) + \phi] - z_1^*) = 0
\tag{14}
$$

The function g has one parameter y^*, and as such, together with its derivative g', both evaluated at the current parameter value y_0^*, they enter Newton's method, which finds the ultimate root. The Newton step is the ratio of $g(y_0^*)$ and $g'(y_0^*)$ (*cf.* Equation (15)). The loop continues until the difference between old and new parameter estimates no longer falls below a defined threshold.

$$
y^* = y_0^* - \frac{g(y_0^*)}{g'(y_0^*)}
\tag{15}
$$

Once the y^* value is known, the z^*-coordinate is computed from Equation (2), and lastly, the x^* can be retrieved from the 3D line equation as $x^* = x_1^* + \frac{z^* - z_1^*}{z_2^* - z_1^*} \cdot$

$(x_2^* - x_1^*)$. The final step brings the locally-determined coordinates to the global ones with $(\omega, \Phi, \kappa; \mathbf{T})^{GCS \rightarrow LCS}$.

4. Imaging System

The imaging setup is comprised of three dSLR cameras (Canon 60D, 20-mm focal length) and three continuous illumination sources (1250 W). The spatial resolution of the videos matched the full HD (1920 × 1080 pix), acquiring a maximum of 30 fps in progressive mode. The video files were lossy compressed with the H.264 codec and saved in a .mov container.

The mean object to camera distance amounted to 10 m, resulting in the average image scale of 1:500. The cameras were rigidly mounted on a mobile bridge (*cf.* Figures 10 and 11) and connected with each other, as well as with a PC, via USB cables to allow for: (i) remote triggering, (ii) coarse synchronization. Fine-alignment of the video frames was possible with the help of a laser dot observed in all cameras. The laser worked in a flicker mode, at a frequency lower than that of the video acquisition, and casually moved over the floating platform's surface, both at the start and the finish of each acquisition. No automatization was incorporated at this stage; instead, the alignment was conducted manually.

Figure 10. Red circles point to three cameras placed on a moving platform, across the model basin. The blue rectangles point to three scale bars. The remaining scale bars are symmetrically arranged on the opposite side of the basin.

Despite the USB connections, the videos were stored on the memory cards. No spatial reference field was embedded in the vicinity of the system; instead, the calibration and orientation was carried out with the moved reference bar method [70].

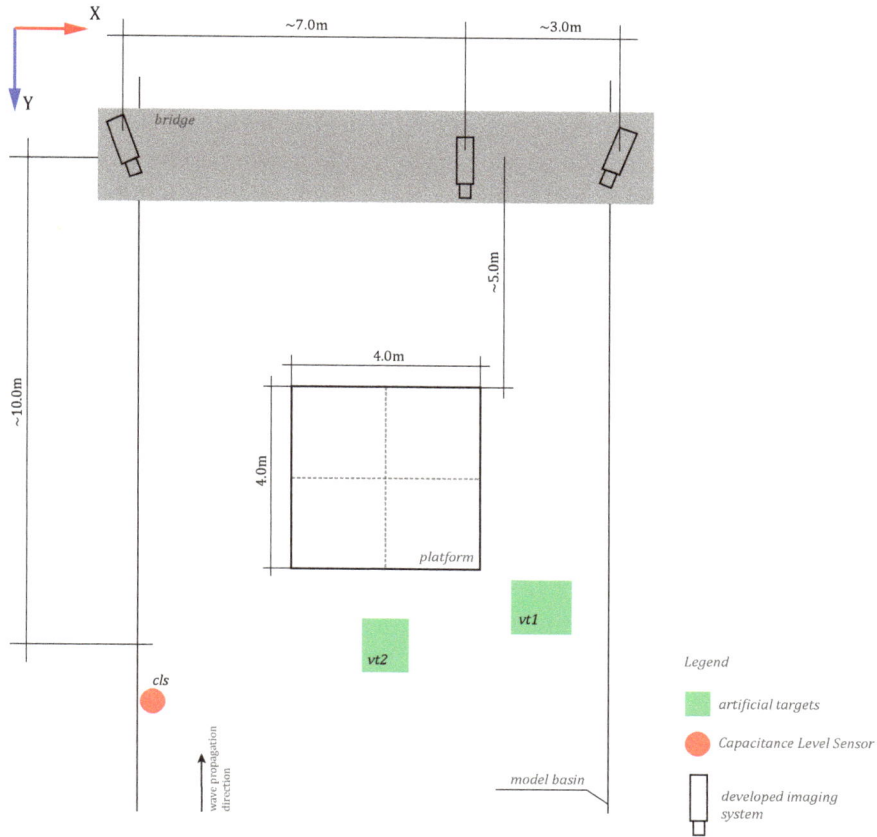

Figure 11. The top view of the measurement workplace. Cameras on top of the drawing are placed on a mobile bridge. The platform occupies the scene's central part and is located *ca.* 5 m away from the bridge. The green rectilinear shapes correspond to the zones of the artificial targets (used in Method 1, and in the evaluation of Methods 1 and 2; see Section 3.1 and Section 5), while the red circle is the capacitive level sensor (used in the evaluation of Methods 1 and 2; see Section 5).

5. Experiments

5.1. Evaluation strategy

Results achieved with Method 1 (*m1*) and Method 2 (*m2*) are confronted with the responses of a capacitive level sensor and validating physical targets (*cf.* Figure 11). The capacitive level sensor was mounted on a rod-like probe and sensed the variations in electrical capacity within the sensor. Given the dielectric constant of the liquid, this information can be directly transformed to the changes in the water level, in which the probe is normally immersed. Because the sensor samples the changes in a singular spot, it provides merely information on the amplitude and frequency of the water level oscillations; likewise validating the targets. The instantaneous wavelength, thereby, remained unknown, as no direct means to judge the accuracy of the calculated wavelength existed. Indirectly, the correctness of all wave parameters, including the wavelength, can be estimated by confronting the response of a number of points distributed along the wave propagation ($vt1, vt2, cls$) with their responses predicted from the model.

In the adjustment, $m1$ adopted three clusters, whereas $m2$ tracked up to seven highlights, corresponding to four to six ground control points (*i.e.*, the lamps' endpoints). The distribution of measured and validating points is displayed in Figure 12. Numerical results of the five measurement series are summarized in Table 1, with the graphical representations provided in Figures 13–18. The third and fourth measurement series was evaluated twice with varied image observations (specular highlights; *cf.* Figure 12). The adopted evaluation strategy is as follows.

5.1.1. Accuracy 1: Validating Targets ($vt1, vt2$)

With validating targets, we refer to points that were not treated in the adjustments aiming at finding the wave parameters. They were measured in images and independently intersected in 3D space. The Z-response of all validating targets is confronted with the value predicted from the devised wave model. The validation takes place in the *LCS*. Figures 13 and 14 illustrate the results projected onto the traveling sine wave. The red corresponds to the response from the target; the blue is the model outcome. The normal probability plots test and confirm that the residuals follow the normal distribution.

5.1.2. Accuracy 2: The Capacitive Level Sensor (cls)

To compare the data collected by the capacitive level sensor and the image-based measurement system, temporal synchronization and data resampling had to take place. The start of the capacitive level sensor (cls) data collection was conveyed to the cameras audiovisually, by switching the light on and by emitting vocal sound. This allowed for rough temporal synchronization. To fine align the two signals, cross-correlation was carried out. The frequency of cls data collection was double the frequency of the camera recording; therefore, to equalize the acquisition rates, every other sample of the cls device was discarded. Figure 18 illustrates the results of the comparisons.

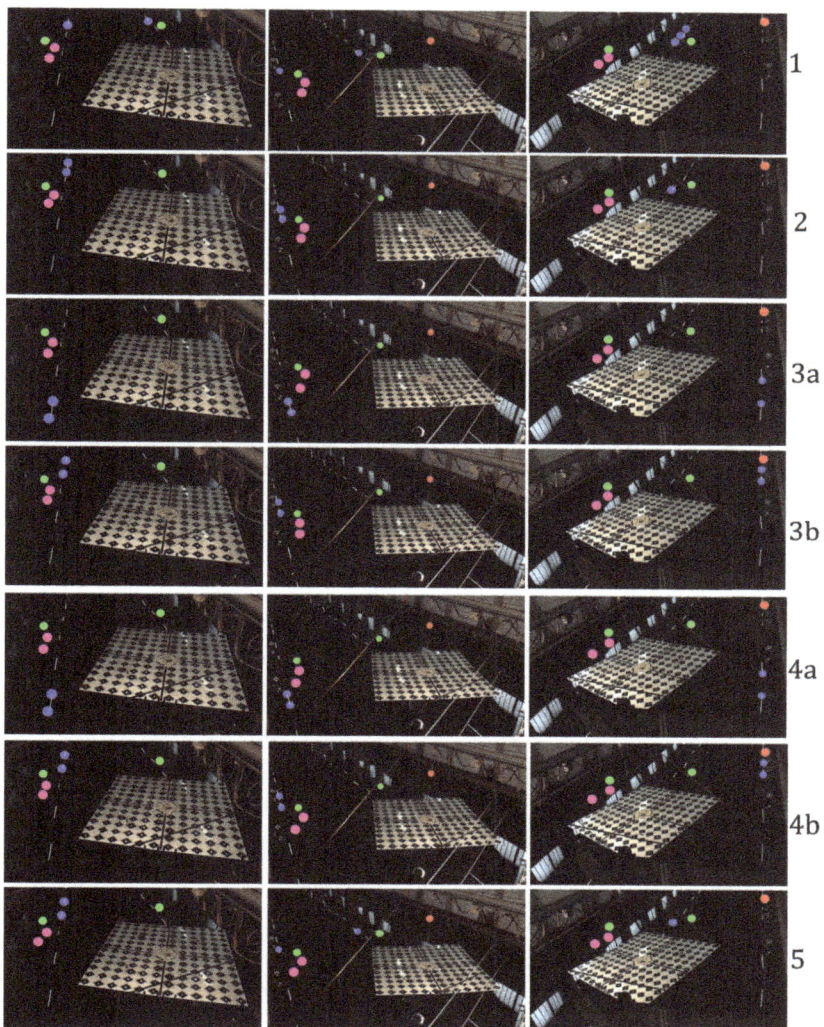

Figure 12. Distribution of observed highlights (blue) used in *m2*, artificial targets used in *m1* (magenta), validating targets (green) and the capacitive level sensor (red) in three camera views, in all measurement series.

(a) $A = 12.9mm, T = 72.1, \phi = 1.0609, \lambda = 8542.8mm$

(b) $A = 12.5mm, T = 64.6, \phi = 0.7812, \lambda = 9401.6mm$

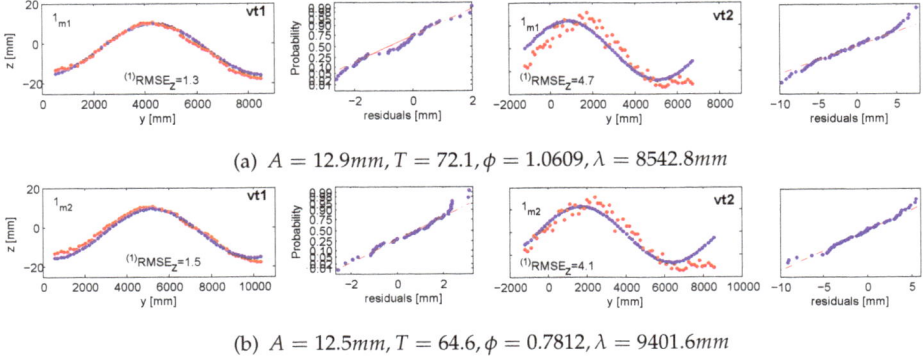

Figure 13. Accuracy 1 validation results in the first measurement series for $m1$ in (a) and $m2$ in (b), compared against the response of $vt1$ and $vt2$. Ground truth in red; fitted wave model in blue. The normal probability plots are in even columns.

Table 1. Mean precision ($\sigma_{x,y,z}$) and accuracy ($^{(1)}RMSE_z$, $^{(2)}RMSE_z$) of five measurement series evaluated with Method 1 and Method 2.

ID	Duration [frames]	$\sigma_{x,y,z}$ [mm]	$^{(1)}RMSE_Z$ [mm] $vt1/vt2$	$^{(2)}RMSE_Z$ [mm] cls
1_{m1}	70	- / 3.2 / 3.3	1.3 / 4.7	3.2
1_{m2}		1.5 / 6.5 / 0.2	1.5 / 4.1	4.0
2_{m1}	70	- / 3.3 / 3.4	2.3 / 5.5	2.3
2_{m2}		1.5 / 12 / 0.5	3.0 / 3.6	2.7
3_{m1}	83	- / 1.2 / 1.2	1.5 / 3.9	3.3
$3a_{m2}$		0.1 / 2.8 / 0.2	2.4 / 5.7	5.3
$3b_{m2}$		0.3 / 5.3 / 0.2	1.8 / 3.4	4.4
4_{m1}	70	- / 1.1 / 1.1	1.6 / 4.9	4.9
$4a_{m2}$		0.1 / 2.6 / 0.2	2.0 / 6.2	6.4
$4b_{m2}$		0.4 / 7.3 / 0.3	2.2 / 3.5	4.0
5_{m1}	100	- / 3.7 / 3.8	2.3 / 3.9	4.8
5_{m2}		1.7 / 8.7 / 0.2	2.2 / 2.8	3.7

200

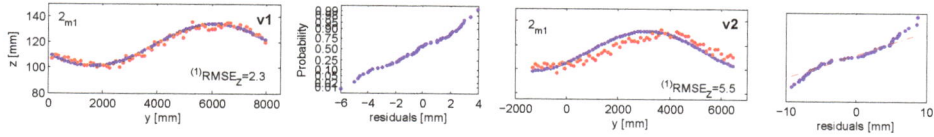

(a) $A = 16.7$ mm, $T = 76.5$, $\phi = -0.3932$, $\lambda = 8915.7$ mm

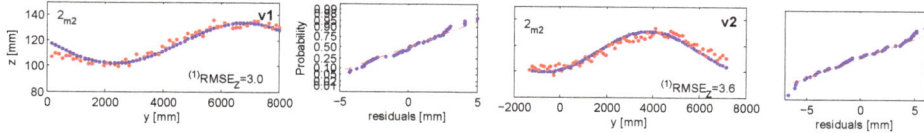

(b) $A = 15.9$ mm, $T = 70.3$, $\phi = 2.3295$, $\lambda = 9349.4$ mm

Figure 14. Accuracy 1 validation results in the second measurement series for $m1$ in (**a**) and $m2$ in (**b**), compared against the response of $vt1$ and $vt2$. Ground truth in red; fitted wave model in blue. The normal probability plots are in even columns.

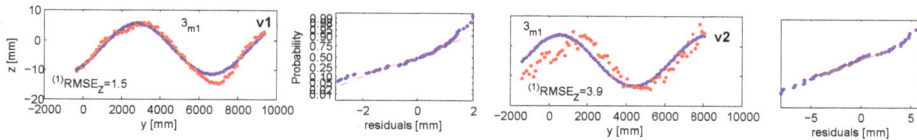

(a) $A = 8.6$ mm, $T = 65.7$, $\phi = 0.0108$, $\lambda = 7658.4$ mm

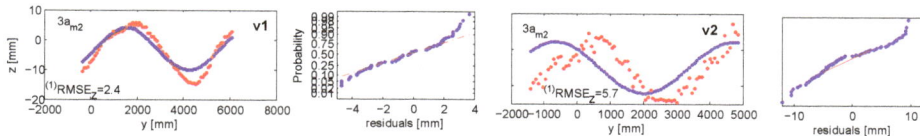

(b) $A = 7.1$ mm, $T = 69.246$, $\phi = 0.7794$, $\lambda = 5353.8.4$ mm

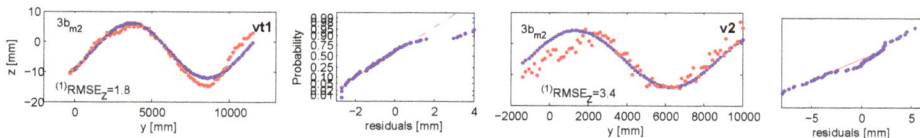

(c) $A = 9.1$ mm, $T = 69.5$, $\phi = 0.7756$, $\lambda = 9874.2$ mm

Figure 15. Accuracy 1 validation results in the third measurement series for $m1$ in (**a**) and $m2$ in (**b**) and (**c**), compared against the response of $vt1$ and $vt2$. Ground truth in red; fitted wave model in blue. The normal probability plots are in even columns.

201

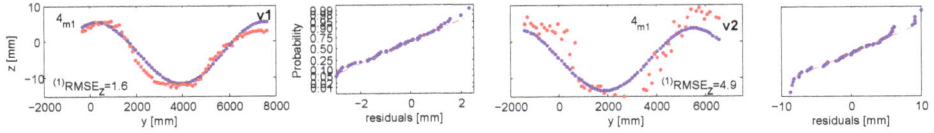

(a) $A = 8.5$ mm, $T = 63.0, \phi = -0.5305, \lambda = 7240.8$ mm

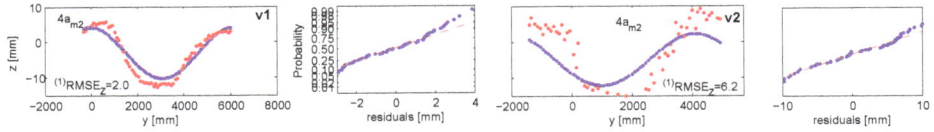

(b) $A = 7.1$ mm, $T = 70.0, \phi = 0.8025, \lambda = 6154.5$ mm

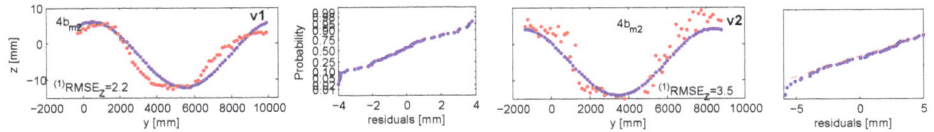

(c) $A = 9.1$ mm, $T = 70.0, \phi = 0.7821, \lambda = 9876.9$ mm

Figure 16. Accuracy 1 validation results in fourth measurement series for $m1$ in (**a**) and $m2$ in (**b**) and (**c**), compared against the response of $vt1$ and $vt2$. Ground truth in red; fitted wave model in blue. The normal probability plots are in even columns.

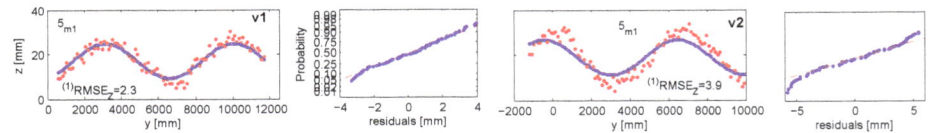

(a) $A = 7.7$ mm, $T = 61.0, \phi = 2.8754, \lambda = 6900.4$ mm

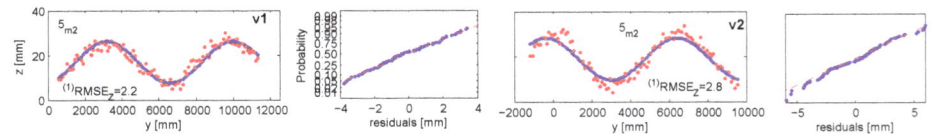

(b) $A = 9.3$ mm, $T = 62.0, \phi = 0.7838, \lambda = 6775.6$ mm

Figure 17. Accuracy 1 validation results in the fifth measurement series for $m1$ in (**a**) and $m2$ in (**b**), compared against the response of $vt1$ and $vt2$. Ground truth in red; fitted wave model in blue. The normal probability plots are in even columns.

202

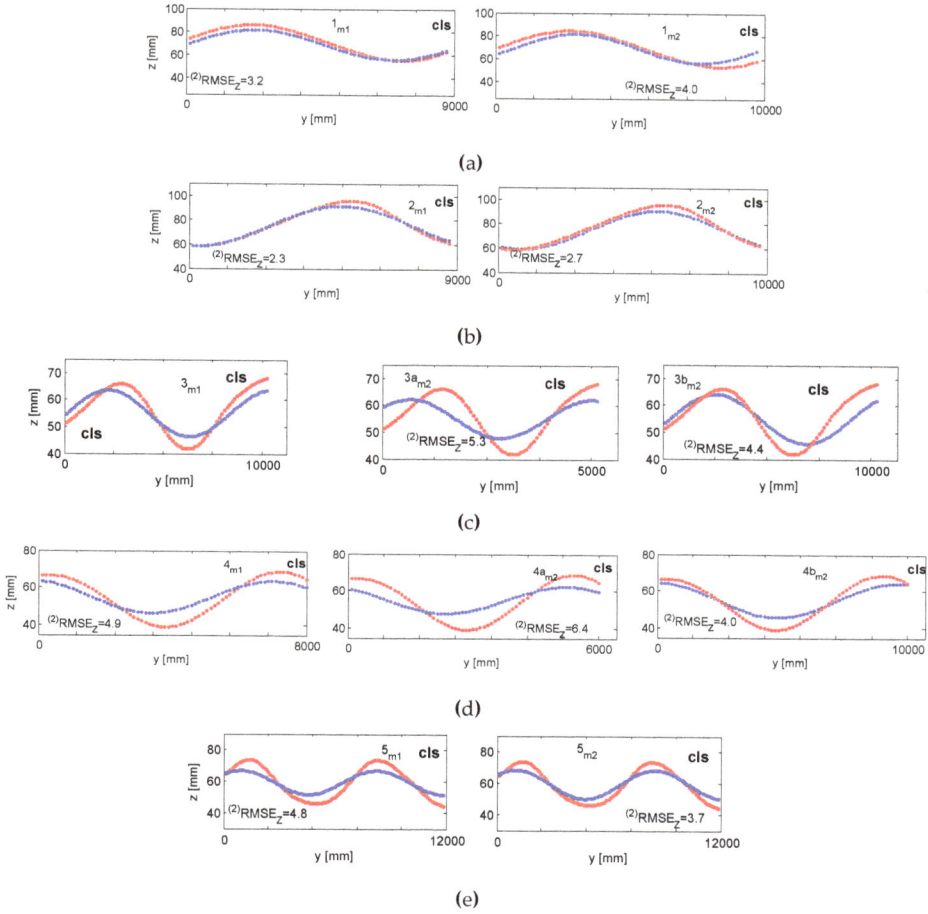

Figure 18. Accuracy 2 validation results for $m1$ and $m2$ in the 1–5 measurement series depicted in **a**–**e**, respectively. In red, the *cls* response; in blue the $m1$, $m2$ responses. All comparisons are carried out at the position of point *cls*.

5.2. Discussion

The results achieved were confronted with the responses of a capacitive level sensor and two validating targets, all of which provided single-point responses, and were placed in various positions across the water basin. In an overall assessment, the specular method (Method 2) proved superior with respect to the diffuse method (Method 1).

203

5.2.1. Accuracy

Method 1 performs well locally, when validated on points in the vicinity of the cluster pair ($vt1$), but as soon as it is confronted with distant points ($vt2$), modeling errors significantly grow; compare, e.g., Figure 13a at $vt1$ and $vt2$. Method 2 employs the entire water field in the computation and therefore has a global scope with the absence of extrapolation effects, and the modeling errors are more consistent, yet have a slightly higher magnitude. It shall be noted that the $vt1$ and cls placed on either end of the basin were under the influence of the principal sine wave, as well as the waves reflected from the basin side walls. The platform floating in the middle also disturbed the principal wave shape. The third and fourth measurement series evaluated with Method 2 on the highlights observed at the top of the basin (series $3a_{m2}$ and $4a_{m2}$) and around the platform (series $3b_{m2}$ and $4b_{m2}$) proved that the wave, having faced an obstacle, decreases its amplitude and wavelength. Compared the significant deviations of the blue/red curves in Figures 15b and 16b at $vt2$, as well as Figures 18c,d at the cls, as opposed to Figures 15c and 16c and Figures 18c,d respectively. This is of high importance in interpreting the behavior of the platform. To do that, one must know the form of the water body just before it hits the platform and not some distance before that interaction, since it no longer corresponds to the real force put on that object.

Evaluation results on cls suggest that the wave form changed spatio-temporally. It was systematically attenuated with increasing distance from the generating source. The cls was mounted closer to the wave maker than $vt1$, $vt2$, other artificial targets or the highlights and, consequently, measured higher wave amplitudes; compare the subfigures of, e.g., Figure 18c or d. Wave superposition effects (the principal and reflected waves) could contribute to higher amplitudes, as well.

5.2.2. Precision

Precision measures should not be interpreted as a single quality measure. As the evaluation proved, they are too optimistic when confronted with the accuracy measures. Moreover, the covariance matrices in Method 1 return a standard deviation homogeneous in all coordinates, while Method 2 manifests large uncertainty in the y-coordinate. This is due to the simplified and rigorous modeling of Methods 1 and 2, respectively. Method 2 simultaneously treats the reconstruction and the modeling task, whereas Method 1 performs merely the modeling, with no special treatment of the preceding steps, other than the known *a priori* standard deviation expressed in the weight matrix.

The inferior precision on the y-coordinate in Method 2 is a side effect of suboptimal network design, with a good base across the model basin (x-coordinate) and practically no shift along the y-axis (see the definition of the coordinate system in Figure 11). In spite of no parallax on the z-coordinate, the precision figures along that axis are satisfying, due to the introduced water surface model.

5.2.3. Wave Parameters

The wave parameters calculated with Method 1 and Method 2 differ, most seriously for the wavelength parameter. The differences are more pronounced at very long waves with small amplitudes (Series 3 and 4) and less evident at shorter waves or high amplitudes (Series 1, 2 and 5). In cases of small amplitude to wavelength ratios, the stability of the solution ought to be brought into discussion. Nonetheless, the subject has not been given further insight within this work.

6. Conclusions

Measurement of a difficult surface was approached with the collinearity equation, treating it as diffuse, and with the piecewise linear equations, when restituted solely from the specular reflections visible on its surface, but coming from a set of lamps hung from the ceiling. The accuracies obtained on physical targets floating on the surface, counted for the entire acquisition length, were more favorable for the latter method, falling between 1 and 3 mm.

The concept of using the lamps' reflections for metrology was partially driven by the fact that lamps in ship testing facilities appear predominantly in the same configurations. They provide a calibration field at no labor cost, of high reliability, making the methodology universal and easy to re-apply. The diffuse method, on the contrary, necessitates extra work to establish a net of targets to be placed on the water. The points are then prone to sinking, partial submergence, occlusions or drift.

The superiority of the specular over the diffuse approach is in full-field *versus* single-point shape modeling. To install a net of points that spans a large area is infeasible; therefore, one is constrained to local observations. On the contrary, the number of specular highlights is a multiplication of every point in the calibration field by the number of cameras within the imaging system. Their allocation over the measurement volume is steerable by the camera placement. If, however, bound to using the diffuse approach and aiming at full-field data collection, eliciting the water shape over extended surfaces may be possible through the adoption of patches of nets in strategic areas. In both conditions, large field modeling demands very careful planning, especially for complex-shaped surfaces. The model definition must contain just enough parameters, whereas the observations ought to deliver enough data for their recovery.

A noteworthy aspect of the specular approach is the magnifying effect present on the water surface. Observed highlights undergo an apparent motion under the deforming surface shape. The motion magnitude and trajectory are known from the law of reflection. This depends on the camera to surface relation and the relation of the point at which reflection is observed (in its real position) to its reflection on the surface (virtual position). By modifying the distance between the reflecting surface and the calibration field (real points), the motion magnitude is changed proportionally. Put differently, very small surface deformations can render large highlight displacements for a sufficiently distant calibration field.

An important issue to consider when doing least squares adjustment, true for the diffuse and specular methods, is the initial approximations of all unknowns. Unless their values are known well enough, the success of the adjustment is put under question. At small amplitudes, the longer the wave is, the more precise must be the approximations. If approximations are imprecise, divergence or convergence to an incorrect solution is highly probable. Reliability measures output from the covariance matrices of the adjustment may serve to evaluate the credibility of the results. However, this has not been investigated within this work.

The model of the wave shape assumes single-frequency oscillations. In large field observations, this assumption is often violated, as has been observed in the presented work. If the assumption is violated and the model becomes insufficient to describe the phenomena, one may still: (i) use the simple model to observe the surface locally, or (ii) extend it to involve wave time-varying components, eventually modeling its shape with a wave being the sum of two elementary waves.

Acknowledgments: Acknowledgments: This research was possible thanks to the support of the Austrian Research Promotion Agency (FFG) and the Doctoral College Energy Systems 2030 (ENSYS2030) at Vienna University of Technology.

Author Contributions: Author Contributions: Ewelina Rupnik concieved and designed the experiments, implemented the methods, and analyzed the data; Josef Jansa and Norbert Pfeifer supervised the research, including on-site experiments, development, and critical analysis.

Conflicts of Interest: Conflicts of Interest: The authors declare no conflict of interest.

Appendix A

$$
\begin{aligned}
\mathbf{f}^{surf} = {} & f_{0,i}^{surf} + \frac{\partial f^{surf}}{\partial A} dA + \frac{\partial f^{surf}}{\partial T} dT + \frac{\partial f^{surf}}{\partial \lambda} d\lambda + \frac{\partial f^{surf}}{\partial \phi} d\phi + \frac{\partial f^{surf}}{\partial y_i^*} dy_i^* + \frac{\partial f^{surf}}{\partial z_i^*} dz_i^* = \\
& f_{0,i}^{surf} + \frac{\partial f^{surf}}{\partial A} dA + \frac{\partial f^{surf}}{\partial T} dT + \frac{\partial f^{surf}}{\partial \lambda} d\lambda + \frac{\partial f^{surf}}{\partial \phi} d\phi + \\
& \underbrace{\sum_{K \in \{X_i, Y_i, Z_i, \omega, \Phi, \kappa, \mathbf{T}^{GCS \to LCS}\}} \frac{\partial f^{surf}}{\partial y_i^*} \cdot \frac{\partial y_i^*}{\partial K} dK + \frac{\partial f^{surf}}{\partial z_i^*} \cdot \frac{\partial z_i^*}{\partial K} dK}_{\frac{\partial f^{surf}}{\partial X}, \frac{\partial f^{surf}}{\partial Y}, \frac{\partial f^{surf}}{\partial Z}, \frac{\partial f^{surf}}{\partial \omega}, \frac{\partial f^{surf}}{\partial \Phi}, \frac{\partial f^{surf}}{\partial \kappa}, \frac{\partial f^{surf}}{\partial \mathbf{T}^{GCS \to LCS}}}
\end{aligned}
\tag{A1}
$$

$$\mathbf{f}^{refl} = f_{0,i}^{refl} + \frac{\partial f^{refl}}{\partial x_O^*}dx_O^* + \frac{\partial f^{refl}}{\partial y_O^*}dy_O^* + \frac{\partial f^{refl}}{\partial z_O^*}dz_O^* +$$

$$\frac{\partial f^{refl}}{\partial x_R^*}dx_R^* + \frac{\partial f^{refl}}{\partial y_R^*}dy_R^* + \frac{\partial f^{refl}}{\partial z_R^*}dz_R^* +$$

$$\frac{\partial f^{refl}}{\partial N_{x^*,i}}dN_{x^*,i} + \frac{\partial f^{refl}}{\partial N_{y^*,i}}dN_{y^*,i} + \frac{\partial f^{refl}}{\partial N_{z^*,i}}dN_{z^*,i} = \qquad \text{(A2)}$$

$$f_{0,i}^{refl} + \sum_{j\in\{O,R\}}\sum_{K\in\{\omega,\Phi,\kappa,\mathbf{T}^{GCS\to LCS}\}}\frac{\partial f^{refl}}{\partial x_j^*}\cdot\frac{\partial x_j^*}{\partial K}dK + \frac{\partial f^{refl}}{\partial y_j^*}\cdot\frac{\partial y_j^*}{\partial K}dK + \frac{\partial f^{refl}}{\partial z_j^*}\cdot\frac{\partial z_j^*}{\partial K}dK +$$

$$\sum_{j\in\{x^*,y^*,z^*\}}\sum_{K\in\{X_i,Y_i,Z_i,\omega,\kappa,\mathbf{T}^{GCS\to LCS}\}}\frac{\partial f^{refl}}{\partial N_{j,i}}\cdot\frac{\partial N_{j,i}}{\partial K}dK$$

$$\mathbf{f}^{copl} = f_{0,i}^{copl} + \frac{\partial f^{copl}}{\partial x_O^*}dx_O^* + \frac{\partial f^{copl}}{\partial y_O^*}dy_O^* + \frac{\partial f^{copl}}{\partial z_O^*}dz_O^* +$$

$$\frac{\partial f^{copl}}{\partial x_R^*}dx_R^* + \frac{\partial f^{copl}}{\partial y_R^*}dy_R^* + \frac{\partial f^{copl}}{\partial z_R^*}dz_R^* +$$

$$\frac{\partial f^{copl}}{\partial N_{x^*,i}}dN_{x^*,i} + \frac{\partial f^{copl}}{\partial N_{y^*,i}}dN_{y^*,i} + \frac{\partial f^{copl}}{\partial N_{z^*,i}}dN_{z^*,i} = \qquad \text{(A3)}$$

$$f_{0,i}^{copl} + \sum_{j\in\{O,R\}}\sum_{K\in\{\omega,\Phi,\kappa,\mathbf{T}^{GCS\to LCS}\}}\frac{\partial f^{copl}}{\partial x_j^*}dK + \frac{\partial f^{copl}}{\partial y_j^*}\cdot\frac{\partial y_j^*}{\partial K}dK + \frac{\partial f^{copl}}{\partial z_j^*}\cdot\frac{\partial z_j^*}{\partial K}dK +$$

$$\sum_{j\in\{x^*,y^*,z^*\}}\sum_{K\in\{X_i,Y_i,Z_i,\omega,\kappa,\mathbf{T}^{GCS\to LCS}\}}\frac{\partial f^{copl}}{\partial N_{j,i}}\cdot\frac{\partial N_{j,i}}{\partial K}dK$$

where:

$$N_{x^*,i} = \frac{\partial f^{surf}}{\partial y_i^*}\cdot\frac{\partial y_i^*}{\partial X_i} + \frac{\partial f^{surf}}{\partial z_i^*}\cdot\frac{\partial z_i^*}{\partial X_i}$$

$$N_{y^*,i} = \frac{\partial f^{surf}}{\partial y_i^*}\cdot\frac{\partial y_i^*}{\partial Y_i} + \frac{\partial f^{surf}}{\partial z_i^*}\cdot\frac{\partial z_i^*}{\partial Y_i} \qquad \text{(A4)}$$

$$N_{z^*,i} = \frac{\partial f^{surf}}{\partial y_i^*}\cdot\frac{\partial y_i^*}{\partial Z_i} + \frac{\partial f^{surf}}{\partial z_i^*}\cdot\frac{\partial z_i^*}{\partial Z_i}$$

References

1. Kohlschütter, E. Stereophotogrammetrische Arbeiten, Wellen und Küstenaufnahmen. *Forschungsreise S.M.S Planet*, 3; Verlag von Karl Siegismund: Berlin, Germany, **1906** (in German).

2. Laas, W. Messung der Meereswellen und ihre Bedeutung für den Schiffbau. In *Jahrbuch der Schiffbautechnischen Gesellschaft*; Springer: Berlin, Germany, 1906; pp. 391–407 (in German).

3. Waas, S. Entwicklung eines Verfahrens zur Messung kombinierter Höhen- und Neigungsverteilungen von Wasseroberflächenwellen mit Stereoaufnahmen. M.Sc. Thesis, Ruperto-Carola University of Heidelberg, Heidelberg, Germany, 1988 (in German).
4. Kiefhaber, D. Optical Measurement of Short Wind Waves–From the Laboratory to the Field. Ph.D. Thesis, Ruperto-Carola University of Heidelberg, Heidelberg, Germany, 2014.
5. Nocerino, E.; Ackermann, S.; Del Pizzo, S.; Menna, F.; Troisi, S. Low-cost human motion capture system for postural analysis onboard ships. In Proceedings of Videometrics, Range Imaging, and Applications XII; Munich, Germany, 14–15 May 2011; pp. 80850L–80856L.
6. Stojic, M.; Chandler, J.; Ashmore, P.; Luce, J. The assessment of sediment transport rates by automated digital photogrammetry. *Photogr. Eng. & Rem. Sens.* **1998**, *64*, 387–395.
7. Godding, R.; Hentschel, B.; Kauppert, K. Videometrie im wasserbaulichen Versuchswesen. *Wasserwirtschaft, Wassertechnik* **2003**, *4*, 36–40; (in German).
8. Chandler, J.; Wackrow, R.; Sun, X.; Shiono, K.; Rameshwaran, P. Measuring a dynamic and flooding river surface by close range digital photogrammetry. In Proceedings of ISPRS Int. Arch. Photogram. Rem. Sens. Spat. Inform. Sci., Beijing, China, 3–11 July 2008; pp. 211–216.
9. Mulsow, C.; Maas, H.G.; Westfeld, P.; Schulze, M. Triangulation methods for height profile measurements on instationary water surfaces. *JAG* **2008**, *2*, 21–29.
10. Adams, L.; Pos, J. Wave height measurements in model harbours using close range photogrammetry. In Proceedings of 15th Congress of the Int. Soc. for Photogram. Rem. Sens., Rio de Janeiro, Brazil, 17–29 June 1984.
11. Redondo, J.; Rodriguez, A.; Bahia, E.; Falques, A.; Gracia, V.; Sánchez-Arcilla, A.; Stive, M. Image analysis of surf zone hydrodynamics. In Proceedings of International Conference on the Role of the Large Scale Experiments in Coastal Research, Barcelona, Spain, 21–25 February 1994.
12. Aarninkhof, S.G.; Turner, I.L.; Dronkers, T.D.; Caljouw, M.; Nipius, L. A video-based technique for mapping intertidal beach bathymetry. *Coast. Eng.* **2003**, *49*, 275–289.
13. Santel, F. Automatische Bestimmung von Wasseroberflächen in der Brandungszone aus Bildsequenzen mittels digitaler Bildzuordnung. Ph.D Thesis, Fachrichtung Geodäsie und Geoinformatik University Hannover, Hannover, Germany, 2006, (in German).
14. Jähne, B.; Klinke, J.; Waas, S. Imaging of short ocean wind waves: A critical theoretical review. *J. Opt. Soc. Am.* **1994**, *11*, 2197–2209.
15. Kiefhaber, D. Development of a Reflective Stereo Slope Gauge for the measurement of ocean surface wave slope statistics. M.Sc. Thesis, Ruperto-Carola University of Heidelberg, Heidelberg, Germany, 2010.
16. Wahr, J.; Smeed, D.A.; Leuliette, E.; Swenson, S. Seasonal variability of the Red Sea, from satellite gravity, radar altimetry, and in situ observations. *J. Geophys. Res. Oceans* **2014**, *119*, 5091–5104.

17. Schiavulli, D.; Lenti, F.; Nunziata, F.; Pugliano, G.; Migliaccio, M. Landweber method in Hilbert and Banach spaces to reconstruct the NRCS field from GNSS-R measurements. *Int. J. Remote Sens.* **2014**, *35*, 3782–3796.

18. Rupnik, E.; Jansa, J. Off-the-shelf videogrammetry—A success story. In Proceedings of ISPRS Int. Arch. Photogram. Rem. Sens. Spat. Inform. Sci., Munich, Germany, 14–15 May 2014; Volume 43, pp. 99–105.

19. Nayar, S.K.; Ikeuchi, K.; Kanade, T. Surface reflection: Physical and geometrical perspectives. *IEEE Trans. Pat. Anal. Mach. Intel.* **1991**, *13*, 611–634.

20. Park, J.; Kak, A.C. 3D modeling of optically challenging objects. *IEEE Comp. Graph.* **2008**, *14*, 246–262.

21. Maresca, J.W.; Seibel, E. Terrestrial photogrammetric measurements of breaking waves and longshore currents in the nearshore zone. *Coast. Eng.* **1976**, doi:10.9753/icce.v15.%25p.

22. Piepmeier, J.A.; Waters, J. Analysis of stereo vision-based measurements of laboratory water waves. In Proceedings of Geoscience and Remote Sensing Symposium, Anchorage, Alaska, USA, 20–24 September 2004; Volume 5, pp. 3588–3591.

23. Cobelli, P.J.; Maurel, A.; Pagneux, V.; Petitjeans, P. Global measurement of water waves by Fourier transform profilometry. *Exp. Fluids* **2009**, *46*, 1037–1047.

24. Bhat, D.N.; Nayar, S.K. Stereo in the Presence of Specular Reflection. In Proceedings of the 5th International Conference on Computer Vision (ICCV), Boston, Massachusetts, USA, 20–23 June 1995; pp. 1086–1092.

25. Wells, J.M.; Danehy, P.M. Polarization and Color Filtering Applied to Enhance Photogrammetric Measurements of Reflective Surface. In Proceedings of Structures, Structural Dynamics and Materials Conference, Austin, TX, USA, 18–21 April 2005.

26. Black, J.T.; Blandino, J.R.; Jones, T.W.; Danehy, P.M.; Dorrington, A.A. *Dot-Projection Photogrammetry and Videogrammetry of Gossamer Space Structures*. *Technical Report NASA/TM-2003-212146*; NASA Langley Research Center: Hampton, VA, USA, 2003.

27. Lippmann, T.; Holman, R. Wave group modulations in cross-shore breaking patterns. *Coastal Eng.* **1992**, doi:10.9753/icce.v23.

28. Cox, C.; Munk, W. Measurement of the roughness of the sea surface from photographs of the sun's glitter. *J. Opt. Soc. Am.* **1954**, *44*, 838–850.

29. Schooley, A.H. A simple optical method for measuring the statistical distribution of water surface slopes. *J. Opt. Soc. Am.* **1954**, *44*, 37–40.

30. Stilwell, D. Directional energy spectra of the sea from photographs. *J. Geophys. Res.* **1969**, *74*, 1974–1986.

31. Ikeuchi, K. Determining Surface Orientation of Specular Surfaces by Using the Photometric Stereo Method. *IEEE Trans. Pat. Anal. Mach. Intel.* **1981**, *3*, 661–669.

32. Healey, G.; Binford, T.O. Local shape from specularity. *Comput. Vis. Graph. Image Process.* **1988**, *42*, 62–86.

33. Sanderson, A.C.; Weiss, L.E.; Nayar, S.K. Structured highlight inspection of specular surfaces. *IEEE Trans. Pat. Anal. Mach. Intel.* **1988**, *10*, 44–55.

34. Halstead, M.A.; Barsky, B.A.; Klein, S.A.; Mandell, R.B. Reconstructing curved surfaces from specular reflection patterns using spline surface fitting of normals. In Proceedings of the 23rd Annual Conference on Computer Graphics and Interactive Techniques, New Orleans, LA, USA, 4–9 August 1996; pp. 335–342.

35. Savarese, S.; Perona, P. Local analysis for the 3rd reconstruction of specular surfaces. *Comp. Vis. Pat. Recog.* **2001**, *2*, 738–745.

36. Bonfort, T.; Sturm, P. Voxel carving for specular surfaces. In Proceedings of the 9th International Conference on Computer Vision (ICCV), Nice, France, 13–16 October 2003; pp. 591–596.

37. Roth, S.; Black, M.J. Specular flow and the recovery of surface structure. *IEEE Comp. Soc. Conf. Comp. Vis. Pat. Recog.* **2006**, *2*, 1869–1876.

38. Adato, Y.; Vasilyev, Y.; Ben-Shahar, O.; Zickler, T. Toward a theory of shape from specular flow. In Proceedings of IEEE 11th International Conference on Computer Vision, Rio de Janeiro, Brazil, 14–21 October 2007; pp. 1–8.

39. Sankaranarayanan, A.C.; Veeraraghavan, A.; Tuzel, O.; Agrawal, A. Specular surface reconstruction from sparse reflection correspondences. In Proceedings of IEEE 23rd Conference on Computer Vision and Pattern Recognition, San Francisco, LA, USA, 13–18 June 2010; pp. 1245–1252.

40. Kutulakos, K.N.; Steger, E. A theory of refractive and specular 3d shape by light-path triangulation. *Int. J. Comp. Vis.* **2008**, *76*, 13–29.

41. Ma, W.C.; Hawkins, T.; Peers, P.; Chabert, C.F.; Weiss, M.; Debevec, P. Rapid acquisition of specular and diffuse normal maps from polarized spherical gradient illumination. In Proceedings of the 18th Eurographics conference on Rendering Techniques. Eurographics Association, Aire-la-Ville, Switzerland 2007; pp. 183–194.

42. Stolz, C.; Ferraton, M.; Meriaudeau, F. Shape from polarization: A method for solving zenithal angle ambiguity. *Opt. Let.* **2012**, *37*, 4218–4220.

43. Hertzmann, A.; Seitz, S.M. Example-based photometric stereo: Shape reconstruction with general, varying brdfs. *IEEE Trans. Pat. Anal. Mach. Intel.* **2005**, *27*, 1254–1264.

44. Lensch, H.P.A.; Kautz, J.; Goesele, M.; Heidrich, W.; Seidel, H.P. Image-based Reconstruction of Spatial Appearance and Geometric Detail. *ACM Trans. Graph.* **2003**, *22*, 234–257.

45. Wang, J.; Dana, K.J. Relief texture from specularities. *IEEE Trans. Pat. Anal. Mach. Intel.* **2006**, *28*, 446–457.

46. Meriaudeau, F.; Sanchez Secades, L.; Eren, G.; Ercil, A.; Truchetet, F.; Aubreton, O.; Fofi, D. 3D Scanning of Non-Opaque Objects by means of Imaging Emitted Structured Infrared Patterns. *IEEE Trans. Instrum. Meas.* **2010**, *59*, 2898–2906.

47. Eren, G.; Aubreton, O.; Meriaudeau, F.; Sanchez Secades, L.A.; Fofi, D.; Truchetet, F.; Ercil, A. Scanning From Heating: 3D Shape Estimation of Transparent Objects from Local Surface Heating. *Opt. Express* **2009**, *17*, 11457–11468.

48. Hilsenstein, V. Design and implementation of a passive stereo-infrared imaging system for the surface reconstruction of Water Waves. Ph.D Thesis, Ruperto-Carola University of Heidelberg, Heildeberg, Germany, 2004.

49. Rantoson, R.; Stolz, C.; Fofi, D.; Mériaudeau, F. 3D reconstruction of transparent objects exploiting surface fluorescence caused by UV irradiation. In Proceedings of 17th IEEE International Conference on Image Processing, Hong Kong, 26–29 September 2010; pp. 2965–2968.

50. Ihrke, I.; Kutulakos, K.N.; Lensch, H.P.; Magnor, M.; Heidrich, W. State of the art in transparent and specular object reconstruction. *EUROGRAPHICS STAR* **2008**.

51. Morris, N.J. Shape Estimation under General Reflectance and Transparency. PhD Thesis, University of Toronto, Toronto, Ontario, Canada 2011.

52. Cox, C.S. Measurement of slopes of high-frequency wind waves. *J. Mar. Res.* **1958**, *16*, 199–225.

53. Jähne, B.; Riemer, K. Two-dimensional wave number spectra of small-scale water surface waves. *J. Geophys. Res.: Oceans (1978–2012)* **1990**, *95*, 11531–11546.

54. Zhang, X.; Cox, C.S. Measuring the two-dimensional structure of a wavy water surface optically: A surface gradient detector. *Exp. Fluids* **1994**, *17*, 225–237.

55. Rocholz, R. Spatiotemporal Measurement of Shortwind-Driven Water Waves. Ph.D Thesis, Ruperto-Carola University of Heidelberg, Heidelberg, Germany, 2008.

56. Sturtevant, B. Optical depth gauge for laboratory studies of water waves. *Rev. Scient. Instr.* **1966**, *37*, doi:10.1063/1.1720019.

57. Hughes, B.; Grant, H.; Chappell, R. A fast response surface-wave slope meter and measured wind-wave moments. *Deep Sea Res.* **1977**, *24*, 1211–1223.

58. Bock, E.J.; Hara, T. Optical measurements of capillary-gravity wave spectra using a scanning laser slope gauge. *J. Atm. Ocean. Tech.* **1995**, *12*, 395–403.

59. Murase, H. Surface Shape Recontruction of a Nonrigid Transparent Object Using Refraction and Motion. *IEEE Trans. Pat. Anal. Mach. Intel.* **1992**, *14*, 1045–1052.

60. Morris, N.J.; Kutulakos, K.N. Dynamic Refraction Stereo. *IEEE Trans. Pat. Anal. Mach. Intel.* **2011**, *33*, 1518–1531.

61. Shortis, M.R.; Clarke, T.A.; Robson, S. Practical testing of the precision and accuracy of target image centering algorithms. Videometrics IV, SPIE , Philadelphia, PA, USA, October, 1995, pp. 65–76.

62. Otepka, J. Precision Target Mensuration in Vision Metrology. Ph.D Thesis, Technische Universtät Wien, Wien, Austria, 2004.

63. Wiora, G.; Babrou, P.; Männer, R. Real time high speed measurement of photogrammetric targets. In *Pattern Recognition*; Springer: Berlin Heidelberg, Germany 2004; pp. 562–569.

64. Kraus, K. *Photogrammetry*; In *Advanced Methods and Applications*; Dümmler Verlag: Bonn, Germany, 1997; Volume 2.

65. Rusu, R.B. Semantic 3D Object Maps for Everyday Manipulation in Human Living Environments. Ph.D Thesis, Computer Science department, Technische Universität Muenchen, Munich, Germany, 2009.

66. Point Cloud Library (PCL). Available online: http://pointclouds.org/ (accessed on November 2014)

211

67. Blake, A.; Brelstaff, G. Geometry From Specularities. In Proceedings of International Conference on Computer Vision (ICCV), Tampa, Florida, USA, December 1988; pp. 394–403.

68. MicMac, Apero, Pastis and Other Beverages in a Nutshell! Available online: http://logiciels.ign.fr/?Telechargement,20 (accessed on March 2015).

69. Deseilligny-Pierrot, M.; Clery, I. Apero, an open source bundle adjustment software for automatic calibration and orientation of set of images. *ISPRS Int. Arch. Photogram. Rem. Sens. Spat. Inform. Sci.* **2011**, *38*, 269–276.

70. Maas, H.G. Image sequence based automatic multi-camera system calibration techniques. *J. Photogram. Rem. Sens.* **1999**, *54*, 352–359.

Adjustment of Sonar and Laser Acquisition Data for Building the 3D Reference Model of a Canal Tunnel

Emmanuel Moisan, Pierre Charbonnier, Philippe Foucher, Pierre Grussenmeyer, Samuel Guillemin and Mathieu Koehl

Abstract: In this paper, we focus on the construction of a full 3D model of a canal tunnel by combining terrestrial laser (for its above-water part) and sonar (for its underwater part) scans collected from static acquisitions. The modeling of such a structure is challenging because the sonar device is used in a narrow environment that induces many artifacts. Moreover, the location and the orientation of the sonar device are unknown. In our approach, sonar data are first simultaneously denoised and meshed. Then, above- and under-water point clouds are co-registered to generate directly the full 3D model of the canal tunnel. Faced with the lack of overlap between both models, we introduce a robust algorithm that relies on geometrical entities and partially-immersed targets, which are visible in both the laser and sonar point clouds. A full 3D model, visually promising, of the entrance of a canal tunnel is obtained. The analysis of the method raises several improvement directions that will help with obtaining more accurate models, in a more automated way, in the limits of the involved technology.

Reprinted from *Sensors*. Cite as: Moisan, E.; Charbonnier, P.; Foucher, P.; Grussenmeyer, P.; Guillemin, S.; Koehl, M. Adjustment of Sonar and Laser Acquisition Data for Building the 3D Reference Model of a Canal Tunnel. *Sensors* **2015**, *15*, 31180–31204.

1. Introduction

The use of three-dimensional (3D) data acquisition systems (point clouds or images) for building models of partly-submerged infrastructures is currently undergoing an important development. In the literature, many systems, including industrial solutions, combine underwater and terrestrial sensors to investigate structures, such as dams, harbors or pipelines [1–4]. However, it may be noticed that only a small number of published works consider the accuracy assessment of the produced 3D models, by comparing them to some reference models [5–7].

In this paper, we focus on the construction of an accurate 3D model of the entrances of a tunnel canal, from static acquisitions of point clouds. This model shall be used as a reference for future accuracy assessments in the context of the development of an embedded acquisition system devoted to the full 3D modeling of canal tunnels (*i.e.*, including both their underwater and above-water parts). Indeed, conventional mobile mapping systems cannot be used for positioning a barge, because global navigation satellite systems (GNSS) do not work, neither in tunnels nor at their entrances, which are most of the time bordered by narrow embankments (see Figure 1 and Figure 2-left), so innovative solutions must be proposed. Potential application concerns, in France, 31 tunnels currently in use, representing 42 km of underground waterways: the maintenance of these structures is a necessity, not only for preserving the historical heritage they represent, but also for protecting goods and persons. In the context of a partnership between Voies Navigables de France (VNF, the French operator of waterways), the Centre d'Études des tunnels (CETU) and the Cerema, in collaboration with the Photogrammetry and Geomatics Group at INSA-Strasbourg (institut national des sciences appliquées), an image acquisition prototype, embedded on a barge, has been devised for imaging the tunnel vaults and side walls (see Figure 1). During this project, solutions to geo-reference data precisely in the tunnel have been proposed and evaluated [8]. This system is going to be equipped with a multibeam echosounder to provide 3D views of the underwater parts of tunnel canals.

Figure 1. Modular on-board mobile image recording system at the experimental site.

Figure 2. Constraints that apply to recording systems in tunnel canals. (**Left**) Global navigation data are not available in tunnels, nor at their entrances, because satellites are masked, hindering conventional mobile mapping; (**Center**) The turbidity of water prevents using optical imaging devices; (**Right**) The canal is shallow and narrow, so robust sonar processing algorithms are needed.

A 3D reference model will be necessary to assess the accuracy of the model of the whole tunnel provided by the mobile recording system under development. We have chosen to build this model from separate, static acquisitions of the under- and above-water parts of the tunnel entrances. For the above-water parts, point clouds have been collected using a 3D terrestrial laser scanning (TLS) system. A previous evaluation of the resulting 3D terrestrial model has shown that its accuracy is 1.7 cm [8]. Since the water turbidity (see Figure 2-center) excludes the use of optical sensors, the acquisitions of the underwater parts of the canal have been performed using a 3D mechanical scanning sonar (MSS) from static positions, in a similar way to a TLS [9,10]. This emerging technology may provide more accurate models than mobile systems. However, unlike the processing of TLS data, the registration and geo-referencing of MSS point clouds are complex, and several challenges need to be solved. MSS data are intrinsically noisy, and the narrowness and shallowness of the canal (Figure 2-right) induce artifacts due to water surface and sidewall reflections. Therefore, robust methods must be sought to alleviate these difficulties and reconstruct an accurate 3D model of the underwater part.

Aligning the 3D underwater model with the TLS one to form the global reference model is the second challenging problem, because, at every scanning position, the location and the orientation of the sonar device are not directly available. In this contribution, we propose to generate directly the full 3D model of the canal tunnel by co-registering the above- and under-water point clouds. Such an approach ideally requires an overlap between the TLS and MSS point clouds. In our case, there is no overlap, but we can exploit some geometric primitives (planes, lines, silhouette of the waterline), which are common to the above- and under-water parts of the structure. Moreover, we partially immersed wooden ladders on each canal bank. The robust fitting of the rungs and stiles of the ladders in both point clouds, using

Maximum likelihood-type estimators (M-estimators), provides additional constraints for modeling the canal tunnel.

The paper is organized as follows. We first review related works in Section 2. Then, in Section 3, we introduce TLS and MSS data recording systems. Sonar data processing is described in Section 4. Section 5 is dedicated to the construction of the full 3D model by co-registering the above- and under-water models. In Section 6, we comment on the experimental results. Section 7 concludes the paper and proposes future directions for this work.

2. Related Work

For the purpose of assessing the accuracy of 3D acquisitions of civil engineering structures, the 3D reference model has to be more accurate than the model under inspection. Under certain circumstances, construction plans are available, and an as-built model may be used. For example, in [11], the evaluation of 3D reconstructions by an underwater SLAM (simultaneous localization and mapping) method was performed with a computer-aided design (CAD) model of a ship hull. Most tunnel canals, however, were bored during the 19th and 20th century (e.g., our test structure, Niderviller's tunnel, located near Strasbourg (France), was bored between 1839 and 1845), and their construction plans, even limited to the headwalls, are not available or not accurate enough, so other solutions must be sought.

The alternative for building ground-truth models is to survey the object with a very accurate measurement device. For the above-water parts, geo-referenced points or point clouds may be used. In [6], the accuracy of data collected by a boat-based mobile laser scanning system is evaluated by using a set of reference spheres, positioned by GNSS. In [12], a set of geo-localized TLS point clouds provided reference data with centimetric precision for a mobile mapping system verification. A similar technique was reported in [8].

In the case of partly-immersed structures, there are two possibilities. The first one is to immerse an artificial reference object that can be surveyed beforehand, using a static TLS, for example. This technique was recently used in [7] to evaluate CIDCO (Centre Interdisciplinaire de Développement en Cartographie des Océans) sonar prototypes. In this work, a test bench made of concrete panels with protrusions and extrusions was scanned by a TLS at a millimeter resolution, with a 2-cm resampling before being immersed and surveyed. The second one is to exploit existing structures that can be emptied, such as dry docks, or filled, such as dam reservoirs. For example, in [5], the Blueview company performed acquisitions in dry docks to evaluate a multi-beam sonar equipment. A TLS survey of the empty dry dock was first performed, and sonar acquisitions were made after filling the dock. In [7], a 3D LiDAR model of a dam was acquired before the reservoir was filled, and then, a survey was made using a multi-beam echosounder (MBES), showing only slight, local differences (less than 5 cm) due to the shift of materials during the filling operation. In a similar manner, it is possible to benefit from natural effects, such as tide. For example, in [1], a surface was surveyed at high tide using a sonar system, and a total station was used to assess the acquisition process at low tide.

In our case, it would have been interesting to complement the existing TLS survey of the above-water parts of our test structure, Niderviller's tunnel, after emptying the canal. Unfortunately, such an operation is costly and may even be hazardous due to the age of the tunnel: the canal walls, in poor condition, might crumble. Therefore, canal managers are often reluctant, and drying operations occur very rarely. For example, Niderviller's tunnel and a nearby one, Arzviller's tunnel, were emptied in 2009 (before the beginning of our study); see Figure 3. This was the first time since 1968. Since it is not possible to empty the canal, we have to resort to 3D underwater imaging techniques. Mobile or static systems might be envisioned.

Figure 3. Entrance of Arzviller's tunnel during its emptying in 2009.

In recent years, mobile surveying systems were developed to inspect partially- or totally-immersed open structures, like harbors or dams. Most of these operational systems provide 3D models thanks to MBES for the underwater parts and terrestrial laser scanner (TLS) for out-of-the-water parts. The acquisition is performed in a dynamic way in order to sweep the surveyed structure. The localization of the system is hence of central importance. Most of the time, the associated positioning system combines GNSS and inertial navigation systems (INS); see, e.g., [1,3]. However, these methods are unsuitable for canal tunnels due to the lack of a GNSS signal. Alternatives to GNSS/INS systems were proposed in the robotics and computer vision literature. For example, [13] introduces a SLAM approach to obtain at the same time the 3D model and the mobile localization, thanks to the registration of TLS point clouds acquired with high frequency. In [8], we proposed a simplified visual odometry technique to estimate the position of our mobile mapping prototype along the tunnel. However, all of these methods suffer from drifts that are especially sensible in elongated structures, such as tunnels, and that can only be corrected using reference points, which are more difficult to set up in the absence of a GNSS signal. Other potentially interesting systems were developed in the context of large-diameter pipe inspection. Recently, two devices were proposed: the ABIS (above and below inspection system) system of the ASI Marine Technology society (see [4]) and the HD Profiler System of the Hydromax USA society (see [2]). Both systems acquire laser and sonar data and HD images. The documentation of these commercial solutions is limited. Furthermore, a towed device cannot be used in certain tunnels, which are curved.

In order to avoid the difficulties related to the localization of mobile mapping systems and to obtain an accurate reference model, we choose to perform static acquisitions to build the underwater part of our reference model. Such an approach is classical for TLS data. Thanks to the recent availability of a 3D mechanical scanning sonar (MSS), it is now possible to get a 3D model from sonar acquisition in a static way, as well. The MSS device was used in several underwater surveys, such as the ones introduced in [9,10]. In a TLS-like manner, several scanning positions were carried out to get a full model. The co-registration of point clouds was done, in a local coordinate system, with the iterative closest point (ICP) algorithm.

The positioning of underwater cloud points in the same coordinate system as the TLS data can be performed in direct or indirect ways (see [14]). Both methods require the knowledge of reference points. In the literature, one may find several solutions to define these points underwater. The first method is acoustic positioning based on triangulation. The operating principle is close to GNSS, except, in the water, acoustic waves are used. For example, in [15], buoys equipped with ultra-short baseline (USBL) transceivers, tied up with GNSS receivers, yield the localization of a remotely-operated vehicle (ROV). When the survey is carried out in shallow water, underwater points can be surveyed with terrestrial methods thanks to a long pole equipped with a prism for a total station or a GPS antenna (see [16]). Lastly, in [17], poles with underwater and above-water targets are partially immersed. Thus, under- and above-water acquisitions can be co-registered, because poles create a link between both models. In the present work, we have chosen a similar solution, in which wooden ladders, as well as existing geometric primitives of the structure are used to link the underwater and above-water models.

3. Data Recording

In this section, we introduce the setup that was used to build the 3D reference model of the canal tunnel on the site of Niderviller using two up-to-date scanner devices.

3.1. Scanner Devices

The above-water acquisitions were performed using a Focus 3D X330 TLS, the latest Faro® scanner; see Figure 4 (left), and Figure 5 (top-left). To survey the environment in 3D over a range of 0.6–300 m, a laser beam sweeps the visible surfaces, vertically and horizontally over almost 360°. The distance between the TLS and the objects is measured using the phase difference technique [14].

Figure 4. Schematic representation of acquisition sensors: The Faro® (Lake Mary, FL, USA) Focus 3D X330 TLS (**Left**); The Blueview® (Bothell, WA, USA) BV5000 mechanical scanning sonar (MSS) (**Right**).

The underwater acquisitions were carried out with the Blueview® (Bothell, WA, USA: www.blueview.com) BV5000 MSS (see Figures 4 (right) and 5 (bottom)) operated by a sub-contractor, the Sub-C Marine company. The device is made of a multi-beam echo-sounder with a vertical swath direction and a rotation system with a vertical axis that enables a 360° horizontal scan (see Figure 5 (bottom-right)). Since the swath aperture is 45°, a mechanical system is used to tilt the sensing head, so potentially, a 320° vertical range can be scanned. We used three different tilt angles (0°, −9°, −30°) for this experiment. The acoustic sensor emits a high frequency signal (1.35 Mhz), which offers a good resolution in distance, but at the same time, limits the maximum acquisition range to 30 m. According to the manufacturer's data sheet, the vertical and horizontal spatial resolution (*i.e.*, the distance between a recorded point and its closest neighbor) at 10 m are 16 mm and 30 mm, respectively.

Figure 5. Data acquisitions in Niderviller's tunnel: Photograph of the TLS device (**Top-left**); Zoom on a spherical targets and a ladder fixation on the dock side (**Top-right**); Schematic representation of MSS device positioning (**Bottom-left**) and Rotary scanning (**Bottom-right**).

The comparison between MSS and TLS features (see Table 1) highlights differences that will impact the relative quality of the recorded point clouds. For example, we can note a large difference in spatial resolution that will affect the point densities. However, the main difference concerns the beam width, which has a direct impact on the ability to distinguish two echoes coming from two different targets. For example, the signal footprint on a plane perpendicular to the signal direction 10 m away from the device is a circle of 6 mm in diameter for the Focus 3D X330 and a square of a 175 mm side length for the BV5000.

Note that all differences between point clouds are not imputable to the characteristics of the devices only. For example, the footprint size depends not only on the recorded object to the scanner, but also on the incidence angle of the signal, as we will see in Section 3.3.

Table 1. Manufacturer's specification of the acquisition devices (top lines) and recording resolutions used for the experiment (two bottom lines).

	Faro Focus 3D X330	Blue View BV5000
beam width	2.25 mm + 2 × 0.011°	1°/1°
ranging error	±2 mm (10–25 m)	15 mm
maximum range	330 m	30 m
field-of-view (vertical/horizontal)	300°/360°	45°/360° (320°/360°)
horizontal resolution	0.035° (6 mm at range 10 m)	∼0.09° (16 mm at range 10 m)
vertical resolution	0.035° (6 mm at range 10 m)	0.18° (30 mm at range 10 m)

3.2. Experimental Setup

The experimental acquisitions have been carried out in Niderviller's canal tunnel (see Figure 1). It is straight and lined with stonework. The radius of the vault is 4 m; the width of the canal is 6.6 m; and its water depth is about 2.5 m. It has a pedestrian path on a ledge that was formerly used as a towing path. About 7000 ships, among which a vast majority are pleasure boats, cross the tunnel annually (according to 2012 statistics). Albeit that the acquisitions took place during a low-traffic period, the recording time was constrained, and it was not possible to scan the full length of the tunnel (475 m). Therefore, the acquisitions were focused on its entrances, for this first experimental campaign.

At each entrance, two laser scans have been performed, from each bank of the canal, simultaneously with the underwater acquisition. The resulting model can be complemented using a previously-performed TLS survey of the whole tunnel. To register point clouds, spherical targets have been placed in the shared scanning area, as shown in Figure 5, top. In order to geo-reference the model, the coordinates of sphere centers have been established with traditional surveying methods based on a set of reference points implemented on site, in the French reference coordinate systems RGF 93 (réseau géodésique français) and NGF-IGN 69 (nivellement général de la France operated by l'Institut National d'Information Géographique et Forestière).

Two sonar scans, from positions placed 10 m away from each other, have been performed on each entrance of the tunnel, one inside and one outside of it; see Figure 6. The MSS is attached to a tripod, which is sunk in the canal from a boat, as shown on the bottom left part of Figure 5. The soil consists of a mixture of rocks and mud. It proved to be sufficiently hard to support the weight of the tripod, which kept stable during all acquisitions (accidental motions of the instrument may be detected at the closure of a scanning rotation). Setting up the MSS took about 15 min by scan. The spatial resolution was adjusted, so the acquisitions themselves lasted 10 min by rotation. Three acquisitions, with different tilt angles, were used. The quality of the obtained signals was checked on site. Overall, each acquisition required about one hour.

Two wooden ladders (3.60 m high and 0.32 m wide) were partly immersed in the water, so they were recorded by both the MSS and the TLS. Ladders are ordinary objects, easily available and easy to set up on site, that are well suited to the application. Their length and geometric characteristics (parallelism, orthogonality, known inter-rung distances) help the registration of the underwater and above-water models (see Section 5). Wooden ladders were preferred to metallic ones because they yield cleaner sonar echoes. The ladders are weighted and attached to the dock side using a weighted frame, as shown in Figure 5 (top-right).

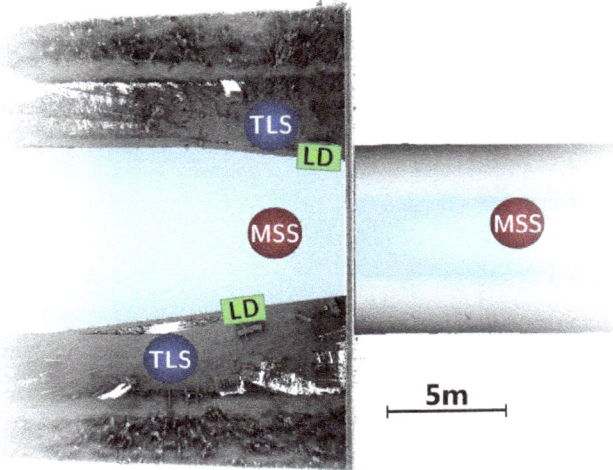

Figure 6. Illustration of the acquisition setup seen from above, featuring TLS and MSS scanning positions, as well as the location of ladders (LD).

3.3. Remarks

A few remarks can be made in relation to the acquisition context. Of course, unlike TLS acquisitions, the underwater recording cannot be supported by visual control. To make a decision about the scanning positions, for example, we could only rely on above-water elements, which raises issues for both recording and interpreting sonar data.

Another remark is that the elongated shape of canal tunnel yields unfavorable incidence angles for sonar acquisition. This influence is visible in the acoustic images, shown in Figure 7. One can see that the vertical line, which corresponds to the footprint of the swath on the sidewalls of the canal, is much wider for a grazing incidence than for an almost perpendicular acquisition. According to a theoretical model of the acquisition setup and to the datasheet of the BV5000, we may estimate the horizontal width of the beam footprint (see Figure 8) by intersecting the emission beam model with a plane representing the sidewall. We see that, in the case of canal tunnels, this length can easily reach more than 0.5 m.

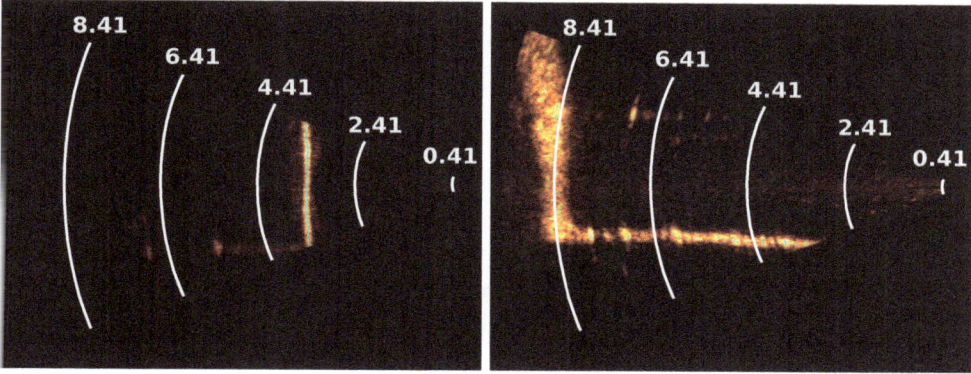

Figure 7. Multi-beam echo-sounder swaths under two different incidence angles. (**Left**) Almost perpendicular incidence (see the red line in Figure 8); (**Right**) Grazing angle (see the green line in Figure 8). The swath is wider in the second case. Distances are given in meters.

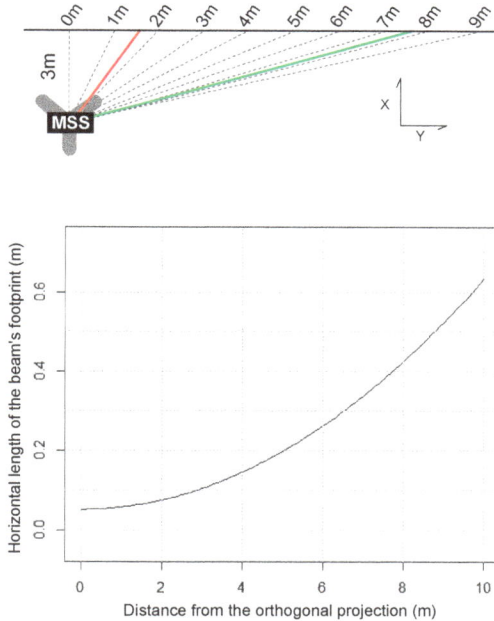

Figure 8. Top view of the theoretical acquisition setup (**Top**); Estimation of the sonar horizontal footprint size as a function of the distance from the orthogonal projection (**Bottom**).

4. Sonar Data Processing

The first observation of raw sonar data highlights measurement errors and also the noisy nature of MSS point clouds. These elements must be handled. Furthermore, to obtain a full 3D model, underwater and above-water point clouds have to be adjusted by registration.

4.1. Sonar Measurement Errors

The raw sonar output is an acoustic image constructed from the received echoes in each beam. More specifically, from each beam, the echo with the highest intensity is used to estimate the 3D point. However, the acoustic image brings out some measurement errors. In most cases, these errors have a lower intensity than the recorded object, so they do not appear in the output point cloud, but some artifacts may remain.

Some measurement errors in sonar data may come from the scanner device itself. A list of error sources is reported in [5], along with their consequences on data recording and possible adjustments to alleviate them. Typical problems that may occur are: platform motion, tilt offset errors, insufficient coverage or incorrect sound speed. Some corrections can be applied in post-processing, but in some cases, new scans need to be taken. These rough errors are checked immediately after scanning. However, slight errors may be insensible in the raw point cloud.

Other errors are due to the configuration in which the MSS scans are operated: shallow water and confined environment. The most visible errors are due to signal reflections on the water surface. In some cases, "phantom" objects may be observed above the water surface (see Figure 9). However, such reflections can be detected by visual inspection of the profiles, and the artifacts are then easily deleted from the point cloud. It is more difficult to detect surface reflections when the sidewalls of the canal are planar. They make the vertical position of the waterline more difficult to estimate. This is another justification of using ladders to help with geo-referencing the underwater point cloud. Once the MSS vertical position is known, surface artifacts can be deleted. Reflections may also occur on the sidewalls or on the raft. In the latter case, phantom objects are observed underneath the soil level, so they can be discarded.

Furthermore, some errors unavoidably arise due to the presence of objects in the canal, such as the boat hull, cables, fishes or suspended particles. All of these errors are deleted manually, but most of them could be removed automatically, because the canal shape is roughly known.

224

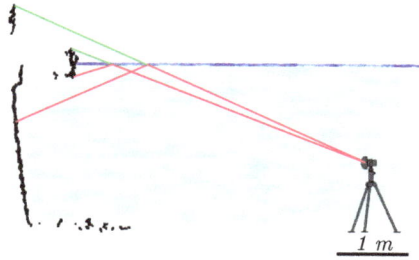

Figure 9. Signal reflection on the water surface may cause artifacts.

4.2. Denoising and Meshing

The first processing step consists of removing the significant noise exhibited by sonar acquisitions. The proposed solution exploits the meshing phase of the reconstruction process. Indeed, most of the time in surveying applications, surface reconstruction is performed to obtain a simplified digital model of the recorded structure, and this operation is generally the last step in the modeling chain. Here, we use it as a pre-processing step, since it provides a visual control on the result, which highlights errors and guides denoising.

We note that TLS point clouds have a negligible noise level, so their processing only involves outlier removing, and all points may be employed as triangle vertices to reconstruct the surface. This is not the case for MSS clouds, for which artifacts may occur when all points are used for meshing; see Figure 10 (left). Removing those artifacts reduces to denoising and can be done using two methods:

- One is to select evenly-spaced points as mesh vertices for triangulation (Figure 10 (center)), based on a minimum distance criterion. The price to pay is that details may be lost.
- Another way is to compute the nearest surface to points using robust estimators (Figure 10 (right)). For this purpose, new points are interpolated. However, the risk is to obtain an over-smoothed model.

Meshing of point clouds may be carried out using specialized software; see, e.g., [18] for a review. We use 3DReshaper® (Genay, France: www.3dreshaper. com), which has the ability to mesh with both previously-mentioned techniques and also to combine them to perform successive refinements of the model using the point cloud. Thus, the underwater model reconstruction is performed in a coarse-to-fine manner. The process starts with a large-scale mesh made by selecting points according to a minimum distance criterion. Then, points are picked again in the cloud or computed by interpolation to progressively increase the mesh resolution. Point selection involves either a distance-to-mesh or a maximum surface deviation criterion. The parameters are empirically tuned by an operator, and the process requires a trade-off between details and noise.

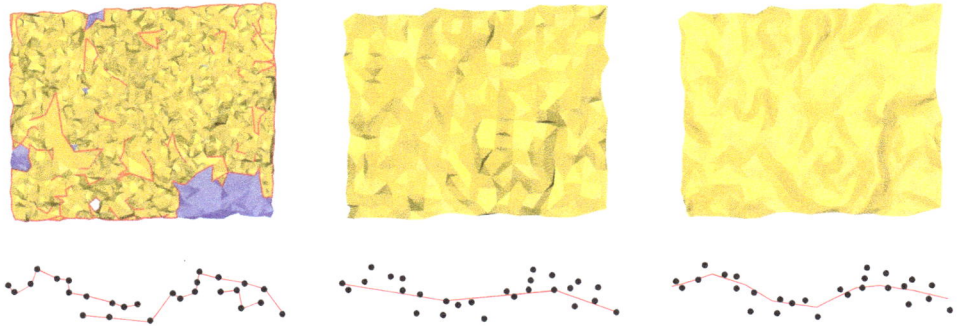

Figure 10. Methods for meshing point clouds: Using all points (**Left**); Using selected points (**Center**); Meshing using interpolated points (**Right**).

We note that this step of the process requires many manual operations, like outlier removal or correction of mesh reconstruction mistakes. While such interventions may be supported by photographs or other physical measurements for TLS data, this is not the case for underwater data, except in particular situations. Hence, the construction of the underwater model from MSS data involves an important part of interpretation.

However, the example in Figure 11 shows that visually-correct results may be achieved using this method. We note that the filtered MSS surface shown on the rightmost part of the figure was obtained without any knowledge of the underwater structure: the photograph was found in VNF archives after the experiment.

4.3. Underwater Registration

The co-registration of MSS data aims at gathering all records in a single point cloud corresponding to the underwater part of the tunnel. For this purpose, the position between scans must be estimated. Since, in our setup, the position and orientation of the underwater scanner cannot be directly measured, we have to resort to the indirect method. Cloud-to-cloud registration seems to be the easiest technique to implement, but it also raises several issues. Some are due to the nature of the technology itself: MSS data are very noisy, and the resolution is rather coarse, so finding correspondences is difficult. Other difficulties are related to the elongated shape of the canal and to our experimental setup: the farther the recorded point, the smaller the grazing angle. Therefore, points in overlapping zones have a poor accuracy, which also influences the quality of registration. In these conditions, it is very difficult to solve the longitudinal ambiguity, *i.e.*, to accurately estimate the translation along the tunnel axis. Immersing geometric reference objects (e.g., ladders) or decreasing the distance between scanning positions to increase the overlap quality are possible solutions to this problem.

Figure 11. (Left) Photograph on the northern tunnel entry, during its emptying for maintenance in 2009 (source: Voies Navigables de France (VNF) archives); **(Right)** Visualization of the MSS denoised model (in grey) superimposed on the image. The orientation is done manually, thanks to the masonry block and the cofferdam grooves, which are visible in the model.

5. Model Alignment and Geo-Referencing

In this section, we introduce the method we propose for registering sonar data on laser data to form the full 3D reference model. We recall that this model will be used for assessing the accuracy of the models provided by a mobile mapping system for canal tunnels, currently under development. Comparing 3D models can be done in any arbitrary coordinate system. However, our test site is equipped with a geodetic model, so the model can be geo-referenced without additional complexity.

5.1. Registration and Geo-Referencing Method

In general, there are two ways of registering and geo-referencing point clouds [14]. The first one is direct: it requires the knowledge of the position of the scanner. The latter can be obtained *a priori*, by placing the device at a point of known coordinates, or *a posteriori*, by surveying its position using conventional techniques, but both solutions are difficult to put into practice for underwater acquisitions. The second one is indirect, *i.e.*, point clouds themselves are used to be registered and geo-referenced. Registration and geo-referencing of point clouds can be based either on targets or on clouds.

Target-based registration requires anticipating and placing targets in the field of view. Their geometry and scale depend on the spatial resolution and precision of the scanner. Spheres are usually used for TLS recording because the determination of their centers can be made very accurately. Of course, the quality of the registration also depends on the distribution and number of targets. Furthermore, when sphere centers are known in coordinates in a defined system, point cloud geo-referencing can be deduced straightforwardly. In our application, we use this method, which is very usual in laser scanning, for geo-referencing the TLS point clouds and form the above-water part of the model. Spherical open frames, proposed by the Blueview company (patent pending), may also be used as targets for MSS data; see [5]. No such targets were available for our acquisition campaign, but we experimented with ladders instead.

In general, cloud-based registration algorithms involve two steps. First, homologous points between scans are found. Then, these correspondences are used to estimate the geometric transformation (rotations, translations) between sets of points. The most popular algorithm is the ICP method, introduced by [19], which iterates these steps to minimize the discrepancy between the first set of points and the geometric transformation of the second set of points. This method requires a certain overlap between point clouds and, also, a first estimation of the transformation. An alternative technique is based on the detection of homologous geometrical entities between scans and finding the best way to align them. These entities can be planes, spheres, cylinders or lines. As for target-based methods, geo-referencing of the point cloud is a by-product of the registration step if the coordinates of some of the features are known.

Four observations can be made that form the basis of the proposed registration and geo-referencing method:

- some elements of the recorded scene can be approximated by geometrical entities, and certain ones of these are surveyed at the same time by the underwater and terrestrial scanners;
- targets (ladders) create a link between both environments;
- the projection of the waterline on the structure is the only contact element between both models;
- the silhouette of the waterline features many salient elements that can be used to align both models in a horizontal plane.

Following the above remarks, we implement a three-step registration method, depicted in Figure 12. First, the orientation angles are corrected thanks to the Procrustes method [20] using common geometric primitives. Then, the fitting of ladders on both under- and above-water point clouds enables estimating the vertical translation vector. Last, the 2D silhouette of the waterline is extracted from both models to estimate the horizontal translation.

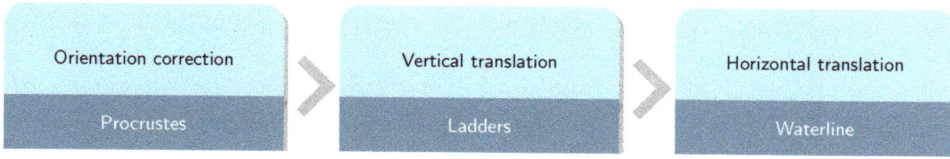

Figure 12. Flow chart of the registration of a sonar point cloud on a laser point cloud.

5.2. Orientation Correction

The first step aims at correcting the orientation angles of the underwater model. To this end, common geometrical entities are manually fitted on certain parts of both models. For example, canal banks are surveyed under and above the water and approximated by planes. The normals to these common planes should be collinear. Using several such normals, the underwater model orientation can be computed, in the form of a 3×3 rotation matrix, $\mathbf{Q}$. Other primitives, such as the directions of salient lines (e.g., cofferdam groove corners; see Figure 13) can also be taken into account: a minimum of two non-collinear vectors are required to estimate the rotation matrix.

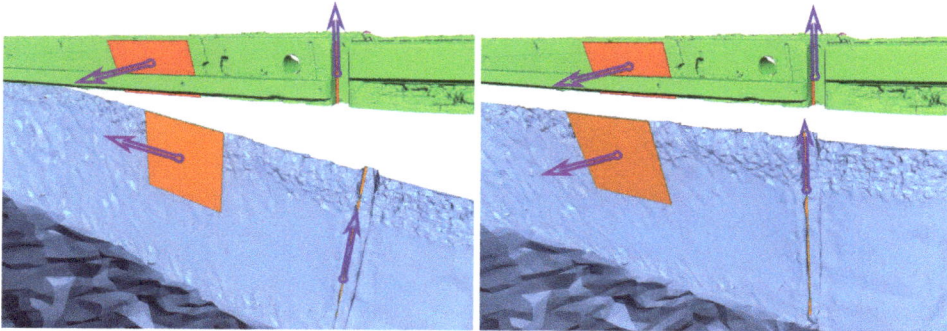

Figure 13. Original orientation of the TLS (in green) and the MSS (in blue) models (**Left**); Aligned models (**Right**).

A property of the rotation matrix that must be taken into account in the computation is its orthogonality. To perform the alignment, we use the solution described in [20] to the so-called "orthogonal Procrustes problem":

$$\min \| \mathbf{A} - \mathbf{BQ} \|_F^2 \quad \text{subject to} \quad \mathbf{Q}^T\mathbf{Q} = \mathbf{I} \tag{1}$$

where $\| \cdot \|_F$ denotes the Frobenius norm and where $\mathbf{A}$ and $\mathbf{B}$ are, respectively, the direction vectors of the elements extracted from the TLS and the MSS models. The

229

algorithm computes the singular value decomposition (SVD) of the $\mathbf{B}^T\mathbf{A}$ product, $\mathbf{U}^T(\mathbf{B}^T\mathbf{A})\mathbf{V} = \Sigma$, to get the rotation matrix $\mathbf{Q}$, as:

$$\mathbf{Q} = \mathbf{U}\mathbf{V}^T \tag{2}$$

Once $\mathbf{Q}$ is estimated, the orientation of the underwater model can be corrected; see Figure 13 (right).

5.3. Vertical Translation

After the orientation of the underwater model has been corrected, the next step is to correct the vertical translation between models. For this purpose, we use the information provided by ladders that were immersed before the acquisitions, in such a way that they are visible in both point clouds; see Figure 14.

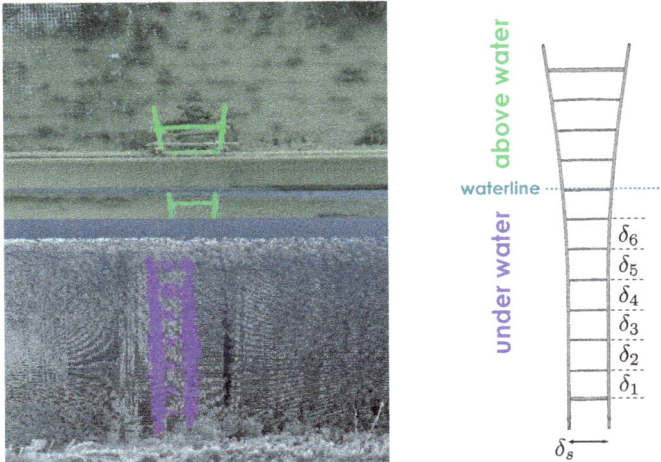

Figure 14. (**Left**) Segmentation of a ladder in the TLS model (**Top**, in green) and in the MSS model (**Bottom**, in blue). The upper part of the ladder is partly masked because it was placed behind a boom, (**Right**) Reference TLS survey of the ladder.

5.3.1. Basic Ideas

The principle of the method is to use the distance between the rungs of the ladders to compute the vertical difference between models. These inter-distances, denoted by δ_i, are given by a reference TLS survey. Figure 14 (right), shows the ladder that was used in this experiment. It is straight in its lower part and flared at the top. It was placed behind a protection rail (boom) during the survey, so its upper part is partly masked in the TLS point cloud (Figure 14 (left)). Note that one rung, that was just below the water surface, was not used. Indeed, only a few points were

230

distinguishable in the point cloud due to surface reflection artifacts. This way, only the straight part of the ladder is used.

In order to estimate the vertical gap between models, it is first necessary to adjust a set of lines on both ladders in the point clouds. The proposed method is split into three steps, as described on Figure 15 and illustrated in Figure 16a–c.

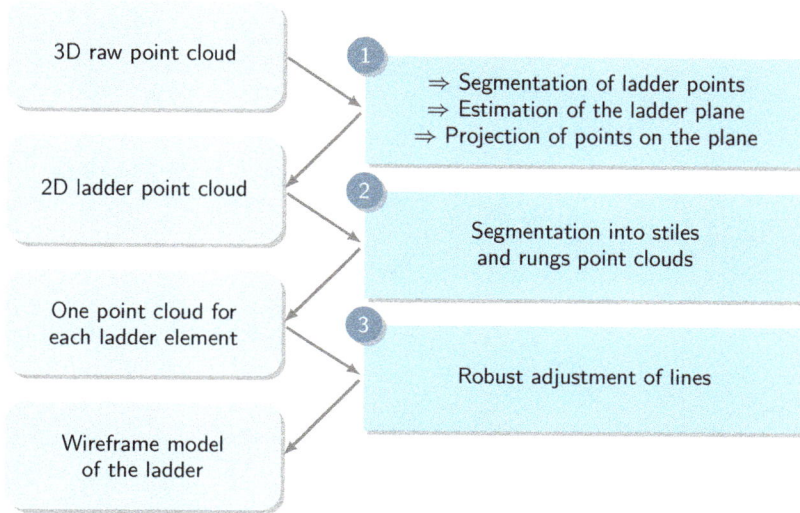

Figure 15. Flow chart of the ladder adjustment method.

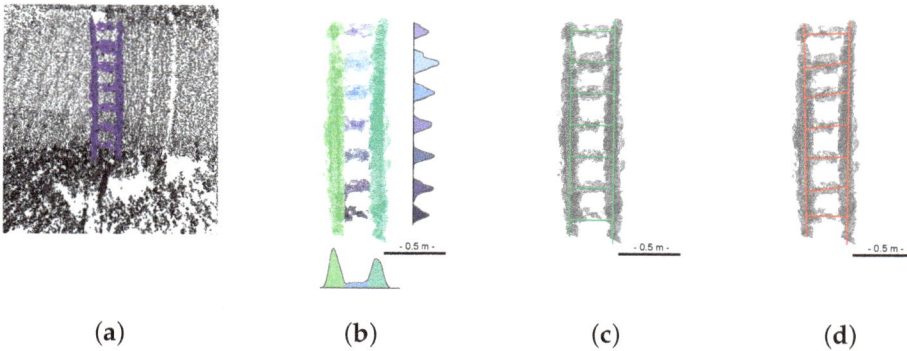

(a)   (b)   (c)   (d)

Figure 16. Illustration of the ladder fitting process: Extraction of ladder points from raw data (**a**); Segmentation into separate point clouds (**b**); Robust fitting of ladder using M-estimation and structural priors (**c**); Non-robust fitting using commercial software (**d**).

231

While the TLS point cloud is not very noisy, there are many outliers in the MSS point cloud (Figure 14 (left)). This makes the extraction of the lower part of the ladder a rather challenging task, which requires the use of robust regression techniques. In this paper, we use M-estimators [21], which, in place of the usual sum of squared residuals, minimize a function of the form:

$$J(\theta) = \sum_i \rho(r_i) \tag{3}$$

where θ is the vector of model parameters, ρ is a non-quadratic potential or penalty function and r_i is the residual, *i.e.*, the difference between the observation and its prediction by the model. In the half-quadratic framework (see, e.g., [22]), it is shown that minimizing J is equivalent to minimizing:

$$J^\star(\theta, b) = \sum_i b_i r_i^2 + \Psi(b_i) \tag{4}$$

where Ψ is a convex penalty whose expression can be related to ρ and b_i is an auxiliary variable, whose role is both to down weight outliers and to linearize the estimation problem. Indeed, $J^\star$ is quadratic with respect to r (hence, for θ in linear regression) when b is fixed. It is convex with respect to b when r is fixed, and the minimum is obtained for $b = \rho'(r)/2r$. Such properties suggest a deterministic algorithmic strategy that consists of alternately fixing each variable and minimizing with respect to the other. The resulting algorithms are iterative and perform a series of weighted least-squares (LS) estimations; see Equation (4). The weights b are adjusted at each iteration according to the value of the residuals. Moreover, $\rho'(r)/2r$ is a decreasing function, such that $b \simeq 1$ for small residuals and $b \rightarrow 0$ for large residuals. Hence, inliers are considered as in ordinary LS, while outliers are given a small weight, which reduces their influence on the estimation.

5.3.2. Robust Projection and 2D Segmentation

The first step of the method is to manually extract the ladder points x_i (for $i = 1, ..., N$) from the MSS point cloud. Then, we suppose that this set of 3D points may be approximated by a plane passing through an origin μ and spanned by two orthogonal unitary vectors e_1 and e_2 ($\|e_1{}^\top e_2\|^2 = 0, \|e_1\|^2 = \|e_2\|^2 = 1$):

$$x_i \simeq \mu + E a_i \tag{5}$$

where $E = (e_1, e_2)$ and a_i are the coordinates of the orthogonal projection of x_i on the plane, which are given by $a_i = E^T(x_i - \mu)$. In the LS framework, the i-th residual is given by $r_i = \|\mu + \mathbf{Ea_i} - \mathbf{x_i}\|$, and the parameters of the model, namely the origin and the basis vectors, are estimated by minimizing:

$$J_{LS}(\mu, \mathbf{e_1}, \mathbf{e_2}) = \sum_{i=1}^{N} \|\mu + \mathbf{Ea_i} - \mathbf{x_i}\|^2 \qquad (6)$$

The solution of this orthogonal regression problem (which is akin to principal component analysis, or PCA; see, e.g., [23]) is, for the origin:

$$\mu = \frac{1}{N} \sum_{i=1}^{N} \mathbf{x_i} \qquad (7)$$

i.e., the sample mean of the point cloud, and for the basis vectors, the solution of:

$$\mathbf{CE} = \lambda \mathbf{E} \quad \text{with} \quad \mathbf{C} = \frac{1}{N} \sum_{i=1}^{N} (\mathbf{x_i} - \mu)(\mathbf{x_i} - \mu)^{\top} \qquad (8)$$

In other words, the basis vectors are given by the two eigenvectors of the sample covariance matrix $\mathbf{C}$ that correspond to its two largest eigenvalues, λ_1 and λ_2. In the robust half-quadratic framework, the augmented energy is given by:

$$J^{\star}(\mu, \mathbf{e_1}, \mathbf{e_2}, b) = \sum_{i} b_i \|\mu + \mathbf{Ea_i} - \mathbf{x_i}\|^2 + \Psi(b_i) \qquad (9)$$

The optimization of $J^{\star}$ is performed by an iterative reweighted PCA algorithm. Each iteration alternates between: computation of the auxiliary variables $b_i = \rho'(r_i)/2r_i$ (with $r_i = \|\mu + \mathbf{Ea_i} - \mathbf{x_i}\|$, for $i = 1, ..., N$), computation of the weighted mean:

$$\mu = \sum_{i=1}^{N} b_i \mathbf{x_i} / \sum_{i=1}^{N} b_i \qquad (10)$$

and diagonalization of the weighted covariance matrix:

$$\mathbf{C} = \sum_{i=1}^{N} b_i (\mathbf{x_i} - \mu)(\mathbf{x_i} - \mu)^{\top} / \sum_{i=1}^{N} b_i \qquad (11)$$

The complete algorithm is given in Appendix A. Once the origin and basis vectors are estimated, all points of the ladder cloud are projected onto the plane, and the rest of the process is performed in two dimensions. To avoid introducing new notations, data points will be denoted by $\mathbf{x_i}$ thereafter, but from now on, they designate 2D projections in the ladder plane.

The robust plane estimation provides axes that follow the stile and rung directions quite well. Then, the distribution of coordinates along both axes shows peaks that correspond to ladder elements. These distributions, shown in Figure 16b, are approximated using Parzen-window density estimation with Gaussian kernels (see, e.g., [23]), at two different resolutions. The rough location of peaks is determined using a kernel resolution (bandwidth) of 1 cm. Then, the intervals between peaks are analyzed using a bandwidth of 1 mm: the list of local maxima of the distribution is iteratively filtered by non-maximum suppressions, until zero or one local minimum remains in the list. This analysis follows the spirit of the fine-to-coarse histogram analysis technique proposed by Delon *et al.* [24]. If no secondary peak exists within the interval, then the limits of the distribution are set by traversing the list of local minimum from the extremities of the interval until the distribution goes below an arbitrarily small value. Otherwise, the threshold is set at the first local minimum that goes below the height of the secondary peak. This method is successively applied to the horizontal and vertical coordinates, so the stiles are first separated from the ladders, then the rungs are extracted individually. Many other thresholding techniques might be used, but this one is rather simple and gives satisfactory results. Moreover, the segmentation stage is not a very sensitive step, since the estimation of the ladder model is performed in a robust way.

5.3.3. Modeling Ladders

The last step entails approximating the ladder by a set of lines, which takes into account the geometrical features of ladder. The structure of the ladder can be defined by the following properties:

- the rungs are parallel;
- the stiles are parallel;
- the rungs and stiles are orthogonal.

The orthogonality can be exploited as shown in Figure 17, by applying a 90° rotation to the stile point clouds. Then, only one direction has to be estimated. When the second condition is not satisfied (e.g., for the top part of the flared ladder), the stiles are simply not considered in the estimation.

In [25], we proposed a robust regression technique, that will be recalled below, to model the ladders using the three above assumptions. In the present paper, we propose to use an even more constrained ladder model by introducing *a priori* information about distances, namely:

- the inter-rungs distances, δ_j,
- the inter-stile distance, δ_s,

which may be estimated from a TLS survey of the ladder; see Figure14.

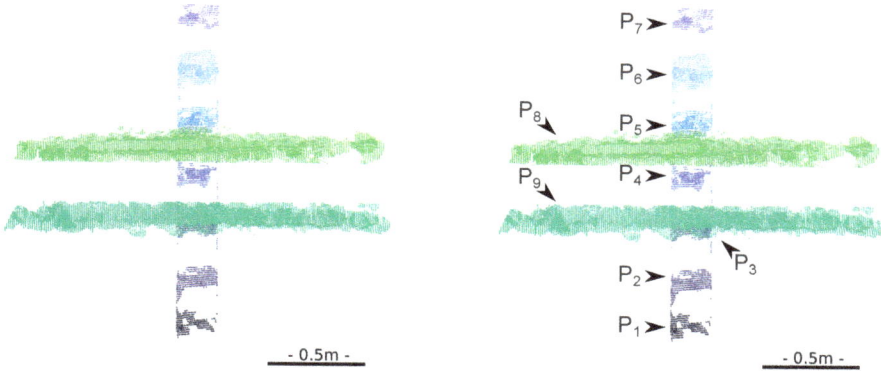

Figure 17. (Left) The 2D ladder point cloud is first split into two stile point clouds and seven rung point clouds; **(Right)** The stiles are rotated, and only one direction has to be adjusted on the data.

The linear regression analysis can be performed using either affine or orthogonal regression. Affine regression involves models of the form $y_i = \alpha x_i + \beta$ and can easily be adapted to the simultaneous robust fitting of multiple lines [26,27]. In our case, the slope α is the same for all components thanks to the parallelism constraint. It would be possible to introduce the inter-distance prior by modifying the model as (for the j-th rung):

$$y_i = \alpha x_i + \beta + \sqrt{1 + \alpha^2} \sum_{k=1}^{j} \delta_k \qquad (12)$$

(see Figure 18-left). However, this leads to a non-linear relationship with respect to α. If the axes provided by the robust estimation of the ladder plane are exactly vertical and orthogonal with respect to the ladder, then $\alpha = 0$, and the problem is reduced to the regression of one intercept, β. However, we prefer to avoid such an assumption by considering the orthogonal regression framework.

5.3.4. Orthogonal Simultaneous Fitting of Rungs and Stiles

The model underlying orthogonal linear regression is similar to Equation (5), except that a single vector (the one that supports the straight line) is considered: $\mathbf{x_i} \simeq \mu + \mathbf{e}a_i$; and that a_i is a scalar. The robust orthogonal linear regression algorithm is then straightforwardly adapted from the one summarized in Appendix A.

235

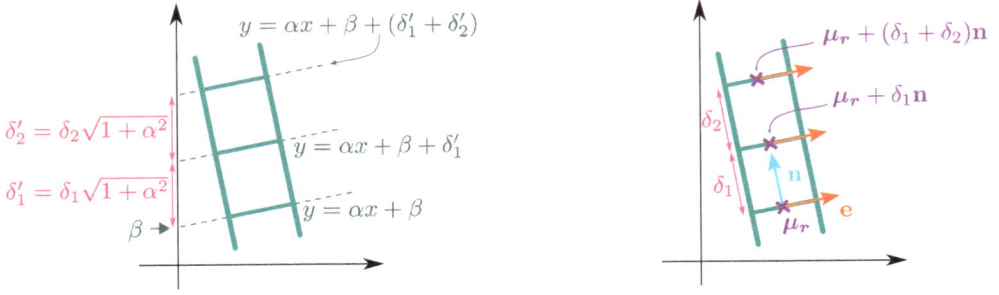

Figure 18. Simultaneous affine (**Left**) and orthogonal (**Right**) regressions.

Let us denote by $P_j = \{\mathbf{x_{ji}}\}_{i=1...N_j}$ each of the R point clouds corresponding to rungs and the $S = 2$ point clouds corresponding to the side rails, with $\sum_{j=1}^{R+S} N_j = N$. Due to the parallelism and orthogonality conditions, the orthogonal simultaneous fitting algorithm is reduced to the estimation of a single direction, $\mathbf{e}$ and $R + S$ centroids, $\boldsymbol{\mu}_j$. In other words, the underlying model is:

$$\mathbf{x_{ji}} \simeq \boldsymbol{\mu}_j + \mathbf{e}a_{ji} \tag{13}$$

The solution of the associated LS problem is given by the sample means, for the centroids:

$$\boldsymbol{\mu}_j = \frac{1}{N_j} \sum_{i=1}^{N_j} \mathbf{x_{ji}} \tag{14}$$

and for the common direction $\mathbf{e}$, by the first eigenvector of the global covariance matrix:

$$\mathbf{C} = \frac{1}{N} \sum_{j=1}^{R+S} \sum_{i=1}^{N_j} (\mathbf{x_{ji}} - \boldsymbol{\mu}_j)(\mathbf{x_{ji}} - \boldsymbol{\mu}_j)^\top \tag{15}$$

The robust counterpart of this algorithm is derived by alternating computations of the weights, b_{ji}, and weighted LS estimations of $\boldsymbol{\mu}_j$ and $\mathbf{e}$; see Appendix B.

In Table 2, we show the distances between rungs that were estimated from TLS and MSS data with the proposed method and a commercial software (3DReshaper®). Note that our method fit the rungs and styles simultaneously, while they must be dealt with separately with the commercial software. Moreover, the proposed algorithm exploits orthogonality and parallelism constraints. The TLS data have been obtained in laboratory conditions and can then be considered as almost noiseless. In such a favorable situation, both methods perform well, and we observe differences of 1 mm maximum. The distances obtained by our method will be considered as a reference in what follows. Unlike TLS data, MSS data are very noisy and contain outliers. As shown in Figure 16d, the performance of the commercial software collapses in that case, and since the rungs are not parallel in the resulting model, inter-rung distances cannot be evaluated. In contrast, our method is robust on this dataset. The fourth column of the table shows that the maximum difference with the reference is 28 mm and that most errors are less than 10 mm (which also corresponds to the mean absolute difference, the median absolute difference being 5 mm). One may note that these results are better than the ones obtained in [25]. This is due to the fact that in [25], the reference measurements were made by hand, and a constant inter-distance (of 280 mm) was considered. Using a TLS survey provides a better reference.

Table 2. Comparison of inter-rung distance (δ_i) estimations obtained with our orthogonal simultaneous fitting algorithm (without distance prior) and a commercial solution (3DReshaper®) from laser and sonar point clouds of the ladder. Distances are in millimeters.

	laser (Low Noise, No Outliers)		sonar (Strong Noise + Outliers)	
	Proposed Solution (Reference)	Commercial Software	Proposed Solution	Commercial Software
δ_1	279	279 (0)	276 (−3)	n/a
δ_2	282	281 (−1)	266 (−16)	n/a
δ_3	283	284 (1)	287 (3)	n/a
δ_4	278	279 (1)	275 (−3)	n/a
δ_5	280	279 (−1)	287 (7)	n/a
δ_6	262	261 (−1)	234 (−28)	n/a

This algorithm can be complemented by a second stage (which was not present in [25]), in which the distance priors are fully taken into account. Since the inter-rung distances δ_j are known (third column of Table 2), the centroids of the rungs are related, as illustrated in Figure 18 (right), so only two parameters, μ_r and $\mathbf{e}$, must be estimated. In fact, a third parameter, μ_s, must be taken into account because the stiles are independent from the rungs in terms of translation, albeit they share the same orientation, up to a 90° rotation. This stage is initialized as follows. A straight line is adjusted on the first R centroids, μ_j's, to obtain a first estimate of the axis of symmetry of the ladder. The orthogonal projection of μ_1 onto this axis defines μ_r. For the stiles, μ_s is defined as the orthogonal projection of μ_{R+1} on the line of direction $\mathbf{e}$ passing through the means of μ_{R+1} and μ_{R+2}. Then, the algorithm alternately updates $\mathbf{e}$, μ_r and μ_s. The complete algorithm is given in Appendix C.

Figure 16c shows an example of robust ladder fitting using the robust orthogonal simultaneous technique with distance priors. One may see that, despite the strong noise level, the ladder is well approximated. It is then possible to correct the vertical gap that was visible in Figure 14 (left). The result is shown in Figure 19 (left).

5.4. Horizontal Translation

Once the orientation and vertical translation have been correctly estimated, the last operation consists of estimating the horizontal translation vector.

The 2D silhouette of the waterline along the structure can be extracted on both the TLS and MSS model by intersecting 3D meshes with the plane that corresponds to the water surface (Figure 19 (center)). Finally, we apply a 2D ICP algorithm to estimate the remaining 2D translation vector between both models. After this step, one may see that the continuity of the ladder is restored (Figure 19 (right)).

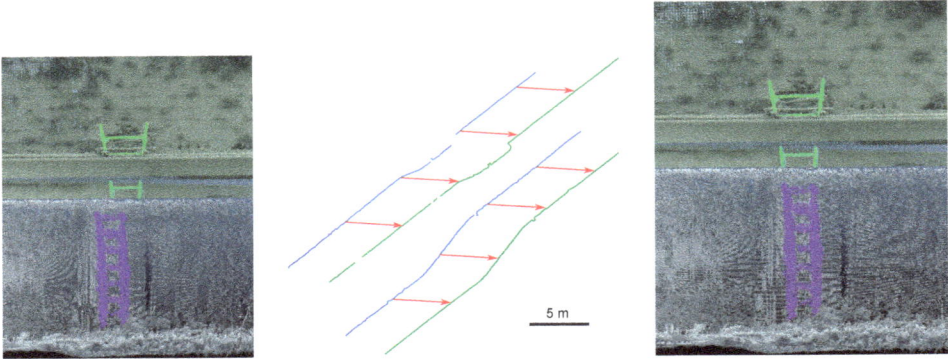

Figure 19. (**Left**) Correction of the vertical translation; (**Center**) Silhouette of the waterline in the TLS and MSS models; (**Right**) Correction of the horizontal translation (final result).

6. Experimental Results

The final 3D, full reference model we obtain with the proposed methodology is shown in Figure 20. We propose two complementary renderings: mesh and point cloud visualization. The latter provides visual information (such as measurement shadows) that may disappear in the mesh visualization, which is smoother. Moreover, some elements (ladders, equipments, packaging) were discarded from the mesh. In the point cloud visualization (Figure 20 (top)), it may be noticed that the above-water and underwater models are visually very satisfying. However, the TLS point cloud seems more homogeneous and denser than the underwater one, which has a lower resolution. Moreover, the MSS model appears more and more grainy as the points are away from the scanning position. For example, many imperfections, both within the MSS model and at the intersection of the models, are visible on canal banks at the bottom of the image. The defects are probably due to the wide footprint of the beam at such a distance from the source at a grazing incidence angle, as already seen in Figure 8. In future experiments, the distances between the MSS positions should be reduced to alleviate these issues.

Figure 20. Resulting full 3D geo-referenced model of Niderviller's canal tunnel entrance. (**Top**) Point cloud visualization; (**Bottom**) Mesh visualization. The red disk indicates the location of the MSS position.

The construction of the MSS model required many operator interventions for interpreting the elements either as noise or as detail. This task is very difficult without any visual references about the underwater part of the canal. However, just at the entrance of Niderviller's tunnel, in the alignment of the cofferdam grooves, a detail, which suggests a kind of step, may be seen (Figure 21) and has been conserved as an element of interest. This detail reminds us of the image of Arzviller's tunnel (see Figure 3) where a step is clearly visible between the cofferdam grooves. Unfortunately, we did not find a similar picture of the entrance of Niderviller's tunnel, which could confirm the relevance of this detail in the MSS model.

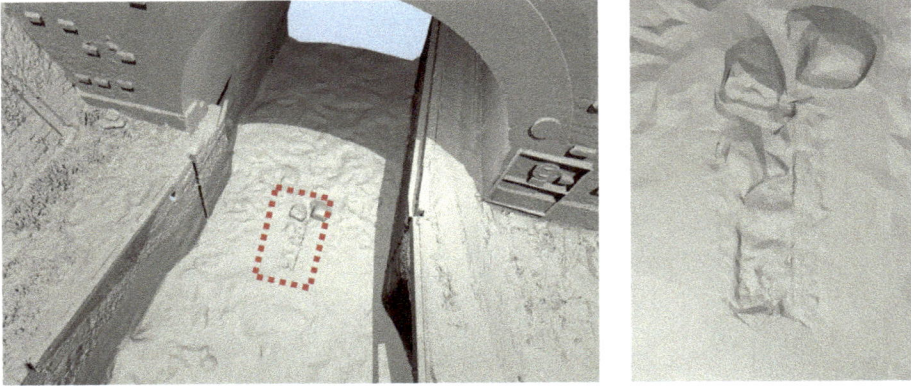

Figure 21. Close-up of the reference model showing two blocks and some structure in the alignment of the cofferdam grooves, which seem similar to the one observed in 2009 at Arzviller (Figure 3).

7. Conclusions and Future Work

In this paper, we have introduced a robust method to build a full 3D model of a canal tunnel. Data have been collected by TLS and MSS devices. The first issue we identified is due to the differences in spatial resolution and beam width between both devices. MSS data are intrinsically noisy and have a much lower resolution than TLS data. In addition, the angular loss of resolution can be rather strong: the oblong shape of the tunnel induces many grazing incidence angles, and the sonar data are rather coarse at large distances from the scanner device.

The first processing step consists of denoising the MSS data by meshing. More specifically, a coarse-to-fine method, which gradually increases the resolution of the mesh, is applied. Of course, differentiating noise from details is a difficult task in the absence of visual or physical reference, and confusion is unavoidable. The interpretation by an operator is required to determine the appropriate trade-off between noise and details. We believe that acoustic and image processing techniques should be explored to devise more automatic and data-driven denoising methods. In particular, moving least squares, bilateral filtering [28], non-local means filtering [29–31] or structure+texture decompositions [32] seem appealing for this task.

A second challenge concerns the co-registration of the point clouds provided by both scanners. The methods for processing TLS data to generate the geo-referenced above-water 3D model are well-known and can be used without any difficulties in our context. On the other hand, handling the MSS point clouds to build the underwater model is much more complicated. In particular, the weak resolution and noisy aspect of MSS data, along with the lack of salient elements along the canal, make classical registration methods, such as ICP, less efficient. To alleviate this difficulty, the experimental setup must be improved by reducing the distance between MSS positions. An interval of about 5 m would be recommended, instead of 10 meters, as in the current experiment. Moreover, additional targets, such as the ladders that are used in our experiment, could be immersed in the canal. They could make the registration easier, by adding references to both point clouds. Our experiments show that the targets must be carefully chosen and placed on-site: for example, the ladders should be wooden and separated from the canal walls; otherwise, their automatic segmentation becomes problematic.

To obtain a full 3D model of the canal tunnel, registering MSS and TSS data is a crucial point. The lack of overlap between point clouds raises difficulties in our application. To solve this challenge, we proposed a three-step procedure in which, first, geometrical entities are exploited to determine the orientation parameters. The second step uses the ladders to estimate the vertical correction: we introduced a robust method based on M-estimation to simultaneously fit lines on the stiles and rungs of the ladder. We assessed the method by comparing the results of the robust fit with real distance measurements. Distance priors are used to increase the precision of the fit thanks to a reference scan of the ladders. The proposed methodology could be adapted to other manufactured targets. Finally, the silhouette of the waterline is extracted in both models, and a 2D ICP algorithm is applied to estimate the remaining horizontal translation vector. Since the TLS point cloud is bound to a geodetic system, the geo-referencing of the model comes as a by-product of our method, without additional complexity.

This experimentation provides an initial overview of underwater acquisition in canal tunnels and yields promising results. Improvements of the model quality may be expected from a better experimental setup (closer scanning positions, more numerous targets). More automatic and data-driven filtering techniques should help with enhancing the data quality and reducing manual interventions. Furthermore, an experiment in a controlled environment, like a dry dock, as proposed in [5], would allow a fine assessment of the model accuracy. One may foresee that the progress of technology will improve the performances of the acquisition devices. Subsequently, we may expect more accurate models using the proposed methodology. The obtained models should be used as a reference for future acquisitions, either with dynamic underwater acquisitions systems (for assessing mobile mapping solutions) or in static ones (to evaluate the tunnel deformation).

Acknowledgments: This work was funded by the Cerema under PhD Grant 2014-1. The MSS acquisitions were subcontracted to the Sub-C Marine Company (Feyzin, France: www.subcmarine.com). The authors thank them for their availability and their cooperation. Many thanks also to VNF's staff at Niderviller.

Author Contributions: All authors conceived and designed the experiments; the first and last two authors performed the acquisitions; the first two authors contributed analysis tools; the first four authors analyzed the data and wrote the paper.

Conflicts of Interest: The authors declare no conflict of interest.

Appendix

A. Robust Regression of Planes and Lines

1. Initialize μ^0 the centroid and $\mathbf{E}^0$ vectors defining the plane, and set the iteration index to $t = 1$.

   - $\mu^0 = \frac{1}{N} \sum\limits_{i=1}^{N} \mathbf{x_i}$
   - $\mathbf{E}^0$ is the eigenvectors of $\mathbf{C}^0$ that match their two largest eigenvalues, where with $\mathbf{C}^0 = \frac{1}{N} \sum\limits_{i=1}^{N} (\mathbf{x_i} - \mu^0)(\mathbf{x_i} - \mu^0)^\top$

2. Compute residuals: $r_i^t = \|(\mu^{t-1} + \mathbf{E}^{t-1}\mathbf{a_i^t}) - \mathbf{x_i}\|$ with $\mathbf{a_i^t} = \mathbf{E}^{t-1^\top}(\mathbf{x_i} - \mu^{t-1})$.

3. Compute auxiliary variables $b_i^t = \frac{\rho'(r_i^t)}{2r_i^t}$ and their sum $S^t = \sum\limits_{i=1}^{N} b_i^t$.

4. Perform a weighted LS estimation to get μ^t and $\mathbf{e^t}$

   - $\mu^t = \frac{1}{S^t} \sum\limits_{i=1}^{N} b_i^t \mathbf{x_i}$
   - $\mathbf{E^t}$ is the eigenvectors of $\mathbf{C^t}$ corresponding to its two largest eigenvalues. with $\mathbf{C^t} = \frac{1}{S^t} \sum\limits_{i=1}^{N} b_i^t (\mathbf{x_i} - \mu^t)(\mathbf{x_i} - \mu^t)^\top$

5. If $\frac{\|\mu^t - \mu^{t-1}\|^2}{\|\mu^{t-1}\|} > \epsilon$ increment t and go to 2, else $\mu^{IRLS} = \mu^t$ and $\mathbf{e}^{IRLS} = \mathbf{e^t}$.

This algorithm can be straightforwardly adapted to the case of line estimation. In that case, only one vector (the direction of the line), $\mathbf{e}$, is sought. The expression of the residual becomes $r_i^t = \|(\mu^{t-1} + a_i^t \mathbf{e}^{t-1}) - \mathbf{x_i}\|$ with $a_i^t = \mathbf{e}^{t-1^\top}(\mathbf{x_i} - \mu^{t-1})$. The estimation of the means does not change, and $\mathbf{e}$ is given by the eigenvector of $\mathbf{C}$ that matches its largest eigenvalue.

B. Simultaneous Robust Fitting of Lines

1. Initialize μ_j^0 centroids and $\mathbf{e}^0$ the direction vector, and set the iteration index to $t = 1$.

   - $\mu_j^0 = \frac{1}{N_j} \sum\limits_{i=1}^{N_j} \mathbf{x_{ji}}$

- $\mathbf{e^0}$ is the eigenvector of $\mathbf{C^0}$ that matches its largest eigenvalue, where:

$$\mathbf{C^0} = \tfrac{1}{N} \sum_{j=1}^{R+S} \sum_{i=1}^{N_j} (\mathbf{x_{ji}} - \boldsymbol{\mu}_j)(\mathbf{x_{ji}} - \boldsymbol{\mu}_j)^\top$$

2. Compute residuals: $r_{ji}^t = \|(\boldsymbol{\mu}_j^{t-1} + a_{ji}^t \mathbf{e^{t-1}}) - \mathbf{x_{ji}}\|$ with $a_{ji}^t = \mathbf{e^{t-1}}^\top (\mathbf{x_{ji}} - \boldsymbol{\mu}_j^{t-1})$.

3. Compute auxiliary variables $b_{ji}^t = \dfrac{\rho'(r_{ji}^t)}{2r_{ji}^t}$ and the sums $S_j^t = \sum_{i=1}^{N_j} b_{ji}^t$ and

$S^t = \sum_{j=1}^{R+S} S_j^t$.

4. Perform a weighted LS estimation to get $\boldsymbol{\mu}_j^t$ and $\mathbf{e^t}$

- $\boldsymbol{\mu}_j^t = \tfrac{1}{S_j} \sum_{i=1}^{N_j} b_{ji}^t \mathbf{x_{ji}}$

- $\mathbf{e^t}$ is the eigenvector of $\mathbf{C^0}$ corresponding to its largest eigenvalue.

with $\mathbf{C^t} = \tfrac{1}{S^t} \sum_{j=1}^{R+S} \sum_{i=1}^{N_j} b_{ji}^t (\mathbf{x_{ji}} - \boldsymbol{\mu}_j^t)(\mathbf{x_{ji}} - \boldsymbol{\mu}_j^t)^\top$

5. If $\dfrac{\|\boldsymbol{\mu}_j^t - \boldsymbol{\mu}_j^{t-1}\|^2}{\|\boldsymbol{\mu}_j^{t-1}\|} > \epsilon$ increment t and go to 2, else $\boldsymbol{\mu}_j^{IRLS} = \boldsymbol{\mu}_j^t$ and $\mathbf{e^{IRLS}} = \mathbf{e^t}$.

C. Simultaneous Robust Fitting of Lines with Distance Prior

1. Initialize $\boldsymbol{\mu}_r^0$ and $\boldsymbol{\mu}_s^0$ centroids and $\mathbf{e^0}$ the direction vector, and set the iteration index to $t = 1$.

   (a) Estimate $\boldsymbol{\mu}_j^{0'}$ centroids and $\mathbf{e^0}$ the direction vector thanks to the simultaneous robust fitting algorithm; see Appendix B.
   (b) Estimate $\boldsymbol{\mu}_r'$ the centroid of the fitted line of $\boldsymbol{\mu}_j^{0'}$ with j={1,...,R} (see Appendix A).
   Thus, $\boldsymbol{\mu}_r^0 = \boldsymbol{\mu}_r' + (\mathbf{n^0}^\top (\boldsymbol{\mu}_1^{0'} - \boldsymbol{\mu}_r'))\mathbf{n^0}$ with $\mathbf{n^0}$ the normal to $\mathbf{e^0}$
   (c) Compute $\boldsymbol{\mu}_s' = \dfrac{\boldsymbol{\mu}_{R+1}^{0'} + \boldsymbol{\mu}_{R+2}^{0'}}{2}$.
   Thus, $\boldsymbol{\mu}_s^0 = \boldsymbol{\mu}_s' + (\mathbf{n^0}^\top (\boldsymbol{\mu}_1^{0'} - \boldsymbol{\mu}_s'))\mathbf{n^0}$
   (d) Compute $\boldsymbol{\mu}_j^0$ centroids from:

   - $\boldsymbol{\mu}_j^0 = \boldsymbol{\mu}_s^0 + \mathbf{n^0} \sum_{k=1}^{j} \delta_k$ for $j = \{1,...,R\}$
   - $\boldsymbol{\mu}_{R+1}^0 = \boldsymbol{\mu}_s^0$ and $\boldsymbol{\mu}_{R+2}^0 = \boldsymbol{\mu}_s^0 + \mathbf{n^0}\delta_s$

2. Compute residuals: $r_{ji}^t = \|(\boldsymbol{\mu}_j^{t-1} + a_{ji}^t \mathbf{e^{t-1}}) - \mathbf{x_{ji}}\|$ with $a_{ji}^t = \mathbf{e^{t-1}}^\top (\mathbf{x_{ji}} - \boldsymbol{\mu}_j^{t-1})$.

3. Compute auxiliary variables $b_{ji}^t = \frac{\rho'(r_{ji}^t)}{2r_{ji}^t}$ and the sums $S_r^t = \sum\limits_{j=1}^{R} \sum\limits_{i=1}^{N_j} b_{ji}^t$, $S_s^t = \sum\limits_{j=R+1}^{R+2} \sum\limits_{i=1}^{N_j} b_{ji}^t$ and $S^t = S_r^t + S_s^t$.

4. Perform a weighted LS estimation to get $\boldsymbol{\mu}_r^t$, $\boldsymbol{\mu}_s^t$ and $\mathbf{e^t}$:

   - $\mathbf{e^t}$ is the eigenvector of $\mathbf{C^0}$ corresponding to its largest eigenvalue, where:

     with $\mathbf{C^t} = \frac{1}{S^t} \sum\limits_{j=1}^{R+S} \sum\limits_{i=1}^{N_j} b_{ji}^t (\mathbf{x_{ji}} - \boldsymbol{\mu}_j^{t-1})(\mathbf{x_{ji}} - \boldsymbol{\mu}_j^{t-1})^\top$

   - $\boldsymbol{\mu}_r^t = \frac{1}{S_r} \sum\limits_{j=1}^{R} \sum\limits_{i=1}^{N_j} b_{ji}^t (\mathbf{x_{ji}} - \mathbf{n^t} \sum\limits_{k=1}^{j})\delta_k$ for $j = \{1, ..., R\}$

   - $\boldsymbol{\mu}_s^t = \frac{1}{S_s} ((\sum\limits_{i=1}^{N_{R+1}} b_{(R+1)i}^t \mathbf{x_{(R+1)i}}) + (\sum\limits_{i=1}^{N_{R+2}} b_{(R+2)i}^t (\mathbf{x_{(R+2)i}} - \mathbf{n^t}\delta_s)))$

5. Deduce $\boldsymbol{\mu}_j^t$ centroids from:

   - $\boldsymbol{\mu}_j^t = \boldsymbol{\mu}_s^t + \mathbf{n^t} \sum\limits_{k=1}^{j} \delta_k$ for $j = \{1, ..., R\}$

   - $\boldsymbol{\mu}_{R+1}^t = \boldsymbol{\mu}_s^t$ and $\boldsymbol{\mu}_{R+2}^t = \boldsymbol{\mu}_s^t + \mathbf{n^t}\delta_s$

6. If $\frac{\|\boldsymbol{\mu}_j^t - \boldsymbol{\mu}_j^{t-1}\|^2}{\|\boldsymbol{\mu}_j^{t-1}\|} > \epsilon$ increment t and go to 2 else $\boldsymbol{\mu}_j^{IRLS} = \boldsymbol{\mu}_j^t$ and $\mathbf{e^{IRLS}} = \mathbf{e^t}$.

References

1. Mitchell, T.; Miller, C.; Lee, T. Multibeam surveys extended above the waterline. In Proceedings of the WEDA Technical Conference and Texas A&M Dredging Seminar, Nashville, TN, USA, 5–8 June 2011; pp. 414–422.

2. Griffiths, J.; Graham, J. *Processing Combined laser, sonar ans HD Imaging for Better Evalution Decisions, No-Dig Show*; North American Society for Trenchless Technology (NASTT): Washington, DC, USA, 2011.

3. Rondeau, M.; Leblanc, E.; Garantc, L. Dam infrastructure first inspection supported by an integrated multibeam echosounder MBES/ LiDAR system. In Proceedings of the CDA Annual Conference, Saskatoon, SK, Canada, 22–27 September 2012.

4. Ingram, E. Hot products preview. *Hydro Rev.* **2014**, *33*, 22–30.

5. Lesnikowski, N.; Rush, B. *Spool Piece Metrology Applications Utilizing BV5000 3D Scanning Sonar*; Technical Report, BlueView Technologies: Seattle, WA, USA, 2012.

6. Vaaja, M.; Kukko, A.; Kaartinen, H.; Kurkela, M.; Kasvi, E.; Flener, C.; Hyyppä, H.; Hyyppä, J.; Järvelä, J.; Alho, P. Data processing and quality evaluation of a boat-based mobile laser scanning system. *Sensors* **2013**, *13*, 12497–12515.

7. Rondeau, M.; Stoeffler, C.; Brodie, D.; Holland, M. *Deformation Analysis of Harbour and Dam Infrastructure Using Marine GIS*; The Hydrographic Society of America (THSOA): Washington, DC, USA, 2015.

8. Charbonnier, P.; Foucher, P.; Chavant, P.; Muzet, V.; Prybyla, D.; Perrin, T.; Albert, J.; Grussenmeyer, P.; Guillemin, S.; Koehl, M. An image-based inspection system for canal-tunnel heritage. *Int. J. Herit. Digit. Era* **2014**, *3*, 197–214.

9. Drap, P.; Merad, D.; Boï, J.M.; Boubguira, W.; Mahiddine, A.; Chemisky, B.; Seguin, E.; Alcala, F.; Bianchimani, O. ROV-3D: 3D underwater survey combining optical and acoustic sensor. In Proceedings of the 12th International Conference on Virtual Reality, Archaeology and Cultural Heritage, Prato, Italy, 18–21 October 2011; pp. 177–184.

10. Sohnlein, G.; Rush, S.; Thompson, L. Using manned submersibles to create 3D sonar scans of shipwrecks. In Proceedings of the IEEEOCEANS Conference, Santander, Spain, 6–9 June 2011; pp. 1–10.

11. Ozog, P.; Troni, G.; Kaess, M.; Eustice, R.M.; Johnson-Roberson, M. Building 3D Mosaics from an Autonomous Underwater Vehicle, Doppler Velocity Log, and 2D Imaging Sonar. In Proceedings of the IEEE International Conference on Robotics and Automation (ICRA), Seattle, WA, USA, 26–30 May 2015; pp. 1137–1143.

12. Vaaja, M.; Hyyppä, J.; Kukko, A.; Kaartinen, H.; Hyyppä, H.; Alho, P. Mapping Topography Changes and Elevation Accuracies Using a Mobile Laser Scanner. *Remote Sens.* **2011**, *3*, 587–600.

13. Papadopoulos, G.; Kurniawati, H.; Shariff, A.S.B.M.; Wong, L.J.; Patrikalakis, N.M. Experiments on Surface Reconstruction for Partially Submerged Marine Structures. *J. Field Robot.* **2014**, *31*, 225–244.

14. Vosselman, G.V.; Maas, H.G. *Airborne and Terrestrial Laser Scanning*; Whittles Publisher: Dunbeath, UK, 2010.

15. Ridao, P.; Carreras, M.; Ribas, D.; Garcia, R. Visual inspection of hydroelectric dams using an autonomous underwater vehicle. *J. Field Robot.* **2010**, *27*, 759–778.

16. Balletti, C.; Beltrame, C.; Costa, E.; Guerra, F.; Vernier, P. Underwater photogrammetry and 3D reconstruction of marble cargos shipwreck. In Proceedings of the International Archives of the Photogrammetry, Remote Sensing and Spatial Information Sciences (ISPRS), Piano di Sorrento, Italy, 16–17 April 2015; pp. 7–13.

17. Menna, F.; Nocerino, E.; Troisi, S.; Remondino, F. Joint alignment of underwater and above the water photogrammetric 3D models by independent models adjustment. In Proceedings of the International Archives of the Photogrammetry, Remote Sensing and Spatial Information Sciences (ISPRS), Piano di Sorrento, Italy, 16–17 April 2015; pp. 143–151.

18. Remondino, F. From point cloud to surface: The modeling and visualization problem. *Int. Arch. Photogramm. Remote Sens. Spat. Inf. Sci.* **2003**, *34*, doi:10.3929/ethz-a-004655782.

19. Besl, P.; McKay, N.D. A method for registration of 3-D shapes. *IEEE Trans. Pattern Anal. Mach. Intell.* **1992**, *14*, 239–256.

20. Golub, G.H.; van Loan, C.F. *Matrix Computations, Johns Hopkins Studies in the Mathematical Sciences*; Johns Hopkins University Press: Baltimore, MD, USA, 2012; Volume 3; pp. 784.

21. Huber, P. *Robust Statistics*; John Wiley and Sons: New York, NY, USA, 1981.

22. Charbonnier, P.; Blanc-Féraud, L.; Aubert, G.; Barlaud, M. Deterministic Edge-Preserving Regularization in Computed Imaging. *IEEE Trans. Image Process.* **1997**, *6*, 298–311.

23. Bishop, C. *Pattern Recognition and Machine Learning*; Information Science and Statistics, Springer: New York, NY, USA, 2006.

24. Delon, J.; Desolneux, A.; Lisani, J.-L.; Petro, A. A Nonparametric Approach for Histogram Segmentation. *Trans. Image Process.* **2007**, *16*, 253–261.

25. Moisan, E.; Charbonnier, P.; Foucher, P.; Grussenmeyer, P.; Guillemin, S.; Koehl, M. Building a 3D reference model for canal tunnel surveying using sonar and laser scanning. In Proceedings of the International Archives of the Photogrammetry, Remote Sensing and Spatial Information Sciences (ISPRS), ISPRS/ CIPA Workshop "Underwater 3D recording and Modeling", Piano di Sorrento, Italy, 16–17 April 2015; pp. 153–159.

26. Tarel, J.; Charbonnier, P.; Ieng, S. Simultaneous robust fitting of multiple curves. In Proceedings of the International Conference on Computer Vision Theory and Applications (VISAPP), INSTICC, Barcelona, Spain, 8–11 March 2007; pp. 175–182.

27. Tarel, J.; Ieng, S.; Charbonnier, P. A constrained-optimization based half-quadratic algorithm for robustly fitting sets of linearly parametrized curves. *Adv. Data Anal. Classif.* **2008**, *2*, 227–239.

28. Tomasi, C.; Manduchi, R. Bilateral Filtering for Gray and Color Images. In Proceedings of the IEEE International Conference on Computer Vision (ICCV), Bombay, India, 4–7 January 1998; pp. 839–846.

29. Buades, A.; Coll, B.; Morel, J.M. A Non-Local Algorithm for Image Denoising. In Proceedings of the IEEE Conference on Computer Vision and Pattern Recognition (CVPR), San Diego, CA, USA, 20–26 June 2005; Volume 2, pp. 60–65.

30. Deschaud, J.E.; Goulette, F. Point cloud non local denoising using local surface descriptor similarity. In Proceedings of the Symposium PCV 2010 (Photogrammetric Computer Vision and Image Analysis), Saint-Mandé, France, 1–3 September 2010; pp. 109–114.

31. Digne, J. Similarity based filtering of point clouds. In Proceedings of the IEEE Conference on Computer Vision and Pattern Recognition Workshops (CVPR), Providence, RI, USA, 16–21 June 2012; pp. 73–79.

32. Aujol, J.F.; Gilboa, G.; Chan, T.; Osher, S. Structure-Texture Image Decomposition–Modeling, Algorithms, and Parameter Selection. *Int. J. Comput. Vis.* **2006**, *67*, 111–136.

Chapter 3:
Sensor Integration and Data Processing

Integrating Sensors into a Marine Drone for Bathymetric 3D Surveys in Shallow Waters

Francesco Giordano, Gaia Mattei, Claudio Parente, Francesco Peluso and Raffaele Santamaria

Abstract: This paper demonstrates that accurate data concerning bathymetry as well as environmental conditions in shallow waters can be acquired using sensors that are integrated into the same marine vehicle. An open prototype of an unmanned surface vessel (USV) named MicroVeGA is described. The focus is on the main instruments installed on-board: a differential Global Position System (GPS) system and single beam echo sounder; inertial platform for attitude control; ultrasound obstacle-detection system with temperature control system; emerged and submerged video acquisition system. The results of two cases study are presented, both concerning areas (Sorrento Marina Grande and Marechiaro Harbour, both in the Gulf of Naples) characterized by a coastal physiography that impedes the execution of a bathymetric survey with traditional boats. In addition, those areas are critical because of the presence of submerged archaeological remains that produce rapid changes in depth values. The experiments confirm that the integration of the sensors improves the instruments' performance and survey accuracy.

Reprinted from *Sensors*. Cite as: Giordano, F.; Mattei, G.; Parente, C.; Peluso, F.; Santamaria, R. Integrating Sensors into a Marine Drone for Bathymetric 3D Surveys in Shallow Waters. *Sensors* **2016**, *16*, 41.

1. Introduction

Bathymetric information is fundamental in all branches of oceanography, paleoclimate studies, and marine geology. It can be supplied by maps that indicate the water body depth as a function of the position (latitude and longitude), similar to topographic maps representing the altitude of the Earth's surface at different geographic coordinates [1].

Most techniques for obtaining these data are difficult to use in shallow waters where bathymetric surveys often entail expensive measurement costs . For most bathymetry acquisition techniques, it is not possible to obtain a better vertical accuracy than 0.5 m at the 95% confidence level. Airborne LiDAR and/or maritime vessels are the only options for surveys with an accuracy requirement of 0.5 m with a 95% confidence level. Other remote sensing techniques can also be used only if the accuracy requirements are relaxed to 2 m, 95% confidence [2].

Airborne laser (or lidar) bathymetry (ALB) is based on a scanning, pulsed laser beam to measure the depths of relatively shallow, coastal waters from the air. It is also named airborne lidar hydrography (ALH) when used principally for nautical charting [3].

The use of maritime vessels capable of carrying out bathymetric measurements is limited by the depth of the waters, so only small crafts are suitable in shallow waters. Because of their reduced dimensions, these vessels are not manned and are categorized as USVs (Unmanned Surface Vehicles) [4,5]. By analogy with avionics applications, they are also called marine drones [6]. Some such drones are also known as Autonomous Surface Crafts (ASCs) and Remotely Operated Vehicles (ROVs). According to [7], ASCs, also called autonomous surface vehicle (ASVs), are a kind of autonomous marine vehicle without the direct operation of humans, while ROVs are controlled by an operator who is not on-board. However, this distinction is not always observed and the terms are sometimes used with no difference in meaning.

In the last few years several specific crafts have been built for surveying in shallow waters, as reported in the literature.

In June 2006, the US Geological Survey Woods Hole Science Center (WHSC) integrated an ASV for hydrographic surveys in shallow waters (1–5 m), which was designed to map seafloor morphology and surficial sediment distribution and thickness. Named the Independently (or) Remotely Influenced Surveyor (IRIS) and designed as a catamaran-based platform (10 feet in length, 4 feet in width, and approximately 260 lbs in weight), this vehicle is equipped with a chirp dual-frequency side scan sonar (100/500 kHz) and seismic-reflection profiler (4–24 kHz), a wireless video camera and single-beam echosounder (235 kHz). IRIS is operated remotely through a wireless modem network enabling the real-time monitoring of data acquisition and navigated using RTK [8].

The ROAZ unmanned surface vehicle was proposed by the Autonomous Systems Laboratory (ASL) from Porto Polytechnic Institute (ISEP) for marine operations. It was designed to work in very shallow rivers and marine coastlines. Because of the possibility of transmitting the entire data collection on-board a base station, the operator receives online feedback on the vehicle's location and performance, as well as side-scan sonar imagery and bathymetry quality [9].

Another example of a craft used for bathymetric surveys in shallow water was developed by the Underwater Robotic Research Group's (URRG) who developed the URG—ASV, a battery-powered vessel [10].

CatOne is an example of a catamaran-robot that can operate in very shallow waters as well as in sensitive ecosystems because of its very low draft and an electric propulsion that guarantees zero pollution emission and low noise. It carries sonar and GPS on-board and can be equipped with other sensors to support different activities such as environment monitoring [11].

The purpose of this research was to create a marine drone, optimized for surveys in very shallow water, and benefitting from previous experiences in this field as noted above. The innovation of this project is twofold. First, the data and video are broadcast directly to several operators, enabling the visualization and the pre-processing of all data in real time, by means of several devices managed by experts from different disciplines (such as an archaeologist, a geophysicist, a topographer or a GIS expert). This feature was implemented in order to carry out interdisciplinary surveys in critical coastal areas In fact, in the two study cases (both in the Gulf of Naples) there are submerged archaeological remains in the survey area. Thus, each expert can verify that the data acquisition is correct from his/her point of view. In addition, in order to also obtain high precision bathymetric data in critical areas, a system of data quality control was implemented, using an inertial platform.

2. Experimental Section

The MicroVeGA drone is an Open Project of USV conceived, designed and built to operate in shallow water areas (0–20 m), where a traditional boat is poorly manoeuvrable. It was engineered by the DIST research group at the University of Naples and was designed to test the procedures and methods of morpho-bathymetric surveys in critical areas. In [7], the initial development phase of MicroVeGA is described.

The drone is a small and ultra-light catamaran that can be assembled in 30 min, with a few draught centimetres, therefore suitable to perform surveys up to the shoreline. It is driven by non-polluting electric motors, and is therefore suitable to perform surveys in marine protected areas. Table 1 lists the characteristics.

Table 1. Technical and physical characteristics of the drone.

Characteristics	Measures
Overall length	135 cm
Width	85 cm
Weight in navigation trim	20 kg
Motors	2 brushless 750 kV/140 W
Operating speed	0.5–2 m/s
Power autonomy	2–4 h

MicroVeGA is an evolving open project that has enabled surveys to be carried out already in its early stage of development (Figure 1). In this paper, two study cases are illustrated, the first created with the MicroVeGA prototype #1 (Figure 1b), the second with the MicroVeGA prototype #2 (Figure 1c).

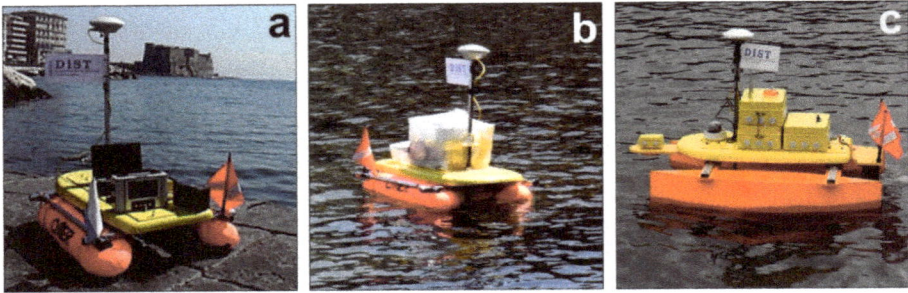

Figure 1. (**a**) Instruments on board of MicroVeGA; (**b**) Prototype #1 of MicroVeGA; (**c**) Prototype #2 of MicroVeGA.

This project is a low risk technology project. The spiral model of project management [12] is divided into smaller sections (Figure 2). Each prototype requires the following phases: requirements, design and refine, build; test, survey and analyse.

The current version of MicroVeGA (*i.e.*, Prototype #2) is remotely controlled by an operator and is equipped with a set of sensors for acquiring morpho-bathymetric high-precision data (see Section 2.2).

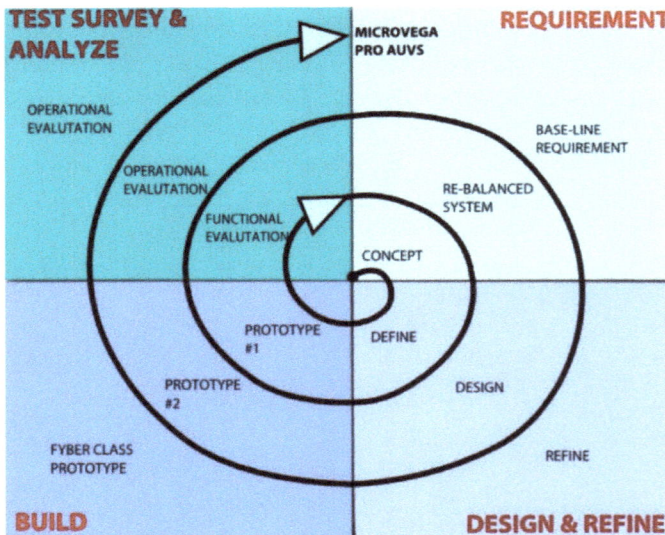

Figure 2. Spiral model of the project management.

2.1. System Architecture

The architecture of the data acquisition system (Figure 3) includes: (i) a base station, with a remote controlled PC and a video terminal; (ii) an on-board computerized system that manages the on-board instrumentation; (iii) a communication system via data link, to connect the UVS with the base station.

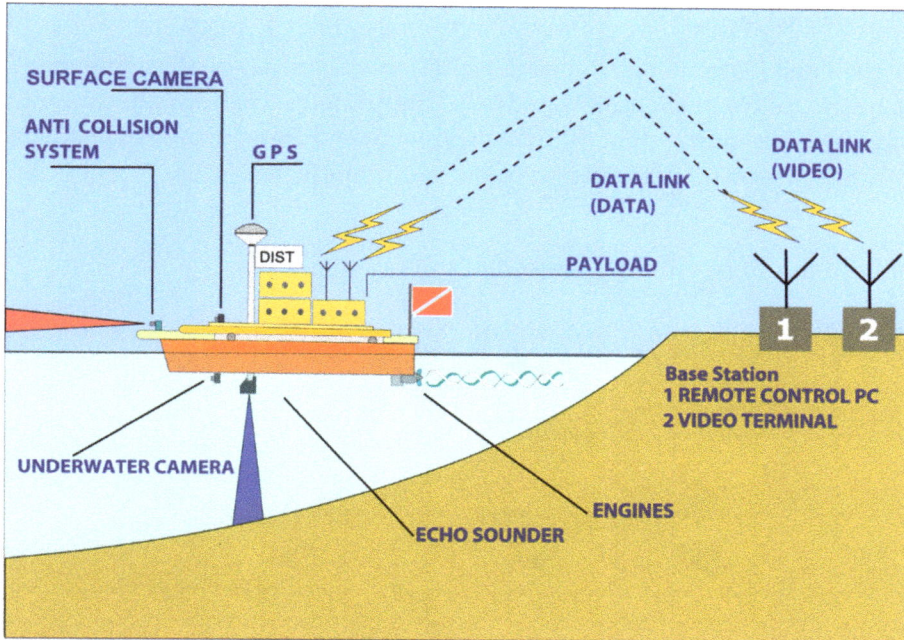

Figure 3. Data acquisition architecture: a base station, with a remote control PC and a video terminal; an on board computerized system that manages the on-board instrumentation; a communication system via data link, to connect the AUVS with the base station.

The operator responsible for the base station manages the mission data by means of TrackStar software by defining the navigation routes and monitoring the mission progress. The TrackStar software (described in Section 2.3 Data Acquisition and Software), implemented by our research group, manages the survey activities and automatically creates a measurement geodatabase.

The data is stored on board in RAW format by a computerized system that acquires and organizes the GPS, echo sounder, inertial platform, and obstacle-detection sensor data. This data is broadcast to the base station by a data link system, after which several operators can simultaneously receive the data in real time.

MicroVeGA data transmission is based on two wireless networks. The first transmits the telemetry data (*i.e.*, position, depth, atmospheric temperature and obstacle detection) from the vessel to Trackstar. The second network transmits the videos of the two on-board cameras to the base station. This information is managed by a specific app, and the images are viewed on a tablet in real time.

2.2. Sensors and Methods for Data Acquisition

The main instruments on-board are: (1) microcomputer; (2) differential GPS system and Single beam echo sounder; (3) integrated system for attitude control; (4) obstacle-detection system (SIROS1) with temperature control system; (5) video acquisition system (both above and below sea level) (Figure 4).

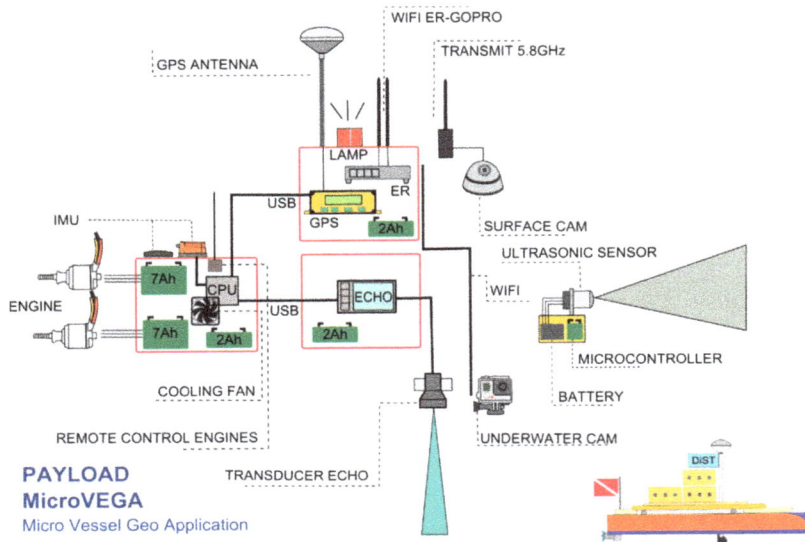

Figure 4. Payload of MicroVeGA drone.

2.2.1. Microcomputer

An OLinuXino microcomputer, with a Linux operating system, and three high-speed serials, manages all the survey phases, the data recording and its wi-fi transmission to the base station. An Arduino microcontroller controls the drone's engines, the temperature measurements, and the management of the obstacle-detection ultrasound systems (Figure 4).

2.2.2. GPS and Single Beam Echo Sounder (SBES)

The GPS receiver (Figure 5b), installed on board MicroVeGA, is the Trimble DSM™ 232 (24-channel L1/L2), which is a robust solution for dynamic positioning tasks in the marine environment. In fact, this device is easily installed and is able to withstand tough environmental conditions, and is thus suitable for surveys in very shallow waters. In addition, the GPS receiver and antenna are modular, and thus it was possible to install on board of MicroVeGA, the antenna vertically with respect to the SBES transducer.

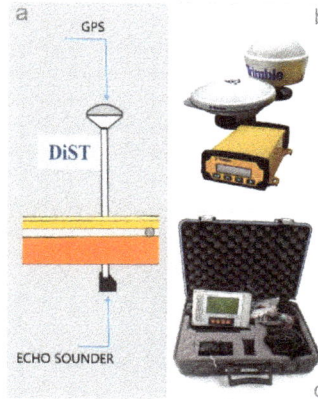

Figure 5. (a) Installation positions of the GPS and echo sounder on drone; (b) Trimble DSM232 GPS; (c) Omex Sonarlite echo sounder.

The Trimble DSM 232 GPS receiver enables the appropriate GPS correction method and accuracy to be selected. In this research, the DGPS option in post-processing was used, using Trimble software.

The SonarLite (Omex) is the SBES installed on-board (Figure 5c). This instrument is optimized for the bathymetric survey in shallow waters, and its transducer is positioned vertically above the GPS receiver in order to remove any offset (Figure 5a).

2.2.3. Inertial Platform Unit (IMU)

The inertial measurement unit used for measuring balance and direction on board of MicroVeGA is the Xsense MTi series G. This device is an integrated GPS and MEMS IMU with a Navigation and Attitude and Heading Reference System processor. It was used on the MicroVeGA drone because of it weighs very little.

The internal low-power signal processor runs a real-time Xsens Kalman Filter (XKF), providing inertial enhanced 3D position and velocity estimates [13,14].

The IMU data are stored in the survey geodatabase and increase the accuracy of the survey since measurements affected by attitude errors are removed [15]. In the case of errors due to pitch and roll, a quality control system that removes all measurements higher than a specific limit d was implemented (Figure 6):

$$d \leqslant spp \tag{1}$$

255

where:

$$d = Z' \sin \beta \tag{2}$$

and Z' = echo sounder measurement; spp = survey parameter precision related to survey scale, depth, survey target; β = angle between Z and Z' = $90° - (90° - \alpha)$; α = pitch or roll.

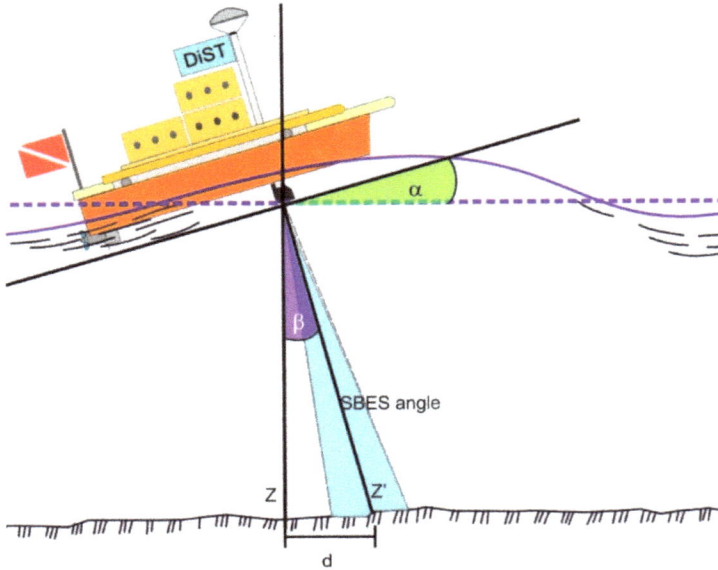

Figure 6. Horizontal error due to pitch or roll.

In the planning phase of the survey, the operator can establish the value of the spp survey parameter, thus defining the horizontal limit d that makes a measurement valid.

In both surveys described below (archaeological survey with rocky seabed and a cartographic scale of 1:1000), the threshold value spp was set equal to 1. As shown in Figure 7, if the measured depth increases, the roll angle becomes even more critical. In fact the same angle of roll (or pitch), equal to 10°, is associated with a valid measurement if the depth is −5 m, while it is associated with an invalid measurement if the depth is greater than −7 m (see also Table 2).

As the weather and sea conditions are essential for the proper execution of a bathymetric survey, surveys are not normally carried out when waves are beyond a certain strength. The validation system is primarily to prevent the storage of incorrect data due to occasional events, such as the passage of a vessel, and thus to improve the quality level of the whole survey.

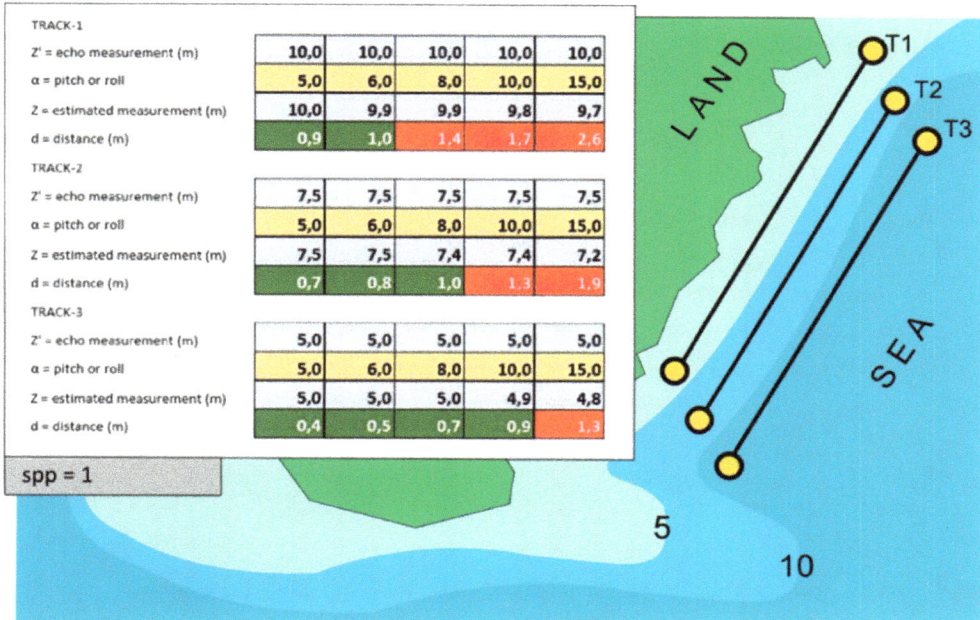

Figure 7. Example of three data acquisitions with spp = 1 constant and with α and Z′ variables.

Table 2. Variation of the distance d with the changing depth (see also Figure 7).

Measurement Parameters	T1	T2	T3
Z′ = Echo Measurement (m)	5.0	7.5	10.0
α = Pitch (or roll)	10.0	10.0	10.0
Z = Estimated Measurement (m)	4.9	7.4	9.8
d = distance (m)	0.9	1.3	1.7

The mission software—Trackstar—manages these calculations in real time highlighting the invalid measurements with a special color scale. This visualization allows the operator to evaluate the areal coverage of the survey, and to decide the possible repetition of a navigation line in real time.

The IMU data are also used to correct the depth with respect to the vertical error due to the wave effect (Figure 8):

$$CWL = Z \pm dZ \qquad (3)$$

where: CWL = clam water level; Z = depth measured by SBES; dZ = vertical error measured by inertial platform.

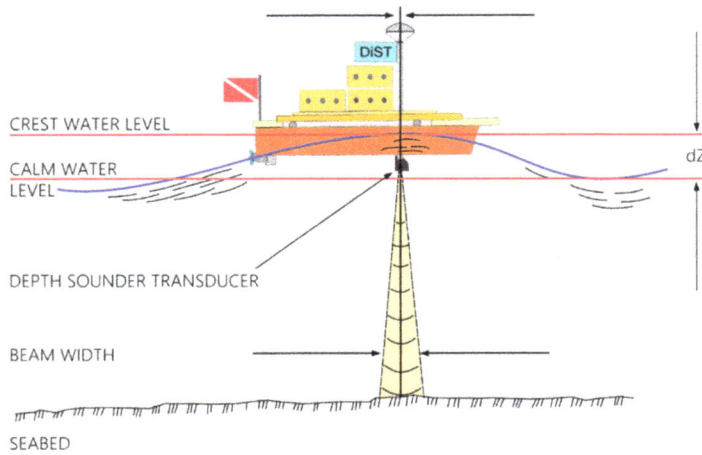

Figure 8. Vertical error in depth measuring due to the wave effect.

2.2.4. SIROS 1 (Obstacle-Detection System—In Italian: Sistema Rilevamento OStacoli)

The system is based on: an Arduino controller; an ultrasonic sensor; a temperature sensor; a servomechanism; an electronic component; and a software application. The main sensor used is the HY-SR05, which is able to detect emerged obstacles in the range of 2–450 cm, with an accuracy of 0.2 cm. The HY-SR05 uses a single output pin on the controller to send a trigger pulse to the sensor, and then another input pin to receive the pulse indicating the object's distance (Figure 9a).

Using a servomechanism, the obstacle-detection system can scan a prow sector of about 160° (Figure 9b). The software controls the ultrasonic sensor and using the servomechanism moves the azimuth of the same sensor in steps of 5°.

The distance detection is a function of the air temperature, and the obstacle-detection system is equipped with a temperature sensor (LM35) that compensates for temperature variations with an accuracy of $\pm 0.5\,^{\circ}\text{C}$, making the obstacle-detection system more efficient.

According to the Laplace law, in the case of the air the speed of sound increases by 0.6 m/s for each increase of $1\,^{\circ}\text{C}$ air temperature:

$$v = 331.3 \frac{m}{s} + 0.606\ T_i \tag{4}$$

where v = sound velocity in the air; 331.5 m/s = the sound velocity at 0 °C; T_i = Air temperature value in a specified measure time.

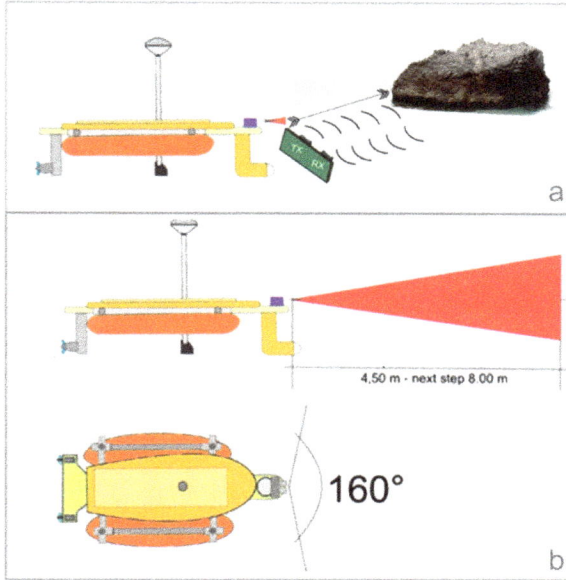

Figure 9. (a) Operation of the ultrasonic sensor; (b) Action range of obstacle-detection system.

Table 3 demonstrates the increasing accuracy of the measurements (dD column), by a comparison between the distance measured at the standard temperature of 20 °C (columns V1 and D1), and the distance measured at the current temperature (columns V2 (t) and D2), showing how the variation in the temperature influences the measured values.

Table 3. Comparison table between the distance measured at the standard temperature of 20 °C (columns V1 and D1), and the distance measured at the actual temperature (columns V2 (t) and D2).

V1 (20°) m/s	T (°C)	V2 (T) m/s	Time (s)	D1 (cm)	D2 (cm)	dD (cm)
343.4	5	334.3	0.010	171.7	167.2	4.5
343.4	10	337.4	0.010	171.7	168.7	3.0
343.4	20	343.4	0.010	171.7	171.7	0.0
343.4	30	349.5	0.010	171.7	174.7	−3.0

The obstacle-detection system, along with the camera's surface, is very useful when there are obstacles, such as scattered rocks, that are not marked on the cartography. This system enables the operator to navigate up to a few centimeters from the docks and piers, and thus is very useful in bathymetric surveys carried out in ports and harbours.

SIROS becomes active when the distance from an obstacle is <400 cm. As soon as this happens, TrackStar displays the progressive distances of the obstacle, thus alerting the operator. Normally the operator decreases the speed and, if necessary changes route. An operator controls the MicroVeGA drone remotely, and thus there are no automatic collision avoidance maneuvers. The only automatic actions of the system are:

- activate alarm visual and sound software management and control,
- activate flashing and sirens on board.

Especially in critical cases, the software automatically stops the engines and activates an alarm (go home command) to warn the operator about the need to stop the mission. Finally, SIROS 1 has a safety navigation system to support the operator in making browsing simpler, safer and fast.

2.2.5. Video Acquisition System

MicroVeGA has a complete system for video data acquisition, above and below sea level. Two GO PRO HERO 3 cameras are installed on-board, one above the water level and the other below. The cameras make a video recording during the whole survey, enabling the operator to check the environmental conditions and to manage the presence of obstacles in real-time. Video data is transmitted to the base station and is recorded on a hard disk.

For performance testing, two methods for transmitting video data from on board to the base station were used. One uses the wi-fi on board a GoPRO camera that (thanks to the Extended Range WiFi positioned on the MicroVeGA) transmits data to the shore. Here any wi-fi device (smart phone, tablet, or PC) can view content in realtime thanks to the app supplied with the GoPRO. The second method uses a 5.8 GHz 100 mW 8 channel video transmitter along with a RC805 5.8 GHz AV Receiver. A small LCD, connected to the receiver, displays real-time video. In the next version of the drone, the second solution will be used, *i.e.*, without the Go-Pro wi-fi, as this will ensure low weight, the high flow rate, and the availability of more transmission channels.

2.3. Data Acquisition and Software

The TrackStar software, developed by our research group, manages the survey activities and automatically creates a measurement geodatabase.

The software displays in real time (Figure 10): the GPS navigation; the deviation of the vessel from the planned line; the SBES bathymetric measurements along the navigation line (bathymetric profiles); the distance from a detected obstacle; and the IMU measurements (pitch, roll, yaw and altitude).

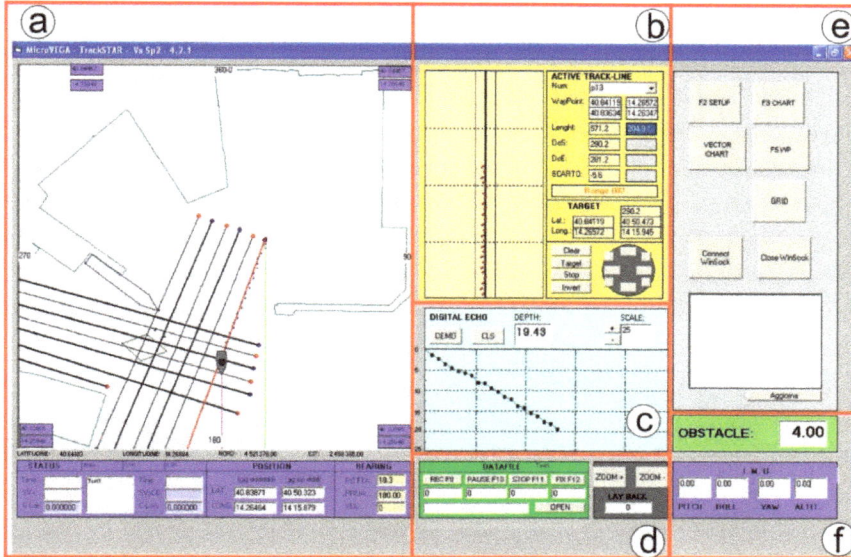

Figure 10. Trackstar desktop: (**a**) real time navigation; (**b**) deviation of the vessel from the planned line; (**c**) real time bathymetric profile; (**d**) real time data recording and datafile creation; (**e**) import of cartography and creation of navigation lines; (**f**) obstacle distance and attitude measurements.

The software also displays the data read from the IMU and, near to an emerged obstacle, shows the distance from the obstacle to the drone, thus facilitating the remote control of operations by the operator. All data from GPS, SBES and IMU are stored in a single datafile in ASCII format. The software was developed in Windows.

3. Results and Discussion

This section describes two cases of the MicroVeGA survey. The main characteristic of these areas is the coastal physiography that prevents any bathymetric surveys with traditional boats. There are also submerged archaeological remains that produce rapid changes in depth values.

The morpho-bathymetric survey carried out in each area was planned in order to obtain a GIS 3D model of the sea floor. The interpolation method used in the post-processing phase was the Inverse Distance Weighted (IDW) interpolation. This interpolator is one of the simplest and most readily available methods for

261

interpolation. It is based on an assumption that the value at an unsampled point can be approximated as a weighted average of values at points within a certain cut-off distance, or from a given number of the closest points [16].

3.1. MicroVeGA Survey 1

The first bathymetric survey of MicroVeGA drone was carried out along the Sorrento Marina Grande coast in the nearshore area (0–3 m depths) between the tufa cliff and coastal protection works using Prototype #1. In Prototype #1 of the drone, the instruments were all contained in a plexiglas non waterproof case and the hulls of the catamaran consisted of two float tubes.

The navigation of the bathymetric survey (Figure 11) had a linear development of about 500 m, with a distance between the navigation lines of about 2 m. In the first instance, the positioning and the morphologic reconstruction were obtained of all the archaeological remains in the area [17], using the GPS, SBES and submerged camera.

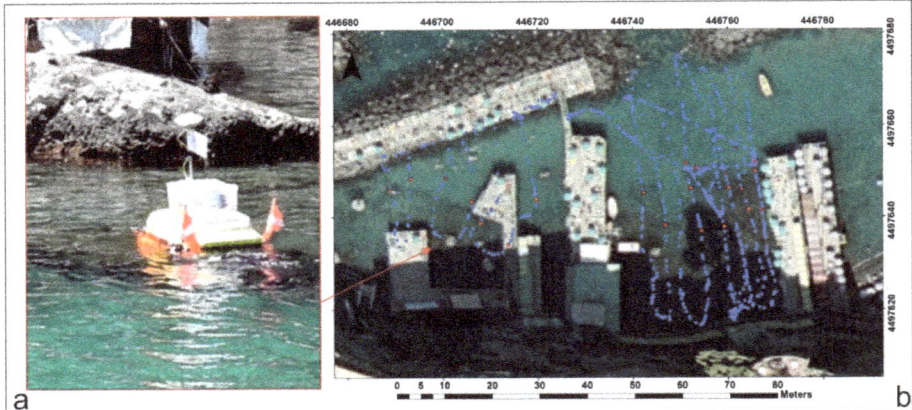

Figure 11. (**a**) MicroVeGA drone prototype #1 used during the survey; (**b**) navigation lines of bathymetric survey in blue and position of archaeological targets located by submerged camera and SBES in red.

In addition, 3D data were processed in the ARCGIS environment, using 3D Analyst. The interpolation of the bathymetric data, through the IDW interpolator, transformed the point measurements into continuous measurements. The final product is a seafloor digital model of the area (Figure 12).

3.2. MicroVeGA Survey 2

The second bathymetric survey of MicroVeGA drone was carried out along the Posillipo Hill (Naples, Italy) coast in the nearshore area (0–10 m depths) of Marechiaro harbour, using Prototype #2. The instruments on board the second prototype were completely contained in a waterproof case and the obstacle-detection system was installed on a waterproofed wooden support on the drone bow, in addition, the catamaran's hulls were made of marine plywood (Figure 13), which widened the hull and lengthened the bearing surfaces side, thus increasing the stability of the drone in navigation by decreasing the pitch and roll movements (Table 4). In addition, the largest volume of the hulls, increasing the displacement, helped to improve the available payload (Table 4).

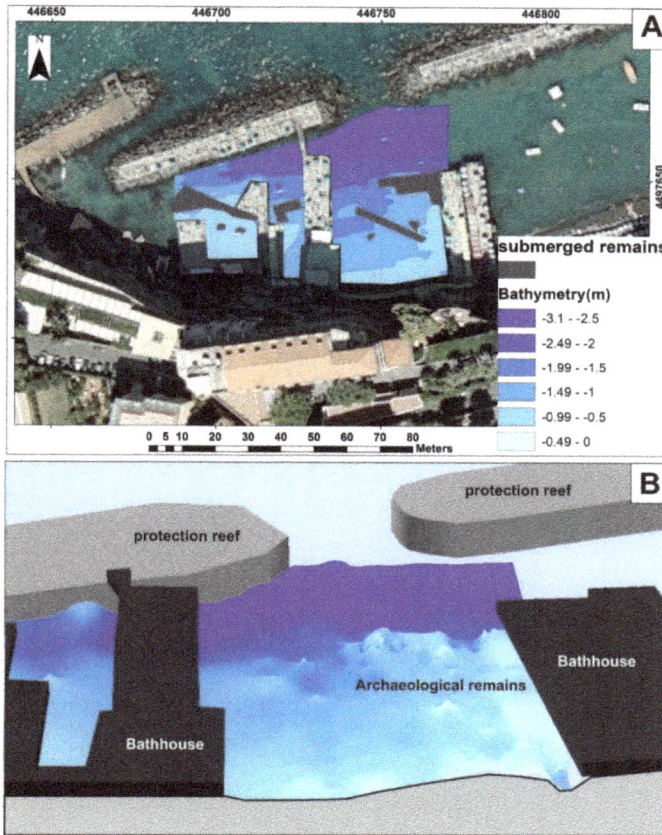

Figure 12. (a) 2D visualization of the sea floor digital model of the study area—Sorrento Marina Grande (Naples, Italy); (b) 3D visualization of the same sea floor.

Figure 13. MicroVeGA in action.

The transverse stability of the hull, in a catamaran like MicroVeGA, increases with the increase in the bearing surface on the water. In fact, while the longitudinal stability counteracts the pitching movements (the "fluctuations" of the vessel from bow to stern), the transverse stability counteracts the rolling motion (the lateral "oscillations" of the vessel). MicroVeGA can be approximated to a rectangular water plane, and the transversal (j_x) and longitudinal (j_y) moments of inertia, as shown in Figure 14, can be calculated as being equal to [18]:

$$j_x = \frac{a \cdot b^3}{12} \tag{5}$$

$$j_y = \frac{b \cdot a^3}{12} \tag{6}$$

Therefore in this version, the increase in the transverse and longitudinal stability increased the navigation safety (Table 4).

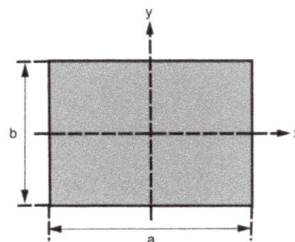

Figure 14. Schema of a rectangular vessel.

264

Table 4. Comparison between physical characteristics of Prototypes #1 and #2.

Prototype	Width (cm)	Length (cm)	J_x	J_y	Payload (kg)
MicroVeGA #1	72	92	0.029	0.047	12
MicroVeGA #2	86	120	0.064	0.124	22

The site of the second survey, was a port in the 1st century AD and several remains of a dock [19] are still present (red dashed line in Figure 15). MicroVeGA passed over these remains thanks to a few centimeters of draught.

The navigation of the bathymetric survey (Figure 14b) had a linear development of about 1500 m, with a distance between the navigation lines of about 5 m. In this survey, the tool that manages the inertial platform measurements eliminated 10% of depth measurement, due to the transition of some boats during the survey.

3D data were processed in ARCGIS, using the Geostatistical Analysis tool. The interpolation of the bathymetric data, through the IDW interpolator, transformed the point measurements into continuous measurements. The final product is a sea floor digital model of the area (Figure 16).

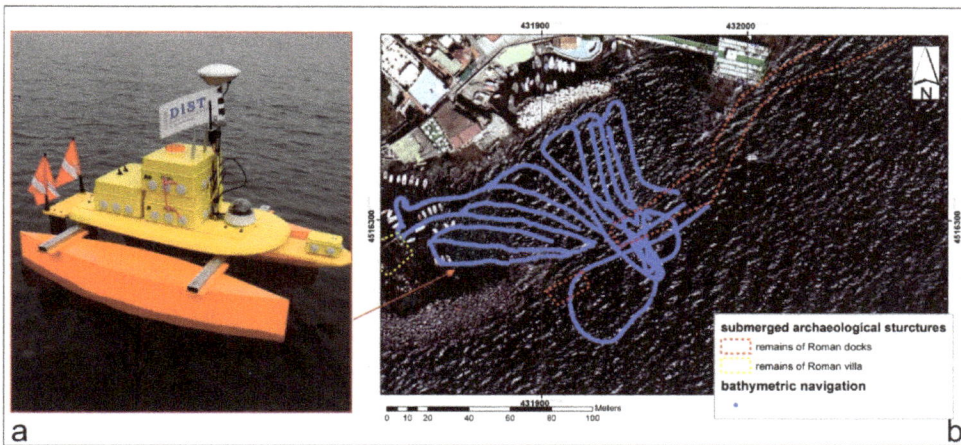

Figure 15. (a) MicroVeGA drone prototype #2 used during the survey; (b) navigation lines of bathymetric survey in blue and submerged archaeological structures in red.

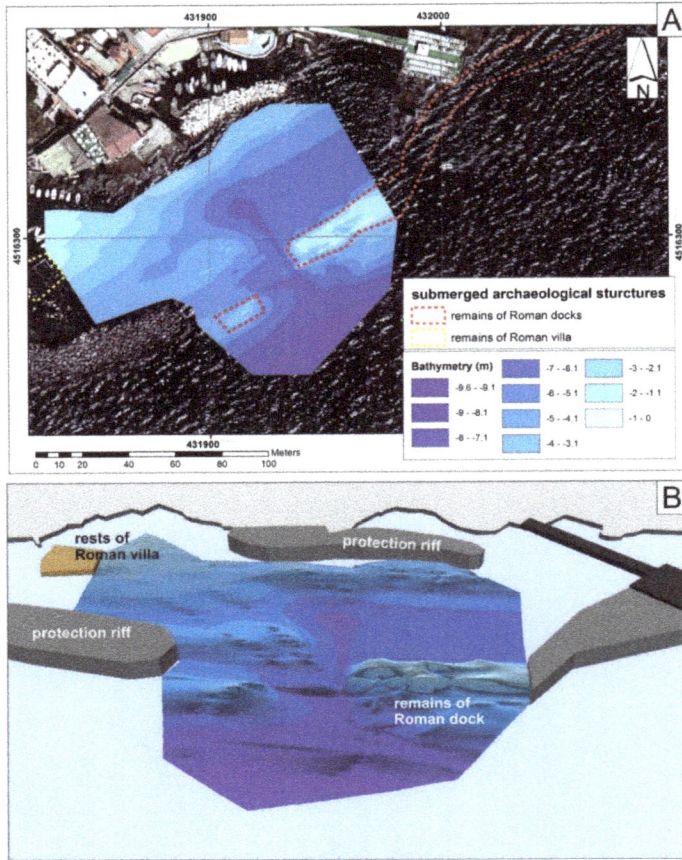

Figure 16. (a) 2D visualization of the sea floor digital model of the study area—Marechiaro harbour along Posillipo Hill (Naples, Italy); (b) 3D visualization of the same sea floor.

4. Conclusions

We have described a prototype of a marine drone optimized for very shallow water, which enables bathymetric surveys to be performed in areas that are not feasible for traditional boats. In the two study cases described in this paper, the various underwater structures would have created many navigation difficulties, if MicroVeGA had not had only a few centimeters of draught.

The experiments performed in the two coastal sites showed that integrating several existing technologies improved the final performance and the quality of the acquired data. The development of a specific software application (Trackstar) improved the accuracy of all the measured data, thus increasing the instruments' performance.

266

Trackstar improves survey accuracy using the inertial platform which extended the survey duration but always guaranteed a high quality control of measurements. In fact, during the planning phase of the survey, we established the survey precision parameter ssp as a function of survey scale, depth and survey target, reducing the attitude errors, as demonstrated in the Marechiaro survey where the effect of the sailing boats was deleted. The control of the speed and the possibility of navigating at a reduced speed also ensured a greater measurement accuracy.

In addition, the safety performance of the operation was improved by integrating the temperature sensor with the ultrasonic sensor, thus increasing the accuracy in the measurements of the distance from the obstacles, as demonstrated in Table 4.

Another important characteristic of this project is the low technology risk philosophy, guaranteed by the spiral model used to manage the drone construction phases. In fact, we had carried out a bathymetric survey in the Sorrento Marina Grande site, already using the first prototype.

Prototypes #1 and #2 provide the basic requirements of practicality and economy. Practicality is clear from the ease of performing the measurements (small footprint, highly portable, ultra lightweight and easy manoeuvrability). Low costs were achieved by assembling and integrating existing systems.

Finally, MicroVeGA is equipped not only with bathymetric sensor but also with an underwater camera which provides an overview of the investigated seabed and the surrounding underwater environment.

In the next (*i.e.*, third) phase of this project, the experience obtained in the current development phases will be used to design morpho—bathymetric surveys in critical areas. Future plans include new survey strategies and an industrial mock up in fiberglass (Figure 17).

Figure 17. An industrial mock up in fiberglass.

Acknowledgments: This paper brings together the results of works performed within the project PRIN 2010-11 financed by MIUR (Ministero dell'Istruzione, dell'Università e della Ricerca) and developed at the University of Naples "Parthenope" (Coordinator: Raffaele Santamaria). The authors would like to thank Alberto Greco, Ferdinando Sposito and PietroRossi for their active collaboration in the engineering of the drone and in all phases of the marine surveys.

Author Contributions: Francesco Giordano took part in the engineering of the drone and coordinated the marine surveys, took part in writing the paper; Gaia Mattei took part in the engineering of the drone and in the marine surveys; she carried out the planning of marine surveys and the data processing, and took part in writing the paper; Francesco Peluso engineered the marine drone and took part in the marine surveys and in writing the paper; Claudio Parente conducted the bibliographic research and supervised the planning of marine surveys and the data processing; he took part in writing the paper; Raffaele Santamaria supervised engineering of the drone, coordinated the marine surveys and supervised the data processing; took part in writing the paper.

Conflicts of Interest: The authors declare no conflict of interest.

References

1. Jawak, S.D.; Vadlamani, S.S.; Luis, A.J. A Synoptic Review on Deriving Bathymetry Information Using Remote Sensing Technologies: Models, Methods and Comparisons. *Adv. Remote Sens.* **2015**, *4*, 147–162.

2. Quadros, N.D. CRSI UDEM2 Project4 Report Stage 2, Final Draft, 2013. Available online: http://www.crcsi.com.au/assets/Resources/695d2af9-e397–4029–8cad-6dfe5510d581.pdf (accessed on 24 December 2015).

3. Guenther, G.C.; Cunningham, A.G.; LaRocque, P.E.; Reid, D.J. Meeting the Accuracy Challenge in Airborne Lidar Bathymetry. In Proceedings of the EARSeL-SIG-Workshop LIDAR, Dresden/FRG, Dresden, Germany, 16–17 June 2000; pp. 16–17.

4. Caccia, M.; Bibuli, M.; Bono, R.; Bruzzone, G.; Bruzzone, G.; Spirandelli, E. Unmanned surface vehicle for coastal and protected waters applications: The Charlie project. *Mar. Technol. Soc. J.* **2007**, *41*, 62–71.

5. Bertram, V. Unmanned Surface Vehicles—A Survey. Available online: http://www.skibstekniskselskab.dk/public/dokumenter/Skibsteknisk/Download%20materiale/2008/10%20marts%2008/USVsurvey_DTU.pdf (accessed on 24 December 2015).

6. Giordano, F.; Mattei, G.; Parente, C.; Peluso, F.; Santamaria, R. MicroVeGA (micro vessel for geodetics application): A marine drone for the acquisition of bathymetric data for GIS applications. *ISPRS Int. Arch. Photogramm. Remote Sens. Spat. Inf. Sci.* **2015**, *1*, 123–130.

7. Zhao, J.; Yan, W.; Jin, X. Brief review of autonomous surface crafts. *ICIC Expr. Lett.* **2011**, *5*, 4381–4386.

8. USGS Woods Hole Coastal and Marine Science Center (2014), IRIS—ASV (Autonomous Surface Vessel). Available online: http://woodshole.er.usgs.gov/operations/sfmapping/iris.htm (accessed on 24 December 2015).

9.   Ferreira, H.; Almeida, C.; Martins, A.; Almeida, J.; Dias, N.; Dias, A.; Silva, E. Autonomous Bathymetry for Risk Assessment with ROAZ Robotic Surface Vehicle. In Proceedings of the IEEE Oceans 2009-Europe, Bremen, Germany, 11–14 May 2009; pp. 1–6.

10.   Hassan, S.R.; Zakaria, M.; Arshad, M.R.; Aziz, Z.A. Evaluation of Propulsion System Used in URRG-Autonomous Surface Vessel (ASV). *Procedia Eng.* **2012**, *41*, 607–613.

11.   Romano, A.; Duranti, P. Autonomous Unmanned Surface Vessels for Hydrographic Measurement and Environmental Monitoring. In Proceedings of the FIG Working Week 2012, Knowing to Manage the Territory, Protect the Environment, Evaluate the Cultural Heritage, Rome, Italy, 6–10 May 2012; pp. 1–15.

12.   Boehm, B.; Hansen, W.J. *Spiral Development: Experience, Principles, and Refinements*; (No. CMU/SEI-2000-SR-008); Software Engineering Institute: Pittsburgh, PA, USA, 2000; pp. 1–35.

13.   Li, W.; Wang, J. Effective Adaptive Kalman Filter for MEMS-IMU/Magnetometers Integrated Attitude and Heading Reference Systems. *J. Navig.* **2013**, *66*, 99–113.

14.   Barshan, B.; Durrant-Whyte, H.F. Inertial navigation systems for mobile robots. *IEEE Trans. Robot. Autom.* **1995**, *11*, 328–342.

15.   Downing, G.C.; Fagerburg, T.L. *Evaluation of Vertical Motion Sensors for Potential Application to Heave Correction in Corps Hydrographic Surveys*; Army Engineer Waterways Experiment Station Vicksburg Ms Hydraulics Lab: Vicksburg, MS, USA, 1987; pp. 1–86.

16.   Mitas, L.; Mitasova, H. Spatial interpolation. Geographical information systems: Principles, techniques. *Manag. Appl.* **1999**, *1*, 481–492.

17.   Mingazzini, P. *Forma Italiae: Latium et Campania*; De Luca Ed: Surrentum, Italy, 1946.

18.   Lodigiani, P. *Capire e Progettare Le Barche. Manuale Per Progettisti Nautici—Aero e Idrodinamica Della Barca a Vela*; Hoepli Editore: Milano, Italy, 2015.

19.   Günther, R.T. *Posillipo Romana*; Electa: Napoli, Italy, 1908.

Underwater Photogrammetry and Object Modeling: A Case Study of Xlendi Wreck in Malta

Pierre Drap, Djamal Merad, Bilal Hijazi, Lamia Gaoua,
Mohamad Motasem Nawaf, Mauro Saccone, Bertrand Chemisky,
Julien Seinturier, Jean-Christophe Sourisseau, Timmy Gambin and Filipe Castro

Abstract: In this paper we present a photogrammetry-based approach for deep-sea underwater surveys conducted from a submarine and guided by knowledge-representation combined with a logical approach (ontology). Two major issues are discussed in this paper. The first concerns deep-sea surveys using photogrammetry from a submarine. Here the goal was to obtain a set of images that completely covered the selected site. Subsequently and based on these images, a low-resolution 3D model is obtained in real-time, followed by a very high-resolution model produced back in the laboratory. The second issue involves the extraction of known artefacts present on the site. This aspect of the research is based on an *a priori* representation of the knowledge involved using systematic reasoning. Two parallel processes were developed to represent the photogrammetric process used for surveying as well as for identifying archaeological artefacts visible on the sea floor. Mapping involved the use of the CIDOC-CRM system (International Committee for Documentation (CIDOC)—Conceptual Reference Model)—This is a system that has been previously utilised to in the heritage sector and is largely available to the established scientific community. The proposed theoretical representation is based on procedural attachment; moreover, a strong link is maintained between the ontological description of the modelled concepts and the Java programming language which permitted 3D structure estimation and modelling based on a set of oriented images. A very recently discovered shipwreck acted as a testing ground for this project; the Xelendi Phoenician shipwreck, found off the Maltese coast, is probably the oldest known shipwreck in the western Mediterranean. The approach presented in this paper was developed in the scope of the GROPLAN project (Généralisation du Relevé, avec Ontologies et Photogrammétrie, pour l'Archéologie Navale et Sous-marine). Financed by the French National Research Agency (ANR) for four years, this project associates two French research laboratories, an industrial partner, the University of Malta, and Texas A & M University.

Reprinted from *Sensors*. Cite as: Drap, P.; Merad, D.; Hijazi, B.; Gaoua, L.; Nawaf, M.M.; Saccone, M.; Chemisky, B.; Seinturier, J.; Sourisseau, J.-C.; Gambin, T.; Castro, F. Underwater Photogrammetry and Object Modeling: A Case Study of Xlendi Wreck in Malta. *Sensors* **2015**, *15*, 30351–30384.

1. Introduction

At the convergence of computer science and the humanities, the Ontology and Photogrammetry; Generalizing Surveys in Underwater and Nautical Archaeology (GROPLAN) project, brings together researchers from academia and industry. The main fields of activity of the GROPLAN project are in underwater and nautical archaeology.

Our central objective is to build an information system based on ontologies. In turn, such an information system will provide a formal framework as well as tools to express and manage digital content and expert knowledge in a homogenous manner. Our second objective is the development of methods for collecting data from sites and the integration of such data into the information system during the acquisition phase. In this paper these two aspects are addressed in the context of the exploration of an exceptional archaeological site—The Phoenician shipwreck off Gozo Xlendi. Resting at a depth of 110 m, it is probably the oldest ancient shipwreck in the central Mediterranean.

The photogrammetric survey was based on an original approach to underwater photogrammetry with scientific assets provided by a partner in the GROPLAN project. A photogrammetric process and the corpus of surveyed objects were combined for ontological formalization. Our approach is based on procedural attachment with the ontology perceived as a combination of the Java class structure. In turn, this manages the photogrammetric survey and the measurement of artefacts. This allows the establishment of reasoning for the ontologies as well as intensive calculations using Java with the same interface. For the sake of transversality, the ontology used to describe the archaeological artefacts from a measurement point of view can be seen as an extension of the CIDOC-CRM ontology used for museo-graphical objects [1,2].

We will also be initiating a process to facilitate the generalisation of this approach to make it available in the field of nautical archaeology. Such an aim will be achieved through several site-specific case studies. The measurement/knowledge relationship will be studied in the scope of nautical archaeology in collaboration with Texas A & M University which has already started working on formalizing ship structures. For a number of years, Texas A & M has also conducted various underwater archaeological excavations in the Mediterranean.

The resolutely interdisciplinary aspect of the GROPLAN project is reflected in the diversity of its partners as well as in the complementary nature of their activities; a computer science lab with an extensively experienced team in close-range photogrammetry, three archaeology departments and two private companies. One of the companies specialises in underwater exploration whereas the other focuses on dimensional control. The key people in this project are computer scientists, photogrammetrists, archaeologists, anthropologists, engineers, and oceanographers.

271

The vast geographic scope of the project, which includes France, Malta and Texas (USA) also highlight GROPLAN's diversity.

1.1. Context

This project deals with the process and the problems of archaeological survey from the perspective of knowledge through the use of photogrammetry. Recent developments in computer vision and photogrammetry make this latter technique a near ideal tool. It could actually be deemed as an essential tool for archaeological survey. Indeed, photogrammetry provides an easy setup remote sensing technique with low implementation costs. This technique also provides denser 3D models when compared to those obtained using laser scanners at ground-level conditions (modern laser scanners can only provide up to 100 K depth points per measuring cycle). In the context of underwater archaeology it is undeniably a must because there is no real alternative.

The main idea of this project is based on the fact that survey, which takes place in the scope of nautical and underwater archaeology relies on a complex well-established methodologies that have evolved over time. The notion of a model as a formalisation of archaeological knowledge is used to guide the survey.

The confluence between the ever-increasing quantities of measured data and knowledge that is progressively formalized (ontologies, semantic networks) raises the issue of the development and automation of survey systems that are able to make the most out of the confrontation of these two aspects.

Measurement data is decoupled from the resulting geometry. On the one hand, the process records all acquired data (measurements, graphical data, annotations, archaeological data or that pertaining to the field of study). On the other hand, after the computation and validation phases of these data the resulting graphical (2D or 3D) images represent the result of a query into the ontologies that manage all collected information.

The measurements obtained through the photogrammetry survey are sparse and these partially fit the theoretical model. A fusion scheme allows the integration between measurements and model. Once instantiated this will produce a graphical representation that is reflective of the needs as expressed by the end-user.

This project is still underway and its first phase focuses on the issues of underwater archaeology through an approach dedicated to a corpus of amphorae. Experiments in underwater archaeology took place on the Phoenician shipwreck in Malta done in collaboration with the University of Malta. The Phoenician shipwreck represents an exceptional site from various points of view. Recently discovered through systematic exploration of the seabed, this deep-water shipwreck lies at a depth of approximately 110 m. From a logistical point of view it is located in a very favourable area on a sandy plateau relatively close to the shore and free of all marine vegetation. The earliest survey available brought to light the upper part of a well-preserved cargo with minimal disturbance. It consists of amphorae, urns and large millstones that explicitly evoke the shape of the ship which measured approximately 12–14 m long and 4–5 m wide. Results from a sub-bottom-profiler survey suggest the existence of a second layer of amphorae. This was confirmed through the photogrammetrical survey performed in July 2014 as part of this project. The first sonar images also confirm this hypotheses and clearly show that the boat and its cargo landed on the seabed before sand rapidly covered the site, thus protecting it from further erosion. The shipwreck is in itself exceptional. Firstly, due to its configuration and its state of preservation which combines to make it particularly well-suited for our experimental 3D modelling project. Exploration of the first layer of amphorae reveals a mixed cargo consisting of items from both Western Phoenicia and the Tyrrhenian-area, which both match the period ranging from between the end of the VIII and the first half of the VII centuries BC. This makes it the oldest known wreck in the western Mediterranean and contemporary to the early days of Carthage and the first Greek settlements in the West. The historical importance of this wreck is highlighted by our work. It is the first time that such technologically advanced techniques have been used on this site. Our fieldwork created a real added-value both in terms of innovation and the international reputation of the project itself.

1.2. Underwater Photogrammetry and Survey

Deep-water shipwrecks have not been widely tackled by researchers, mainly due to a lack of information as well as issues related to accessibility. The lack of information arises because diving beyond 50 m using standard SCUBA equipment with compressed air is prohibited. The depth limit for divers breathing air is specified by several organisations including BSAC in the UK [3] or French law [4]. Diving beyond this limit requires further in-depth training, the use of enriched air and significant facilities on the surface. Furthermore, these deep-sea wrecks are also protected by various natural physio-chemical factors including low light, cooler temperatures and reduced oxygen. Such factors combine to help further preserve such wrecks.

However, threats to deep-water sites are increasing. One major threats stems new forms of trawling that destroy the surface layer of these sites and interfere with their readability. In fact, the protection that has always been afforded to such sites is now something of the past. Trawling nets today can be deployed to depths of up to 1000 m. Consequently, many of these shipwrecks are presently more likely to be damaged before they can be studied, or even observed.

Aside from very limited accessibility by divers, a deep water site cannot be physically reached by the majority of underwater archaeologists, marine biologists or other experts in related marine sciences. It is therefore important, even crucial, to implement better techniques, which are easily deployed and that are able to accurately survey deep-water sites. This represents one of the interests of our research.

The acquisition system used for the photogrammetric survey was installed on the Rémora 2000 submarine made by COMEX. This two-person submarine has a depth limit of 610 m with a maximum dive time of 5 h. Five hours provides more than enough time for the data acquisition phase of the photogrammetry survey. What is of crucial importance to us are the three high-resolution cameras that are synchronized and controlled by a computer. All three cameras are mounted on a bar located on the submarine just in front of the pilot. Continuous lighting of the seabed is provided by a Hydrargyrum medium-arc iodide lamp (HMI) powered by the submarine. The continuous light is more convenient for both the pilot and the archaeologist who can better observe the site from the submarine. The high frequency acquisition frame rate of the cameras ensures full coverage whereas the large scale of acquired images gives the eventual 3D models extreme precision (up to 0.005 mm/pixel for an ortho-photograph).

Deployed in this way the acquisition system entails zero contact with the archaeological site making it both non-destructive and extremely accurate. The on-board processing within the submarine permits the creation of real-time 3D structure estimation of the area covered by the vehicle. This ensures that the pilot can obtain complete coverage of the survey area before the he returns the vehicle to the surface.

Photogrammetry in an underwater context makes it possible to obtain a comprehensive survey of all visible parts of the site without impacting the objects. Moreover, such a task can be accomplished in relatively short time and with a high degree of precision. This approach offers specialists and members of the general public a complete view of a site that is normally not possible due to the turbidity of the marine environment and a lack of light [5]. This aspect of the survey is described in the second section of this paper when discussing the example of the Phoenician shipwreck off Malta.

Our initial focus shall be on an aspect that is often neglected in discussions related to survey techniques: The relationship with knowledge. As archaeological excavations often result in irreversible damage, it is important to ensure that they are accompanied by relevant documentation. Such documentation must take into account the accumulated knowledge gained from the site. This documentation is generally iconographic and textual. Graphic representations of archaeological sites such as drawings, sketches, photographs, topographic renditions, artist impressions and photogrammetric studies are all essential phases or archaeological surveys. However, as highlighted by Olivier Buchsenschutz in his introduction to the conference entitled "Images and archaeological surveys—From proof to demonstration", held in Arles in 2007 [6]: "Even when very accurate, drawings only retain certain observations to support a demonstration, just as a speech only retains certain arguments, but this selection is not generally explicit". In a certain way, this sets the foundation for the further development of this work: Surveys are both a metric representation of the site as well as an interpretation of the same site by the archaeologist.

Surveys are very important components of the documentation and their importance is mostly due to the fact that concepts handled by archaeologists during an excavation are strongly related to space. The very structure of an excavation is based around the notion of a stratigraphic unit. Inherited from a geological approach and subsequently formalised for archaeology by E.-C. Harris [7]. Stratigraphic units are connected to each other through geometrical, topological and temporal relationships and give a structure to the reading of the excavation.

Two families of objects must be distinguished: parts of terrain, or more generally, areas of space that are organised into stratigraphic units and the artefacts that must be positioned in that space. Such an exercise is essential for post-excavation studies that are conducted after the site has been "destroyed". In this paper, we will principally cover the second of the aforementioned objects: artefacts. The survey is therefore based on a priori knowledge, formalized in close collaboration with archaeologists. This knowledge was used during the measurement phase and communicated right through the final representations of the site. Based principally on our knowledge of measured artefacts, this approach used this knowledge to measure the size and localise the object.

Finally, it is imperative to note that archaeological data is, by its very nature, incomplete; mainly because it is heterogeneous and discontinuous as well as being subject to possible updates and revisions (both theoretical and technological). The verification of the final coherence of the survey will be one of the primary objectives of this work.

1.3. Archaeological Experimentation

The shipwreck was first located by Aurora Trust in 2008 during a systematic survey off the coasts of Malta and Gozo that it was conducting with state of the art digital side scan sonar. This ongoing broad survey is authorised by the Superintendence of Cultural Heritage with the aim of creating an inventory of all the underwater ruins located in Malta's territorial waters at depths ranging from 50 to 150 m. Further studies of one sonar target by a remote operated vehicle helped identify the site as a very ancient shipwreck. A collaborative group was gradually created by Timmy Gambin in order to put together a strategy adapted for an in-depth the study of this shipwreck.

The shipwreck is located near a stretch of coastline that is characterised by limestone cliffs that plunge into the sea and whose foundation rests on a continental shelf at an average depth of 100 m below sea level. The shipwreck rests on a practically flat area of this submerged plateau at a depth of 110 m. Sonar images as well as the image of the entire archaeological deposit shows that the cargo remains tightly grouped together and appears to have been only slightly disturbed. An "amphora mound" effect could not be detected indicating the good state of preservation of the transported goods. It is generally accepted that amphora-mounds arise when the hull disintegrates in open water (after sinking) due to the absence of sedimentation processes. This type of phenomenon results in a haphazard spilling and spreading out of cargo. With the site currently under discussion the ship's longitudinal axis and cargo remain visible, spread out over 12 m and easily identifiable. The maximum width of the non-disturbed part is approximately 5 m (see orthophotograph in Figure 1).

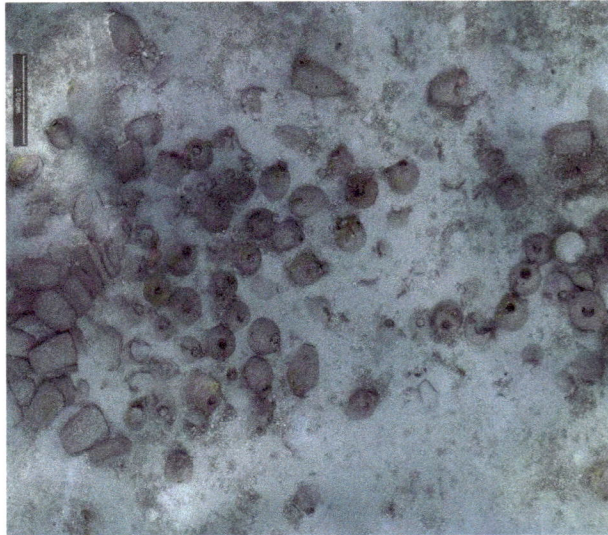

Figure 1. A snapshot taken from a very high resolution orthophoto, the full resolution image can be found in the project's website (http://www.groplan.eu). The overall image resolution is 41,507 × 60,377 pixels at a scale of 0.005 mm per pixel.

Some matters arising from this project may, at first glance, seem distant from the primary interest usually afforded by archaeologists to underwater shipwrecks. However, initial discussions on potential survey strategies brought to light questions involving new requirements. These requirements were progressively established through cross-disciplinary dialogue between collaborators. For the archaeologist, the production of reliable documents from geo-referenced ortho-photographic surveys such as those described here (see Figures 1 and 2 below) provides significant added-value. However, it became apparent that what may seem to be an accomplishment enabling the interpretation of the wreck's visible surface layer proved in practice to be insufficient.

Given that the site being explored is as fragile as it is exceptional, as are other deep water sites, it was imperative that past mistakes, such as haphazard approaches to survey work, be avoided. Preliminary reflections identified two priority actions: (1) the detection, localisation and mapping of deposits; and (2) their characterisation in terms of nature, contents, organisation and time-frame. Thus, the cultural and scientific tasks were closely intertwined. However, a degree of experimentation with available tools was still necessary. This is because such an approach would be needed to quickly provide the desired information despite the constraints that arise from working in a deep-water environment. This is reflective of GROPLAN project's spirit, which seeks developments in the fields of photogrammetry and shape analysis through ontology. In turn, such developments are immediately deployed in the field of archaeological research, management and protection of submerged cultural assets.

Figure 2. Very high resolution orthophoto from the project's website (http://www.groplan.eu). A zoomed-in of the image shown in Figure 1.

Contrary to other archaeological deposits located at depths of less than 50 m, which are often eroded and greatly pillaged, deep-sea wrecks are generally better preserved and offer the chance for a more contextual and spatial approach. As such, the Phoenician wreck offers an interesting and promising site for scientific application and experimentation. This is due to its aforementioned excellent preservation. Its excellent state of preservation can be further deduced through the observation of precisely grouped divisions of located objects as well as their variety. Preliminary observations enabled a precursory characterisation of the objects that were identified using available traditional typological tools. Based on these "traditional" observations, standard in archaeological practice, the groups of objects were identified and tentatively dated. The main idea here is to go (this already established) to an automatic shape recognition experiment. The latter starting from a photogrammetric survey based on the ontological analysis of amphora shapes. The interest of conducting such an experiment is to build a tool that can automatically recognise objects, even if only partially visible, by verifying the relevance of its responses all along the gradual construction of the descriptive arguments as well as by comparing it to an archaeological analysis. The aim is to build a descriptive tree structure that is sufficiently accurate so that a selected shape can be automatically recognised from a repository of known shapes and clearly identified beforehand.

Due to its "mixed" cargo (see Figure 3), which consists of different types of objects, the Phoenician wreck is particularly well suited for this type of approach. In the medium term, once the library of shapes is sufficiently populated, such an approach should permit automatic explorations at greater depths, offering a preliminary reasoned identification of objects from the cargo of deep-sea wrecks, both rapidly and with few resources.

(a)   (b)

Figure 3. Examples of the wreck images taken by the ROV Super-Achille deployed during the photogrammetric survey carried out from the submarine Rémora 2000. The mixed" cargo is visible on both (**a**) and (**b**) images.

2. Mission Field, Underwater Survey

2.1. The Survey Approach

The photogrammetric system was originally designed to be mounted on a lightweight remotely controlled vehicle, implying an optimised size and weight for the assembly—The system is called ROV-3D [3].

A previous version, presented here, was mounted on the submarine Remora 2000. In the ROV 3D configuration, the operator has direct access to the on-board acquisition unit (UAE) via a high-speed Ethernet link between the inside of the inhabited vehicle and the system on the surface. In this version, a connection was established between the embedded sub system, which was fixed on the submarine and the pilot (inside the submarine).

To facilitate the portability, all the components of the photogrammetric system were arranged in a removable assembly that could be attached to the submarine technical bar or under the remotely controlled vehicle (ROV).

Our setup implements a synchronized acquisition of high and low resolution images by video cameras forming a trifocal system. The three cameras are independently mounted in separate waterproof housings. This implies two separate calibration phases: The first one is carried out on each camera housing in order to compute intrinsic parameters and the second one is done to determine the relative position of the three cameras which are securely mounted on the rigid platform. The second calibration can easily be done before each mission. This calibration phase affects the final 3D model scale. This trifocal system is composed of one high-resolution, full-frame camera synchronized at 2 Hz and two low-resolution cameras synchronized at 10 Hz (see Table 1).

Table 1. Intrinsic parameters for the trifocal system.

	Cam 1, Low Resolution	Cam 2, Low Resolution	Cam 3, High Resolution
Manufacturer		Allied Vision Technologies	
Model	AVT PROSILICA GT1920		AVT PROSILICA GT6600
Focal length (mm)	5.7578		28.72
Sensor size (mm)	6.61×8.789		24×36
Image resolution (px)	1456×1936		4384×6576

The lighting, which is a crucial part in photogrammetry, must meet two criteria: the homogeneity of exposure for each image and consistency between images. This is why we use the HMI light system mentioned earlier.

The trifocal system has two different aims: the first one is the real-time computation of system pose and the 3D reconstruction of the zone of seabed visible from the cameras. The operator can pilot the submarine using a dedicated application that displays the position of the vehicle in real-time. A remote video connection also enables the operator to see the images captured by the cameras in real-time. Using available data, the operator can assist the pilot to ensure the complete coverage of the zone to be surveyed. The pose is estimated based on the movement of the vehicle between two consecutive frames. To do this, homologous points are found on two successive pairs of images in low resolution (four images) and sent to the surface computer. The computation is described in detail later in this paper. The second goal is to perform an offline 3D reconstruction of a high resolution metric model. This process involves the use of the high-resolution images for the production of a dense model (see Figure 4) that is scaled based on baseline distances.

Figure 4. A snapshot of the 3D survey at low resolution for web display through *skechfab*, available on the GROPLAN website (http://www.groplan.eu).

To achieve these goals, the system architecture (see Figure 5) must be able to record a large quantity of data in a synchronous way while performing real-time computations. Within given hardware constraints we have developed a modular architecture on which the different tasks are distributed. The first module is dedicated to image acquisition (two low-resolution cameras and one high-resolution camera), which is an electronic synchronization mechanism which ensures that all shots are correctly lit (see Figure 6). This synchronization mechanism is also used to tag images with homogeneous timestamps and to make it possible to retrieve image pairs (for real-time processing) and image triples (for metric high resolution processing). The three cameras are linked by an Ethernet/IP link to an on-board computer that store all produced images (see Figure 6b). Cameras, synchronization and storage can be configured and controlled by remotely sending UDP commands to the onboard computer.

Visual odometry computation is divided in two modules. The first module takes place on the on-board computer, it is responsible of extracting and matching feature points from the low resolution image pairs in real-time before they are stored. The extracted 2D homologous points are then sent through the Ethernet—TCP/IP link to the on-board computer were a second module is dedicated to the visual odometry calculation and visualization. The system presented here is patented by COMEX and the French National Centre for Scientific Research CNRS [8].

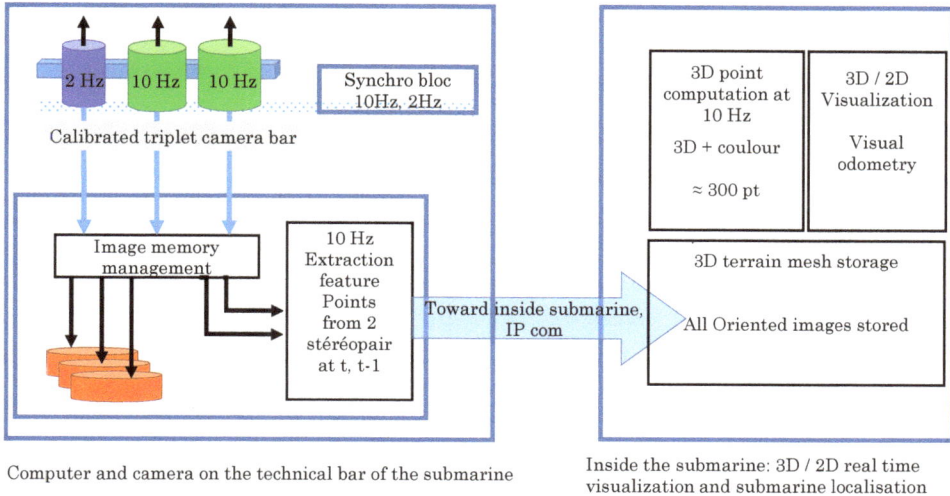

Computer and camera on the technical bar of the submarine

Inside the submarine: 3D / 2D real time visualization and submarine localisation

Figure 5. Synoptic description of the photogrammetric approach.

(a)

(b)

Figure 6. Images of the configuration used installed on the submarine; the technical bar of the Rémora 2000, the three cameras in their vertical cylindrical housing; (**a**) front view; (**b**) back view showing the embedded computer.

2.2. Photogrammetry

To ensure the complete coverage of the study area by the submarine, knowing its position in real-time is crucial. Since rigid transformation links the vehicle and the photogrammetry system, tracking the latter is sufficient for being able to deduce that of the submarine.

The motion of the photogrammetry system, consisting of three high-resolution cameras whose internal and external orientations are theoretically known, must therefore be evaluated. Its assembly should be calibrated before starting the mission. During the mission, the ego-motion of the system is computed on the fly via the embedded computer.

In the literature, the bundle adjustment methods used to refine the pose estimation have proved their effectiveness [9–11] and more generally the multiple views approach [12,13]. Nonetheless, a good approximation of initial values passed as input to the bundle adjustment method is required in order to speed up the convergence.

2.3. Orientation of Images

Let's consider that the camera assembly to be oriented, where M_j is the set of projection matrices, observes a set of points X_i in space and that x_{ij} is the 2D projected point of the ith 3D point on the jth image frame. Therefore, the orientation of the images depends on finding values M_j and X_i that solve the following equation:

$$M_j X_i - x_{ij} = 0 \qquad (1)$$

here the matrix M_j embeds both the rotation and the translation information. The bundle adjustment method proceeds to refine M_j based on forming the Equation (1) as a minimization problem:

$$\min_{M_j X_i} \sum_{i,j} d\left(M_j X_i, x_{ij}\right) \qquad (2)$$

where $d(x,y)$ is the Euclidean distance between two image points x and y. The minimization Equation (2) can be solved iteratively using the following convergence form:

$$J^T J \delta = J^T \epsilon \qquad (3)$$

where J is the Jacobian of the reprojection function (for more information, refer to [9,14]).

Using least squares approach is very costly in terms of computation resources in the given context. Moreover, its cost increases considerably as the number of parameters, *i.e.*, extrinsic parameters of the cameras and the 3D positions of the points, grows, so in order to reduce the calculation time, we have to reduce the number of parameters.

In our application, the photogrammetry system used is a stereo system whose relative pose is fixed and previously determined through a calibration phase. This characteristic allows to reduce the number of parameters linked to the cameras by a factor of 2 (see Xue and Su [15]). In fact, for a stereo pair, the extrinsic parameters of the right camera can be determined using those of the left camera. By taking this fact into account, Xue and Su proposed a method for bundle adjustment that reduces the number of parameters while keeping the information from the observations of the left and right images of the stereo pairs. This reduces the bundle adjustment computation time. The obtained results prove to be sufficient for our application. An illustration of the orientation of several images is shown in Figure 7.

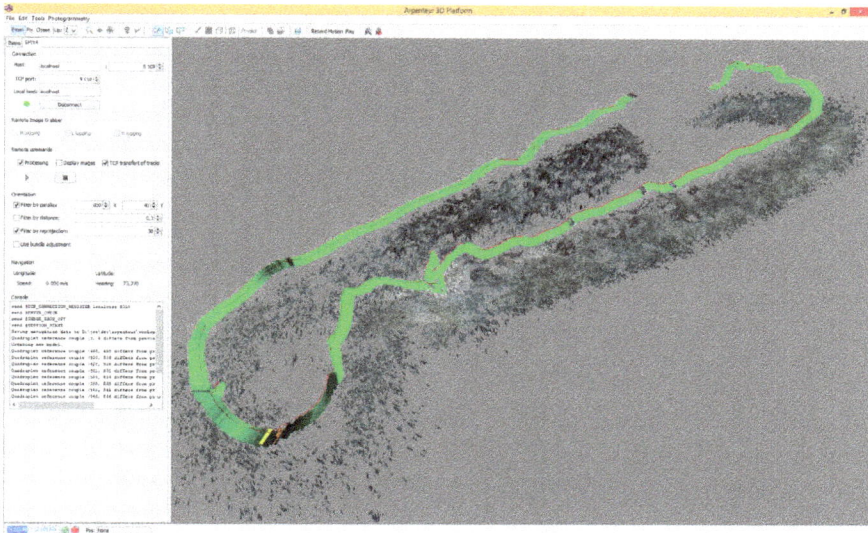

Figure 7. Real-time visual odometry as seen from the on-board computer. This figure shows the 3D point cloud calculated over the surveyed zone and the position of the vehicle for each image pair. The density of the points measured on each image is colour-coded (yellow, green and black) which gives an overview of the quality of the orientation (yellow < 30 points, green between 100 and 200 points, dark green > 300 points).

2.4. Visual Odometry: Rapid Calculation of the Orientation of Images

In the context of relative motion estimation in robotics, Sünderhauf and his team [16] proposed a method that simplifies and accelerates the bundle adjustment. This method consists of working with a window containing a subset of consecutive images instead of all the images. The images in this window are oriented, and the estimated parameters are used for the initialisation of the next orientation, whose window is shifted one image further. Moreover, to compensate for the lack of precision due to small baseline distances of the images taken within the same window, only images whose positions are sufficiently separated (*i.e.*, using a preset threshold distance), are retained. This corresponds to the minimal distance of displacement of the vehicle between two images.

By combining the work of Sünderhauf and that of Xue and Su, we implemented a new algorithm in order to find the orientation of the stereo images taken by the submarine:

- Starting with image I_i, we look for the closest image I_{i+k} so that the displacement between I_i and I_{i+k} exceeds a certain threshold. This threshold is experimentally around one centimeter for a camera shooting at a frequency of 10 Hz. This is determined using odometry.
- Add image I_{i+k} to the window.
- i becomes i + k and we repeat Step 1 until we have three stereo image pairs in the window.
- Apply the bundle adjustment method proposed by Xue and Su using the frames selected in the first step for the initialisation.
- Then, shift the window over one image and repeat the procedure until all images are processed.

2.5. Calculation of the Approximate Orientation of Images for Bundle Adjustment

In order to calculate the bundle adjustment as described in the previous paragraph, it is necessary to calculate the approximate position of the cameras. We have at our disposal a series of 2D points, homologous over four images (two consecutive pairs). The points correspondences have been established on the stereo camera pair at time t and the same at time $t - 1$.

At each step, a set of 2D points, which corresponds to pairs t and $t - 1$ are sent to the on-board computer in order to calculate the relative position of the vehicle. The procedure of calculating the orientation of an image pair at time t with respect to an image pair at time $t - 1$ is as follows:

- The vehicle moves slowly and we consider that the distance travelled between time t and $t - 1$ is slight as well as the change in orientation. Under these conditions, we consider the camera exposures at time t are the same as camera exposures at time $t - 1$. Which is a good approximation due to the slow motion and the high image acquisition rate. Formally:

$$(\mathrm{RT})_{\mathrm{right}(t)} = (\mathrm{RT})_{\mathrm{right}(t-1)} \tag{4}$$

$$(RT)_{\text{left}(t)} = (RT)_{\text{left}(t-1)} \tag{5}$$

- Knowing the relative orientations of the left and right cameras and knowing that these values remain fixed in time, we can obtain the 3D points from the 2D points through triangulation, one time by using the image pair t and another by using the pair $t - 1$. We thus obtain two homologous point clouds calculated at time t and at time $t - 1$ but with the camera exposures for times t and $t - 1$.

If the vehicle was effectively motionless, the two point clouds would be mixed together. In fact, the vehicle's motion causes a displacement of the images which leads to a displacement of the point cloud that corresponds to the image pair $t - 1$ with respect to such that corresponds to the image pair t. The rigid transformation [RT] required for expressing the cameras t in the reference pair $t - 1$ is the rigid transformation required to move the 3D point cloud at time $t - 1$ to the one obtained at time t. Hence, the problem of calculating the orientation of the cameras at time t in relation to time $t - 1$ leads back to the calculation of the transformation used to move from one point cloud to the other. This is possible under our configuration, with small rotation.

Below, we present the method to compute the transformation for passing from the point cloud calculated at time t, denoted P, to the one calculated at time $t - 1$, denoted P'. So we have two sets of n homologous points P = {Pi} and P' = {P'i} where $1 \leqslant i \leqslant n$. n is the size of the point cloud.

We have:

$$P'_i = R \times P_i + T \tag{6}$$

where R is the rotation matrix and T is the translation.

The best transformation minimises the error *err*, the sum of the squares of the residuals:

$$\text{err} = \sum_{i=1}^{n} \| RP_i + T - P'_i \|^2 \tag{7}$$

To solve this problem, we use the singular value decomposition (SVD) of the covariance matrix C, which shows to be robust and have low computation time. We note COMp and COMp' the Centre of Mass of the set of 3D points P and P':

$$C = \sum_{i=1}^{n} (P_i - COM_P) \times (P'_i - COM_{P'})^T$$
$$[U, S, V] = SVD\,(C) \tag{8}$$
$$R = VU^T$$

$$T = -R \times COM_P + COM_{P'} \tag{9}$$

Once the image pair tare expressed in the reference system of the image pair t − 1, the 3D points can be recalculated using the four observations that we have for each point.

A set of verifications are then performed to minimize the pairing errors (verification of the epipolar line, the consistency of the y-parallax, and re-projection residues).

Once validated, the approximated camera exposures at time t are used as input values for the bundle adjustment as described in the previous subsection. An example of reconstructed 3D model based on this method is shown in Figure 8.

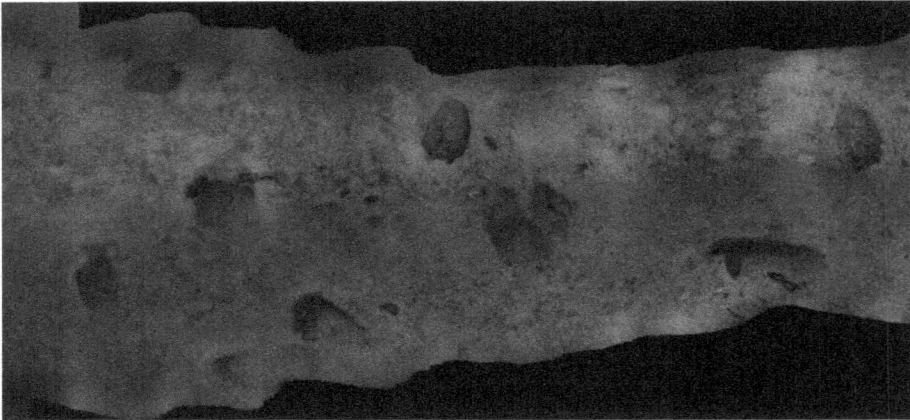

Figure 8. Partial orthophograph produced by the real-time visual odometry orientation. Some small imperfection are visible due to the orientation accuracy, however, the 3D model is visually appealing and acceptable to validate the survey.

2.6. Dense 3D Reconstruction

Following the orientation stage, in our case the orientation of two consecutive stereo pairs, the 3D point cloud that we obtained have low density. In the processing chain implemented here, the orientation of these two pairs is performed in a real-time loop at 10 Hz. It may be useful for operation managers to have a partial 3D model, for localization purposes, while the vehicle performs its survey. To do this, we developed a point cloud densification model based on the image sequence, which is possible as soon as the images are transferred on board (in the future, this densification could be performed using the embedded computer, but the current lack of resources of this machine makes it difficult. In fact, the embedded computer is subject to size, power consumption and temperature restrictions which affect its performance).

This densification is however necessary in order to reconstruct a realistic 3D model. This is done using Multi View Stereo (MVS) methods that produce a dense point cloud using images and the camera parameters. Furukawa [17] proposed a method based on a "patch" based reconstruction (Patch-Based Multi View Stereo or PMVS). They made a model of the surface S, with a random 3D point p from somewhere in the scene, and modelled, using a square section of a plane tangent to S at p, the patch. The 3D position of the patch is determined by minimizing the variance between these projections on the images. The algorithm functions in three steps:

- Initialization of a set of patches by interest points.
- Expansion, which consists of reconstructing new patches around those already identified.
- Filtering to strengthen consistency and remove any erroneous patches.

We have integrated this method in our processing chain (see Figure 9). On the other hand, contrary to PMVS, our developments directly use the images produced by the cameras, without any distortion correction nor rectification, with an adapted algorithm, dedicated to the calculation of epipolar lines.

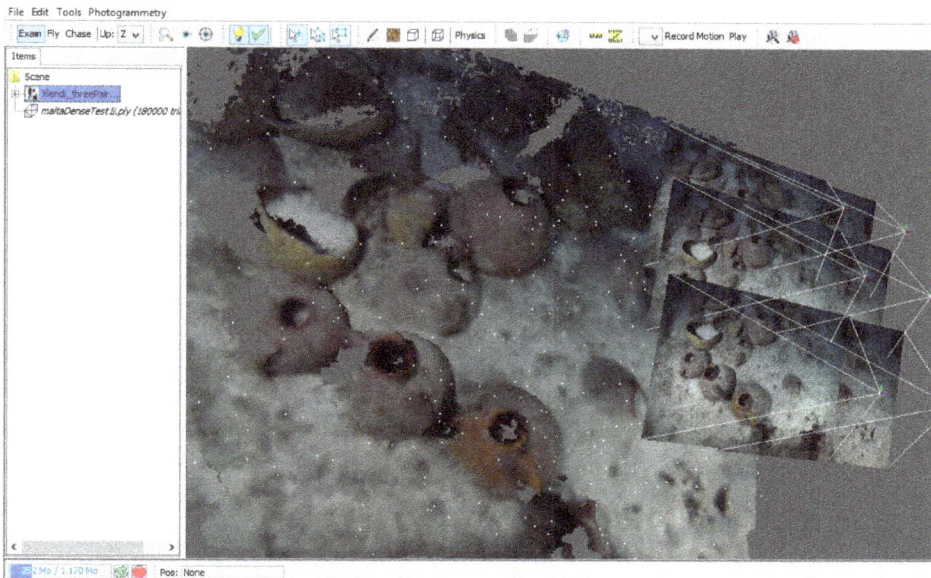

Figure 9. Three consecutive stereo pairs, oriented using the bundle adjustment approach described above, with a local densification of the original images. The visualisation is done inside the 3D tool developed by LSIS lab as a part of the Arpenteur project.

287

In fact, we applied a distortion transformation approach which made it possible to control the final adjustments and the precision of the resulting model. The distortion transformation proceeds by approximating the distorted epipolar line as a curve to simplify the calculation and this curve is modelled by an arc of circle based on the fact that the radial distortion is much greater. It mainly disturbs the projection of the scene on the image (see Figure 10). The following algorithm is used to calculate the equation of the epipolar curve for a point M:

(a)   The M coordinates are corrected for distortion and eccentricity. Let M' denotes the corrected coordinates.

(b)   Using the new coordinates, the equation for the ideal epipolar line l is determined as $l = FM'$ where F where the Fundamental matrix is described in [13]. It is computed using the fixed calibration parameters.

(c)   The two intersection points of l with the window of the image are calculated.

(d)   The centre of mass of these two points is calculated.

(e)   The distortion and eccentricity of Camera 2 are applied to the three points (the two intersection points and their centre of mass).

(f)   The three new points thus obtained are the points that define the arc of circle.

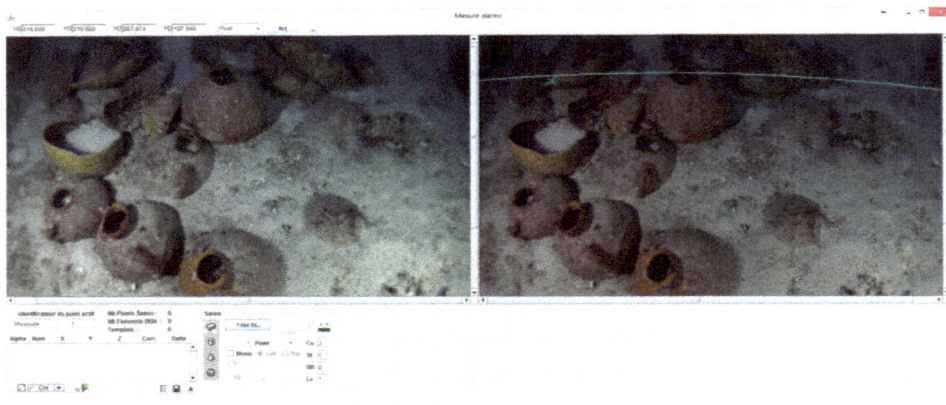

Figure 10. Photogrammetric stereo images taken during odometric navigation while surveying the Xlendi wreck. In the image on the right, one may observe a representation of the epipolar curve clearly showing the strong distortion present in images taken using underwater cameras. Modelling the epipolar line as an arc of circle is essential in calculating the visual odometry in real time as well as in the point cloud densification step (Photograph: GROPLAN consortium, LSIS, COMEX).

2.7. Precision and Control

The assumption of having small baseline distances when computing the orientation in real-time, cannot give excellent results for the entire set of images. In fact, the main problem is limiting common points to only four or six images, *i.e.*, two or three consecutive stereo pairs. Although the visual odometry method presented here is sufficient to ensure the complete coverage of the site in real time, it does not provide enough precision for the final model, especially for our goal which is the automatic recognition of each artefact, and to evaluate its variance with the theoretical model.

We therefore implemented a second step where the homologous points are extracted and matched for all images as well as a global bundle adjustment was performed to ensure the best possible orientation. This was done whilst taking into account the constraints related to the set of three fixed and calibrated cameras.

Two software programs were then used to interface with our system: Agisoft's PhotoScan and Bingo. The use of both programs permitted the control the final adjustments and the precision of the resulting model.

When comparing the results of the two models, the one obtained using odometry and traditional bundle adjustment which take into account all the possible observations of the 3D points, reveals the presence of residues of approximately 5 mm on the (X,Y) plane (which is lateral to the motion), and within 1 cm range depth-wise (cameras pointing direction) on sequences with more than 1000 images. In fact, these data vary in function of the quality of the surveyed terrain. When the seafloor is sandy and low textured, the matched points are less and of lower quality. In the areas where the amphorae were found, the number of detected feature points is more, their quality is better, and the residues between models is less pronounced.

The overall 3D model, obtained using the global bundle adjustment applied on all the high-resolution images is scaled by introducing a stereo base of 0.297 m (value obtained after the triplet calibration, done before the mission in shallow water) as a constraint in the bundle adjustment. At the end, more than 1000 stereo-pairs poses were refined by this constraint so that the residues become less than one millimetre.

3. Ontologies, 3D Pattern Recognition and Modeling Artefact

One of the primary objectives of the GROPLAN project is to provide archaeologists with a set of measurement tools that do not require the presence of a specialist to be used. The goal is to obtain a 3D model of a site with archaeological information already integrated.

The development of such tools depends on the collaboration between experts from various fields of research along with measurement specialists. The transfer of knowledge between all the players involved requires the development of an appropriate knowledge representation.

We opted for a representation based on the notion of an archaeological entity, a notion already used in the Arpenteur project [18,19]. The basic structure is therefore an object-type structure, based on a concept taxonomy describing the archaeological knowledge involved as well as the photogrammetrical knowledge used for the survey. A double formalism is used for the implementation, Java programming language for computing and generating the survey from the images and OWL2 for its implementation by ontology, which is used to manage the coherence of the results. This double implementation allows for an effective procedural attachment and ensures the full use of the two aspects of this double implementation, logical and computational.

3.1. Multidisciplinary Knowledge

The development of measurement tools designed for use by non-specialists revolves around two axes:

- Understanding the needs of the experts in the field.
- Developing measurement methods that meet these needs.

An answer to the problem raised by these two axes is the development of concepts to represent objects that group together the expectations of the experts and that can be measured. These objects can be physical objects or sets of data required by experts for their work and which can be determined during the measurement process.

Starting from this informal description, we can reduce our notion of the field of knowledge to the knowledge-base linked to the objects. For example, in the scope of underwater archaeology, the field of knowledge consists of models of archaeological objects (amphorae, ships' stores and wreckage) that include, among other aspects, metrological values, coherence relationships, dating information, as well as subsequent bibliographical information.

Once the field of knowledge notion is adapted to our needs, we can adjust to the understanding of the expert. We consider an expert in a given field as someone who acts as an interface between a field and others outside of the field. A field can have of course several experts and photogrammetrists themselves are, in the context of such collaboration, to be considered as experts.

The creation of a measurement system based on knowledge requires the collaboration of at least one expert in a field with one expert in measurements. This collaboration is only possible if knowledge representation is formalised and represents the required knowledge coming from the various fields defined.

3.2. Representing Objects in a Given Field

The definition of objects in a given field is based on their formal description. Experts of a given field have a comprehensive knowledge of the precise descriptions of these objects based on heterogeneous information. In fact, this information can be in the form of data sheets, schematics or drawings, spatial data, textual information or literature, bibliographies, geographical information or even classifications. Figure 11 shows one type of data related to the description of amphora: their profile. In the specific case of amphora, the following information is available:

- metrological data (height, maximum diameter, volume, . . .);
- spatial data (position, convex envelope, 3D representation);
- physio-chemical data (type of pottery, colour, container analysis);
- archaeological documentation (chronologies, bibliographies, studies).

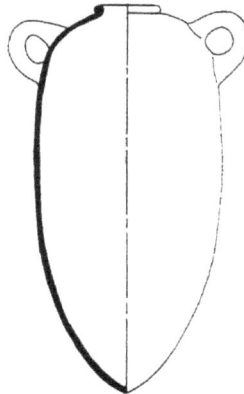

Figure 11. Standardised view of an amphora of type Ramon 2.1.1.1-72. [20].

The media holding this information varies from data sheets filled in by archaeologists, to digital 3D models, as well as electronic field databases. Due to the heterogeneity of the available information and media, a formalism adapted to our problem is a conceptual formalism. Indeed, the definition of concepts is left up to relevant experts and different concepts are assembled into global concepts from a common knowledge viewpoint.

Formalized concepts are developed based on the knowledge from the field of research. The used descriptions as a basis for conceptualisation are expressed by publications, interviews with experts of the given fields, and pre-existing formalisms. In the scope of underwater archaeology for example, conceptualisation is based on the work of archaeologists such as Dressel or Ramon for certain amphorae discovered in the Phoenician wreck and more specifically the work concerning the identification of typologies is done by archaeologists more involved in the specific study of the wreck. Here, the Phoenician shipwreck is studied by Gambin and Sourisseau.

The first step to represent objects in a given field begins with a conceptualisation. Staring with heterogeneous descriptions, experts express an archaeological concept as well as set of relationships linking it to various other concepts that describe, for example, materials or a shape. This set of relationships and concepts can be used to understand the concept of an amphora. This can be then used during the measurement process as well as during the specific study carried out by the experts.

3.3. Notion of a Measurable Item

The conceptual representation of objects from different fields allows us to give expression to objects, or at least a portion of our knowledge of the object, from one field to another. However, it is impossible to develop measurement techniques and adapted tools if each set of concepts from a particular field is not intelligible to experts of another field. In the scope of this work, beyond our study of a cargo of amphorae the implicated fields of study are varied and often independent or only possess slight common knowledge. For example, the study of the ship's structure and its cargo can be carried out by different specialists each having their own specific knowledge. The construction of a representation that can be used by experts from different fields then depends on the establishment of a minimal body of knowledge that is coherent and shared by all experts and which also guarantees the transversality of the concepts employed by using a point of view shared by all. As the context of our work is based on photogrammetrical measurements, all the objects studied are necessarily measurable by photogrammetry. The concepts that characterize the items from all the fields involved (archaeology, architecture, biology) can then be expressed from a measurement point of view and share the notion of a measurable item. We define a "Measurable Item" as an item on which it is possible to carry out measurements. All the experts from any field can then extend the notion of a Measurable Item so that it can be further specified and integrate knowledge from a particular field. In addition to the in-site measurements, the surveyed items are intended to be studied and preserved in a museum. This aspect constraints the use of a dedicated ontology such as CIDOC-CRM. A strong link between the various ontologies is therefore necessary.

3.4. Taxonomy of Measurable Items

As mentioned earlier, all the concepts that characterize the items to be measured are sub-concepts of the measurable item. We can then organise the concepts from a measurement point of view by defining a set of relationships linked to their morphology and based on the information obtained during the measurement process. For example, an amphora can be characterized dimensionally by its maximum diameter, height and the internal and external diameters of its neck, the presence of a "shoulder", as well as the ratio of its maximum diameter and its height. Some of these relationships are specific to amphorae and are used to define their typology, whereas others are common to certain concepts which subsume them either directly or in a more distant manner.

The presence of common relationships allows the concepts to be organised according to a heritage relationship. Each concept B possessing all the morphological relationships of concept A and having additional morphological relationships is a sub-concept of A. In this case, concept A is known as a super-concept, by analogy with the nomenclature of the Item Model. This heritage relationship allows us to define a taxonomy of measurable items.

3.5. Limits of the Taxonomy and Typology

The implemented taxonomy is expressed from a measurement point of view. Although it enables us to represent items coming from various fields in a single framework, it does not permit the specification of items beyond a certain level. In the field of underwater archaeology for example, amphorae are classified according to various typologies, such as for example, Ramon T.2.1.1.1 and Ramon T.2.1.1.2. It is impossible to represent these typologies as concepts of our taxonomy as all amphorae possess the same morphological attributes. The differentiating criterion between amphorae is not the existence of certain attributes, but rather their value, or even the relationship between these values. For example, the relationship between the height and its maximum diameter, or the height (Z side) where the amphora's maximum diameter is located. Our taxonomy is unable to express such a classification because the critical criteria are completely linked to the field of study and their integration is incompatible with the hierarchical relationships that we use. In order to solve this problem, we defined the notion of typology. The typology of a measurable item is a set of value range and default values for the attributes of an item in its field of study. A typology is not a concept in the sense of our taxonomy, but is used to characterize an item using its links with the field of study involved. Typology can be seen as a parameter of the instantiation of the item that specifies the default values of its attributes. These values are set based on the knowledge from the field of study.

The typology of an item is based on the notion of a default morphological relationships and value ranges. It is possible to represent the value ranges using constraints on these relationships. An item is then characterized by a typology only if its morphological relationships are within the defined ranges. In order to verify the validity of the morphological relationship of an item for which we possess a typology, we must define constraints known as intrinsic constraints. For an item to be considered intrinsically coherent, it must meet all the intrinsic constraints related to it.

Just as items are organized into taxonomies, intrinsic constraints are too. An item from a concept C sub-concept of A must meet the intrinsic constraints related to C, but also those related to A if they exist. For example, a basic and general intrinsic constraint for all measured items: "The length of a measurable item is positive". For every "measurable" item, this constraint must be met.

3.6. Completeness of Objects

By their nature, photogrammetrical surveys are incomplete. Since they are performed without any contact, in order to correctly record the artefacts' position upon their discovery only the visible side of the objects lying on the sediment will be measured. The problem with the survey as presented here involves the confrontation of a measured object with its theoretical model. Two types of problems have to be resolved: The first is the incompleteness of the survey in regard to a theoretical model; the second is obtaining a complete theoretical model.

Indeed, the mass production techniques used at that era may encourage us to bring up the issue of an almost industrial constant theoretical model. However, the development of this model still depends on the choice of a paradigm instance. No theoretical blueprints can be of course consulted. Thus, archaeologists specialised in developing typologies publish blueprints and schematics of these typologies based on surveys of studied and compared instances. Furthermore, it should be also noted that a shipwreck of this importance holds a relatively large number of amphorae compared to that of amphorae of this type already identified in the world.

The development of a theoretical model is therefore not trivial and it must be validated for this shipwreck when all the visible amphorae have been recognised and modelled. In order to obtain a reliable theoretical model, we started with exhaustive surveys performed using a laser scanner system for few amphorae carefully removed from the site. An example of the scanning procedure is shown in Figure 12.

In this way we can obtain 3D models for certain amphorae coming from the wreck. Some other models have been already defined in bibliographical references. The confrontation of partial measurements of the amphorae or fragments present on the Xlendi wreck will take into account the origin of the theoretical model with the help of a variability threshold present in the measurements instantiation phase.

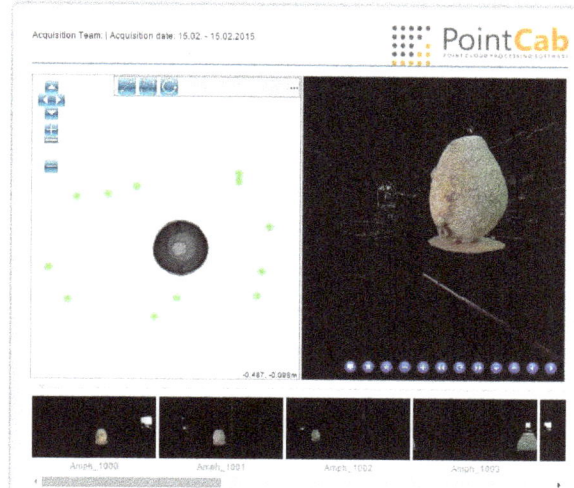

(a)   (b)

Figure 12. An exemplar output of a laser scanner of an amphora removed from wreck during a dig in July 2014; (**a**) a complete 3D model of an amphora; (**b**) the fusion of various captures of the laser scanner in order to obtain the complete 3D model (performed by DeMicoli & Associates with the University of Malta).

3.7. Knowledge Representation in the GROPLAN Project

Knowledge representation within the project is an extension of the conceptual representation. Two distinct aspects of the representation must be taken into account:

- the intrinsic aspect
- the extrinsic aspect

The intrinsic aspect defines the studied items. It consists of all the concepts with their relationships to heritage, characterisation and aggregation as well as the constraints concerning the default values and attributes. In fact, the conceptual representation is used to describe the properties of the entities studied, but does not give them a priori values or even limit the values they may have. In the scope of measurements based on knowledge, many properties can have default values or may even not possess certain values. Formally, an Entity is a set expressed as $E = \left\{ C, V_d, C^I, R, C^E \right\}$, where:

(1)   C is a concept.
(2) V_d is a set of default values for its attributes.
(3) C_I is a set of constraints on its attributes.
(4)   R is a set of relationships between instances of C.

295

(5) C_E is a set of constraints on its relationships of R.

Each item characterized by an entity E is an instance of class C. Components 1, 2 and 3 form the intrinsic part of the entity only apply to one item at a time. While the components 4 and 5 only apply to sets of items; they form the extrinsic part of the entity. The correspondence between a set of items O and an entity is called characterisation. In the scope of the study of amphorae measured by photogrammetry, the entities studied are based on the amphora concept that groups together attributes such as height, rim diameter, body diameter, *etc.*

In addition to the classic conceptual model, the notion of entity is based on the definition of constraints on the attributes. From a formal point of view, an intrinsic constraint applies to one or more attributes in order to set the boundaries of its values (for a numerical attribute), to specify a set of possible values (for a descriptive attribute) or even link values with different attributes.

In the scope of archaeology and more specifically the scope of measuring archaeological items such as amphorae, we can describe the intrinsic constraints based on knowledge compiled in the typologies and in the corpuses already measured. For example, the amphora Ramon 2.1.1.1, described in the work of Ramon [20], have a height of HR2111 $\pm$ Δh cm and a maximum diameter of DR2111 $\pm$ Δd cm. Of course, it is obvious that these constraints are necessary, but insufficient. The values HR2111, DR2111, Δh, Δd fall under archaeological expertise and are generally determined based on existing literature.

An entity characterizes a set O only if the following elements are verified:

- all the items o_i of O are instances of class C;
- the instantiation process to initialize the attributes of o_i with the values of V_d;
- each item o_i of O satisfies the set of intrinsic constraints c_{Ik} of C_I;
- all the constraints on the relationships c_{El} of C_E are satisfied.

Implementing the characterisation of a set of items by an entity means being able to enter values for the attributes of the concepts, to calculate the existence of relationships between the various items as well as evaluate the intrinsic and extrinsic constraints for the items studied.

4. Implementation in Java and Ontologies

The developments in Java are based on the ARPENTEUR platform [21], which includes various photogrammetric tools dedicated to heritage applications. It is designed to be used for photogrammetric measurements and the management of surveyed heritage artefacts. A taxonomy of measurable items is thus defined in agreement with specialists in the field and a photogrammetric measurement process is established for each item, hence, the knowledge of the field thus guides the photogrammetric measurements and ensures the consistency of the result (see the UML diagram in Figure 13). Although this approach is well structured from a software engineering point of view, it has been found to be limited with regards to its reasoning abilities concerning measured items as well as the weakness in representing inter- and intra-entity relationships. In fact, the problem in expressing these relationships rapidly became decisive, first when verifying the coherence of the instances in regards to their theoretical model and then the coherence of the organisation of the entities.

The proposed solution is a "double formalism"; the Java programming language for photogrammetrical computations and for measuring heritage artefacts, and the Web Ontology Language (OWL) in order to define an ontology that describes the concepts involved in the measurement process and the link with the measured items.

4.1. Implementation of the Representation by Entity Using OWL

For several years Web Ontology Language (OWL) has been used as a standard for implementing ontologies (W3C, 2004a). In its simplest form, it enables the representation of concepts (class), instances (individuals), attributes (data properties) and relationships (object properties).

The construction of an ontology in OWL, doubled with the Java taxonomy, is not done automatically. Each concept of the ontology is built so that it can be instantiated in Java but yet does not exactly reflect the Java tree structure. For example, the ScratchMatrix class in Java is inherited from the DenseMatrix class developed by Bjørn-Ove Heimsund in the framework of the Matrix Toolkit Java (MTJ) library. MTJ proposes a native implementation and a Java interface of the library Linear Algebra PACKage (LAPACK) originally written in Fortran, then used in MATLAB for solving systems of linear equations. You can see that the details of the Java implementation are not useful at the level of the ontological description. Nevertheless, it is essential to have the possibility of instantiating a Java matrix using OWL code and reciprocally being able to express an instance of the ScratchMatrix class in OWL (see the Java Figure 13 and OWL tree structures below).

For each concept of the ontology, a procedural attachment method was developed using JENA (an open-source Semantic Web framework for Java), each Java instance having a homologue in the ontology is capable of generating OWL content and possesses a constructor that can accept OWL contents as a parameter.

JENA is currently one of the most complete engines. It implements RDF, RDFS and OWL as well SPARQL queries. Moreover, a forward (Rete), backwards (logic programming) and hybrid chaining engine is available. This engine is used to implement the RDFS semantic and OWL.

Since Version 2 (W3C, 2009), OWL integrates the notion of constraints on the attributes (property restrictions) that are used to restrict possible values and their cardinalities. The OWL framework by itself enables the representation of a part of the entities (a concept C, its attributes, its set of relationships R and the sub-set of the intrinsic constraints restricting attributes cardinalities). The first step in the implementation of an entity concerns the concept itself, its attributes and the relationships it shares:

```
<owl:Class rdf:about=''#Amphorae''/>
<owl:Class rdf:about=''#AmphoraTypology''>
  <owl:equivalentClass><owl:Class>
      <owl:oneOf rdf:parseType=''Collection''>
        <rdf:Description rdf:about=''\#Ramon\_T2111''/>
        <rdf:Description rdf:about=''\#Ramon\_T2112''/>
      </owl:oneOf>
    </owl:Class></owl:equivalentClass>
</owl:Class>
<owl:DatatypeProperty rdf:about=''\#hasHeight''>
  <rdfs:domain rdf:resource=''\#Amphorae''/>

  <rdfs:range rdf:resource=''\&xsd;double''/>
</owl:DatatypeProperty>
<owl:DatatypeProperty rdf:about=''\#hasMaxDiameter''>
  <rdfs:domain rdf:resource=''\#Amphorae''/>
  <rdfs:range rdf:resource=''\&xsd;double''/>
</owl:DatatypeProperty>
<owl:ObjectProperty rdf:about=''\#hasTypology''>
  <rdf:type rdf:resource=''\&owl;FunctionalProperty''/>
  <rdfs:domain rdf:resource=''\#Amphorae''/>
  <rdfs:range rdf:resource=''\#AmphoraTypology''/>
</owl:ObjectProperty>
<owl:ObjectProperty rdf:about=''\#isIntersectingBoundingBox''>
  <rdfs:range rdf:resource=''\#Amphorae''/>
  <rdfs:domain rdf:resource=''\#Amphorae''/>
</owl:ObjectProperty>
```

In this example, an amphora is represented by the concept Amphorae containing the attributes height (hasHeight), belly diameter (hasMaxDiameter) as well as a single typology (hasTypology). The various typologies of amphorae (limited here to Ramon T.2.1.1.1 and Ramon T.2.1.1.2) are represented by the listing AmphoraTypology. An amphora cannot be assigned dynamically to a given typology in OWL; in fact, this is determined by a set of constraints on the attributes that OWL cannot express.

Since 2004, the OWL framework was extended with the Semantic Web Rule Language (SWRL) (W3C, 2004b) that is able to define rules for classes and properties in order to deduct new information from a set of individuals as well as verify its coherence. Several OWL/SWRL inference engines, also known as semantic thinkers, are currently available and offer an acceptable level of performance for managing sets of individuals (Pellet, Hermit, RacerPro, *etc.*). Formally, an SWRL rule is defined as a Horn clause reduced to unary and binary predicates that express a datatype property, an object property or its belonging to a class. In the case of an amphora belonging to the typology Ramon T.2.1.1.1, we can write the following constraint on the metrology:

---

hasTypology(a, Ramon_T2112) ← Amphore(a) ˆ sup(a.height, 90) ˆ inf(a.height, 140)

This clause can be translated into SWRL as follows:

```
<swrl:Variable rdf:ID="amphora"/> <swrl:Variable rdf:about="\#height"/>
<swrl:Imp rdf:about="\#Ramon\_T2111-metrology">
  <swrl:body rdf:parseType="Collection">
    <swrl:ClassAtom>
      <swrl:classPredicate rdf:resource="\#Amphorae"/><swrl:argument1 rdf:resource="\#amphora" />
    </swrl:ClassAtom>
    <swrl:DatavaluedPropertyAtom>
      <swrl:propertyPredicate rdf:resource="\#hasHeight"/>
      <swrl:argument1 rdf:resource="\#amphorae"/><swrl:argument2 rdf:resource="\#height"/>
    </swrl:DatavaluedPropertyAtom>
    <swrl:BuiltinAtom>
      <swrl:builtin rdf:resource="\&swrlb;greaterThan"/>

      <swrl:arguments><rdf:List><rdf:first rdf:resource="\#height"/><rdf:rest>
        <rdf:List><rdf:first rdf:datatype="\&xsd;double">90.0</rdf:first>
          <rdf:rest rdf:resource="\&rdf;nil"/>
          </rdf:List></rdf:rest>
        </rdf:List></swrl:arguments>
    </swrl:BuiltinAtom>
    <swrl:BuiltinAtom>
      <swrl:builtin rdf:resource="\&swrlb;lessThan"/>
      <swrl:arguments><rdf:List><rdf:first rdf:resource="\#height"/><rdf:rest>
        <rdf:List><rdf:first rdf:datatype="\&xsd;double">140.0</rdf:first>
          <rdf:rest rdf:resource="\&rdf;nil"/>
          </rdf:List></rdf:rest>
        </rdf:List></swrl:arguments>
    </swrl:BuiltinAtom>
  </swrl:body>
  <swrl:head rdf:parseType="Collection">
      <swrl:IndividualPropertyAtom><swrl:propertyPredicate rdf:resource="\#hasTypology"/>
        <swrl:argument1 rdf:resource="\#amphorae" />
        <swrl:argument2 rdf:resource="\#Ramon\_T2111" />
      </swrl:IndividualPropertyAtom>
  </swrl:head>
</swrl:Imp>
```

The SWRL rule previously defined has two purposes. First, it assigns an amphora to a typology in function of its metrological attributes; second, it ensures the coherence of the information because if the amphora is already associated to a different typology, the inference will generate an incoherence by assigning it to the Ramon T.2.1.1.1 typology, which contradicts the typology uniqueness.

An entity is therefore completely implementable if limited to the use of OWL2 and SWRL. The table below shows the various OWL/SWRL components involved:

Entity	OWL
Concept	owl:Class
Instance	owl:NamedIndividual
Attribute	owl:DatatypeProperty
Relationship	owl:ObjectProperty
Intrinsic constraint	owl:Restriction/SWRL rule
Extrinsic constraint	SWRL rule

More formally, this representation is based on an OWL2/SWRL sub-set limited to its descriptive part (OWL-DL). The entity and its associated instances can therefore be implemented.

4.2. The Link with CIDOC-CRM

The ontology developed in the framework of the GROPLAN project takes into account the manufactured items surveyed, as well as the method used to measure them; in this case, photogrammetry. The surveyed item is therefore represented from the measurement point of view and has access to all the photogrammetrical data that contributed to its measurement in space. Two ontologies are aligned in this context; one dedicated to photogrammetrical measurement and the geo-localisation of the measured items, whereas the other is dedicated to the measured items, principally the archaeological artefacts, describing their dimensional properties, ratios between main dimensions, and default values.

These ontologies are developed with close links to the Java class data structure that manages the photogrammetric process as well as the measured items. Each concept or relationship in the ontology has a counterpart in Java (the opposite is not necessarily true). Moreover, surveyed items are also archaeological items studied and possibly managed by archaeologists or conservators in a museum. It is therefore important to be able to connect the knowledge acquired when measuring the item with the ontology designed to manage the associated archaeological knowledge. CIDOC CRM is a generic ontology that does not support the items that it represents from a photogrammetric point of view, a simple mapping would not be sufficient and an extension with new concepts and new relationships would be necessary.

This modelling work is based on a previous study that started from the premise that collections of measured items are marred by a lack of precision concerning their measurement, assumptions about their reconstruction, their age, and origin. It was therefore important to ensure the coherence of the measured items and potentially propose a possible revision. For more information, see [22–26].

The extension of the CIDOC-CRM ontology is structured around the concept E22 Man-Made Object. The root of ItemMesurable developed in GROPLAN extends this concept.

The mapping operation is done in Java by interpreting a set of data held by the Java classes as a current identification of the object: 3D bounding box, specific dimension such as maximum diameter or rim diameter in case of amphorae. These attributes are then computed in order to express the right CRM properties.

For example, the amphorae typology, which is strongly connected with the E52 Time-Span, in our point of view, the amphorae typology is linked with some relations between maximum height and maximum diameter (and of course others). This means that E52 instance of our amphora is filled in after a set of computations performed by the Java instance.

Several methodologies can be chosen regarding mapping two ontologies. For example, Amico and his team [27] choose to model the survey location with an activity (E7) in CRM. They also developed a formalism for the digital survey tool mapping the digital camera definition with (D7 Digital Machine Event). We see here that the mapping problem is close to an alignment problem which is really problematic in this case. Aligning two ontologies dealing with digital camera definition is not obvious; a simple observation of the lack of interoperability between photogrammetric software shows the wildness of the problem. We are currently working on an alignment/extension process with Sensor ML which is an ontology dedicated to sensors. Although some work have already been achieved [28,29], but not enough to clearly hold the close range photogrammetry process, from image measurement to artefact representation.

As the link between the data structure of the Java classes (Figure 13) and the GROPLAN ontology (Figure 14) is not trivial, the calculation of the properties of the individuals represented in the ontology is not a simple reproduction of the existing values in the corresponding Java instances. Let's take a simple example: The unit in which the sizes of the measured items are expressed. For CIDOC-CRM, these sizes can be represented by the class E54 Dimension, which possesses a property P91 has unit. It is effectively rigorous to assign a unit to a size. Nonetheless in the scope of the Java class data structure representing the photogrammetrical process as well as the measured items; it's not exactly the same thing. Photogrammetrical measurements use pixels for images. Once the images are oriented, this is done by the class Model (in the context of photogrammetry, a model is a set of images having common points and being oriented in the same reference system). 2D points observed in several images are calculated in 3D in the reference system of the model. It is therefore the model that contains the pixel/site reference transformation and with that the unit in which the 3D measurements are expressed.

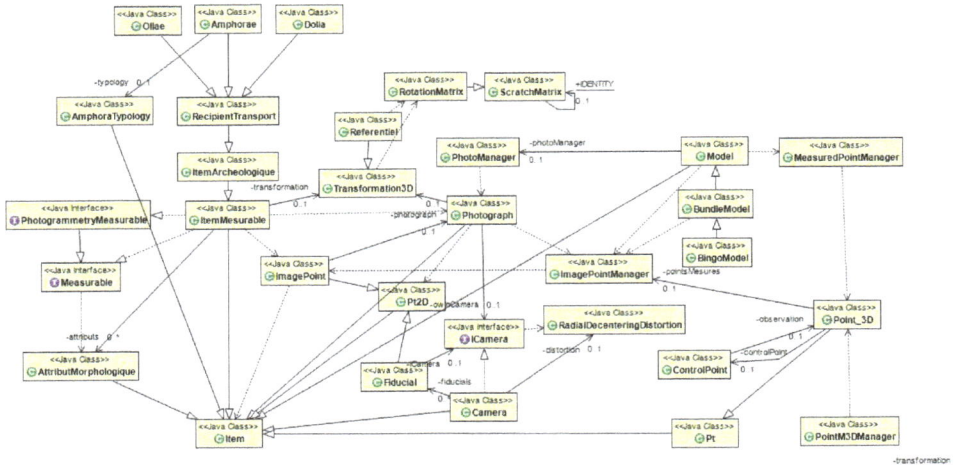

Figure 13. Partial UML diagram of Java classes managing the photogrammetrical process as well as the measured items.

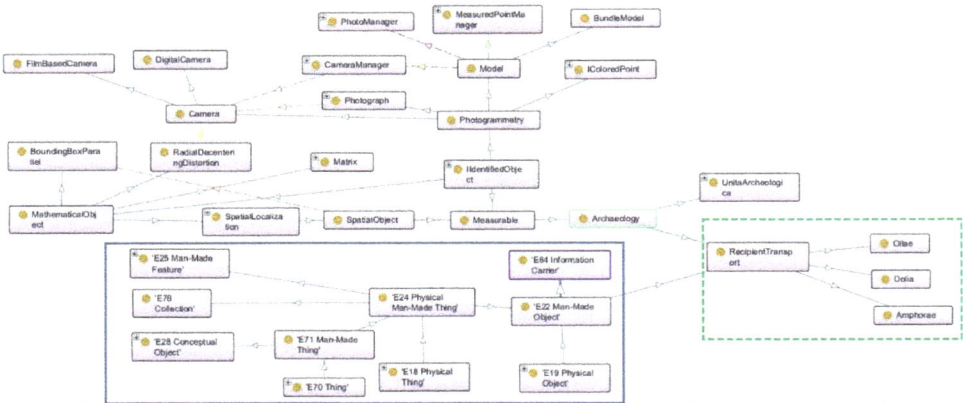

Figure 14. Partial view of the GROPLAN ontology and the link with CIDOC-CRM. (Screen shot taken from the "Protege" software application). In the blue frame, a sub-tree of the CIDOC-CRM ontology.

Measured items do not have a reference to a property unit but towards a photogrammetrical model that contains a set of images in which this item was seen and measured, as well as a reference system defining the unit for expressing 3D measurements. Property P91 is therefore filled in indirectly by the Java instance that possesses a reference to the model that generated it.

4.3. 3D Object Recognition

3D shape retrieval is still an open field of research. Knowing that until now it is difficult to find an automatic method that is efficiently able to identify a 3D model with occlusion. Many existing 3D shape retrieval methods require a complete surface model of a query object in order to find it in the dataset [4,30–32] while others seek to find a signature that is invariant to rotation of the object [33–35]. We are mainly interested to partial shape retrieval methods and we refer the reader to a survey article by Yi *et al.* [36] for more details.

In this work, we present a novel object recognition approach that deals with partial objects. The purpose is to search for a partial object in a training dataset that contains full 3D objects (point clouds). Our algorithm is based on creating a dataset of partial 2D projections that are called level curves, whenever there is an enquiry, we search the dataset in order to find the correspondence. For creating the dataset, we take as input a set of 3D objects. We subsequently produce a set of samples of level curves by using a viewing sphere. The level curves are a set of 2D planar contours that are the projection of points on several perpendicular planes. The viewing sphere contains the target object and represents the base of samples on which the level curves are obtained. Our matching algorithm, used to compare level curves, is based on 2D planar curves alignment by using the intrinsic properties of curves which are the curvature and the arc-length. The properties are used by Sebastian *et al.* [37] for whole-to- whole matching curves.

4.3.1. Related Work

There exist two main approaches that address the problem of partial shape retrieval; the local descriptors-based methods and the view-based methods. Local descriptors-based methods aim to extract the description (or signature) in the neighborhood of surface points whereas the methods based on view generate a set of 2D images of a 3D model from different points of view by projection. Partial shape retrieval ends to compare views. Johnson and Hebert [38] introduced the concept of spin images where they compute a 2D histogram of the 3D points projections on the cylindrical coordinates, Yi *et al.* [36] propose to use this signature with Monte-Carlo sampling on the surface model. Rusu *et al.* [39] propose Fast Point Feature Histograms (FPFH), an optimized method for real time use that characterizes the local geometry of a 3D point and stores the information in 16-bin histograms. Malassiotis *et al.* [40] extract a descriptor from snapshots of the surface over each point using a virtual camera oriented perpendicularly to the surface around the point. Cornea *et al.* [41] used the curve-skeleton of a 3D shape and compare between curves, by using Earth

303

Mover's Distance [42] to evaluate the partial similarity. Sun *et al.* [43] generate a sequence of 2D planar contour by projecting the geodesic circles onto the tangent plane. In this work, we introduced a new approach for partial 3D object retrieval from a database by using level curves. The originality of this work is to use a viewing sphere to extract the contours of the object from several viewpoints in order to create a database with complete information about the 3D objects. Next, the extraction of level curves that present the contours of the object at various levels in order to reduce the problem of matching between two 3D point clouds to a matching between 2D planar curves.

4.3.2. The Approach

3D object models are used in many applications, such as computer vision, computer graphics and Computer-aided design. We can find a large number of databases of 3D models on the Web such as: Stanford 3D Scanning Repository [44], NTU 3D Model database [45] and 3D Keypoint Detection benchmark [46]. In this work, we used some 3D models from [44,46] and by using viewing sphere on each 3D model, we create our dataset where 3D models are represented by level curves obtained from each viewpoint. To find out if a partial model is part of an existing 3D model of our dataset, we have to match level curves of that partial model against all other in our dataset.

As mentioned in the previous section, our approach is based on curve matching, the fact that we believe that the best description of 3D objects are their contours, led us to think of the level curves. Those curves can be extract by slicing out point clouds (3D models) using several planes with a regular step (see Figure 15). Two ways of slicing are possible: using one point of view *i.e.*, one "cutting" plane shifted along the model typically as level curve with horizontal plane in cartography. Or choosing the cutting direction from several point of view settable on a sphere defined around the studied object. Here we consider the latter case as it produce more complete information about the object, despite the larger number of curves that will be created which affects the computation time.

Matching process aims to extract the best part from one curve that matches as whole or just part of the second curve, to solve this problem, we need to find the position where the query curve aligns the best curve from the dataset. To measure the similarity between the query curve against the curves from the dataset, we slide the signature of the query curve on the current curve of the dataset. The small Euclidean distance indicates the position of the best fitting.

To remove the false matches, we added a new step to compute the Euclidean transformation between points on the matched parts. The alignment error is computed by using Root Means Square Error (RMSE), which represents the sum of distances among the points of matched parts. This error and the similarity measure can be used as indications on the quality of the curve matching. Figure 15 on the right shows an example of curve extraction from the complete 3D model of amphora Ramon 2111-73. This curve is matched with the illustrated curves in Figure 16 which shows the problem of the direction of parameterization as it is highlighted in [47]. Therefore, to solve this problem, each curve from the dataset is matched by using the first and second direction of the query curve parameterization, then the best match is kept.

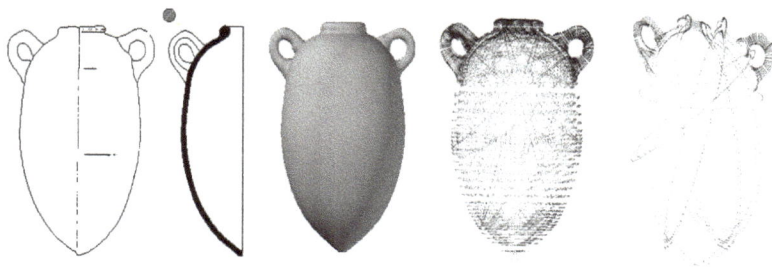

Figure 15. From left to right; the amphora model designed by the archaeologist, then the 3D model computed from the archaeological design. Two illustrations of level curves extraction using several planes.

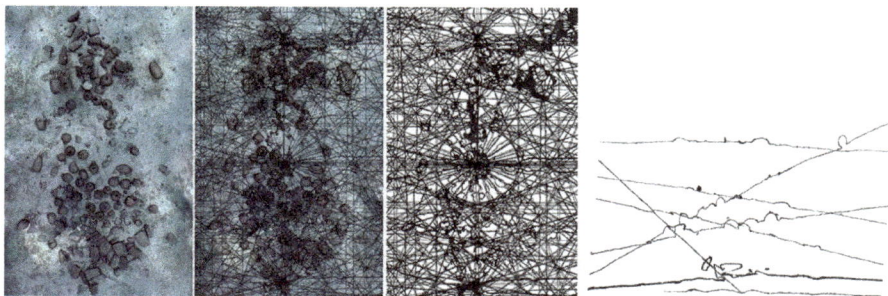

Figure 16. On the left, the ground 3D model computed by photogrammetry, on the right, level curves extraction from several planes with partial trace of the amphorae and other artefacts.

4.3.3. Experiments and Results

We first performed 3D partial object retrieval experiments using the publicly available database [44]. This work has already been published in [48]. In this paper, we work on a real case of the Phoenician wreck as shown in Figures 1 and 3. Figure 4 shows the first step of this work in progress. Our target is to make our automatic matching algorithm reaches the accuracy of the manual matching (which is indeed an effort and time consuming task) as shown in Figure 17. Although the proposed approach is able to correctly detect the position of amphorae, rotation alignment is not accurate enough in some cases due to the small overlap. A possible solution is to extend matching aspects by considering other aspects such as colour and texture information.

Figure 17. First matching using a manual recognition of the typology.

The resulting 3D representation is very useful to archeologists as it furnishes a simple manner to interpret information about the localisation and position of amphorae. This facilitates any further exploration and inspection missions. Furthermore, it provides a fundamental platform upon which an accurate excavation strategy can be developed. Such an underlying strategy has direct impacts on aspects such as excavation techniques, conservation and budget.

On the other hand, when it comes to graphically representing an object of interest, a specialist in one domain may provide a different description of the object based on his own interest. This description may vary from one domain to another. In our context, what we perceive as easy to interpret and visually appealing, may not be the case for a researcher from another field. This is certainly the case with the 3D models produced. Other types of modelling can be further exploited to provide complementary information with respect to the given 3D point cloud-based representation. In this scope, we have achieved an attempt to generate Non-photorealistic rendering (NPR) of the scene. The 3D model obtained preserves the geometry of the original model, whereas the boundaries are highlighted. An example is given in Figure 18, here the model on the bottom right may bring more relevant information to the domain of archeology. Furthermore, the NPR representation can be used at the beginning to identify an area where an amphora is probably located. It can also be used during the process to improve object recognition and matching as an alternative to the current proposed matching scheme. The work presented here is still in progress and the first experiments are yielding very encouraging results. As shown in Figure 18, the highlighted object's boundaries in NPR model represent a discriminative feature to recognise amphorae.

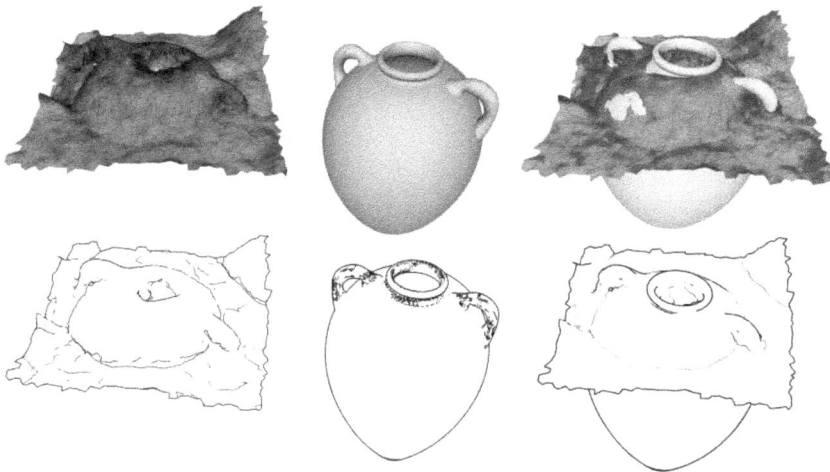

Figure 18. First row: Object recognition based on 3D models using point cloud representation. Second row: Illustration of object recognition using NPR representation.

5. Ongoing and Future Developments

5.1. Pattern Recognition and Measurement

The goal of object recognition is to automatically and efficiently extract the relevant contents of a site; *i.e.*, determine the identity of the objects (of which a model is known a priori) present on the site as well as its spatial orientation.

In the scope of the GROPLAN project, we plan to integrate a pattern recognition tool for exploring underwater archaeological sites. This will help the archaeologist when studying the site, notably during the recognition and orientation phases. There are two types of available data, on one hand a set of models describing the items potentially present on the site, which is currently limited to amphorae, but this database can later be broadened by the user. And on the other hand, a 3D point cloud, measured on the site, obtained through an automatic photogrammetry process and whose density is roughly the same as that used to describe the models of the sought items.

The items present on the site are either damaged or partially covered with sediment; the objective is to help the archaeologist by determining the position and orientation of these items in space and suggesting a class to which they belong. For this aim, we are planning to exploit the 4PCS algorithm proposed by Aiger [49]. This allows the registration of two point clouds and deals with the problem of noisy data. The method is based on extracting all co-planar 4-points sets from a 3D point cloud, that are approximately congruent, under rigid transformation, to a given set of coplanar 4-points. To apply this algorithm in our context, a series of pre-processing steps have to be performed in order to prepare the two input point clouds to match the input requirements constrained by the algorithm in terms of noise and sufficient overlap. This approach has been already used in archaeological context by Reuter in the SeARCH project [50].

Furthermore, inspired by the works of Agarwal and Triggs [51] and Fei-Fei [52], we suggest modelling the appearance and the geometric configuration of the local areas of an item in a descriptor vector. The local areas will be represented by the element closest to a pre-calculated visual vocabulary and the geometric relationships between all the visual word pairs (elements of the visual vocabulary) detected in the model will be encoded in a descriptor vector. A classifier will be used to detect areas containing the item in the site. During the recognition phase, local areas will supply a certain number of matches and they will be used as a verification step to validate their similarity.

We plan to test several types of local descriptors such as Spin Image, 3D shape contexts as well as harmonic shape contexts [53]. In the learning phase, we will use SVM approaches where we will test several variations (SVM decision tree [54], multi-class SVM [55]).

5.2. Application on Nautical Archaeology

The first and most important aspect of photogrammetry on nautical archaeology is cost. Reducing time spent underwater and the amount of equipment deployed on an archaeological survey, independently of its depth, is a major requirement in the discipline. Likewise, so is the development of a theory of knowledge in nautical archaeology. The application of an ontology and a set of logical rules for the identification, definition, and classification of measurable objects is a promising methodology to assess, gather, classify, relate and analyze large sets of data, which also permit the development of broader studies related to the history of seafaring.

Nautical archaeology is a recent sub-discipline of archaeology, which developed after 1960. It is therefore normal that its early steps were concerned with recording methodology and accuracy in underwater environments. Techniques which permit working in specific conditions impose an array of practical constraints. These include reduced time to work and have less light, low visibility, a narrower field of vision, and other conditions such as those derived from surge, current, or depth [56–58]. Since the inception of nautical archaeology, theoretical studies aimed at identifying patterns and attempting to address larger anthropological questions related to culture change. However, the sample sizes were too small to allow generalizations. Few seafaring cultures have been studied and understood well enough to allow a deep understanding of their history, culture, and development. A good and perhaps unique example is the Scandinavian Vikings [59].

Through our work we hope to provide researchers from different marine sciences with the appropriate tools for the facilitation of their work that will in turn contribute to the aforementioned broader studies related to cultural changes.

The study of the history of seafaring is the study of the relations of humans with rivers, lakes, and seas, which started in the Palaeolithic. An understanding of this important part of our past entails the recovery, analysis, and publication of large amounts of data, mostly through non-intrusive surveys.

The methodology proposed in GROPLAN aims at simplifying the collection and analysis of archaeological data, and facilitating the establishment of relations between measurable objects and concepts. It builds upon the work of Steffy, who in the mid-1990s developed a database of shipbuilding information that tried to encompass a wide array of western shipbuilding traditions through time and relate conception and construction traits in a manner that allowed comparisons and the establishment of new relations between objects. Around a decade later Monroy transformed Steffy's database into an ontological representation in RDF-OWL, and expanded its scope to potentially include other archaeological materials [10,60,61]. After establishing a preliminary ontology, completed through a number of interviews of naval and maritime archaeologists, Monroy combined the database with a multi-lingual glossary and built a series of relational links to textual evidence that helped contextualize the archaeological information contained in the database. His work proposed the development of a digital library that combined a body of texts on early modern shipbuilding technology, tools to analyse and tag illustrations, a multi-lingual glossary, and a set of informatics tools to query and retrieve data [10,60–64].

The GROPLAN approach extends these efforts into the collection of data, expands the analysis of measurable objects, and lays the base for the construction of extensive taxonomies of archaeological items. The applications of this theoretical approach are obvious, in that it simplifies the acquisition, analysis, storage, and sharing of data in a rigorous and logically supported framework. From a practical viewpoint, GROPLAN is also advancing the development of lighter, cheaper, and easier to handle equipment packages.

These two advantages are particularly relevant in the present political and economic world context brought about by globalization. The immediate future of naval and maritime archaeology depends on a paradigm change. Archaeology is no longer the activity of a few elected scholars with the means and the power to define their own publication agendas. The survival of the discipline depends more than ever on the public recognition of its social value. Cost, accuracy, reliability (for instance established through the sharing of primary data), and its relation with society's values, memories and amnesias, are already influencing the amount of resources available for research in this area. GROPLAN stands as a pioneer and major contribution to the advancement of not only nautical archaeology, both in shallow and deep water, but its applications extend to land archaeology as well, and tie with the needs of a widening group of stake holders, which include a growing public.

As stated in the beginning of this paper, the main objective of this project is the development of an information system based on ontologies and capable of establishing a methodology to acquire, integrate, analyse, generate and share numeric contents and associated knowledge in a standard, homogenous form. Although still in an early stage of development, the GROPLAN approach has the potential to open a paradigm changing research direction. Even if we consider only questions related with the storage and sharing of primary data, GROPLAN has the potential to advance a set of basic rules of good practice in maritime archaeology. In 2001 the UNESCO Convention for the Underwater Cultural Heritage established the necessity of making all data available to the public [65]. GROPLAN project builds upon this philosophy, proposes a way to share archaeological data, and will undoubtedly change the rules in the field.

Acknowledgments: This work has been partially supported by The National Agency of Research (ANR) of France within the framework of the GROPLAN project.

Author Contributions: All the authors contributed equally to this work. All the authors substantially contributed to the study conception and design as well as the acquisition and interpretation of the data and drafting the manuscript.

Conflicts of Interest: The authors declare no conflict of interest.

References

1. Gergatsoulis, M.; Bountouri, L.; Gaitanou, P.; Papatheodorou, C. Mapping Cultural Metadata Schemas to CIDOC Conceptual Reference Model. In Proceedings of the 6th Hellenic Conference on AI, STEN 210, Artificial Intelligence: Theories, Models and Applications, Athen, Greece, 4–7 May 2010; pp. 321–326.

2. Niccolucci, F.; D'Andrea, A. An Ontology for 3D Cultural Objects. In Proceedings of the 7th International Symposium on Virtual Reality, Archaeology and Cultural Heritage VAST, Nicosia, Chyprus, 30 October–4 November 2006.

3. Watson, J. BSAC, SAFETY TALK—Depth Limits. 2006. Available online: http://www.bsac.com/core/core_picker/download.asp?id=508&filetitle=Depth+limits (accessed on 11 November 2015).

4. Legifrance. Arrêté du 30 Octobre 2012 Relatif Aux Travaux Subaquatiques Effectués en Milieu hyperbare. 2012. Available online: http://www.legifrance.gouv.fr/affichTexte.do?cidTexte=JORFTEXT000026762149& categorieLien=id (accessed on 11 November 2015).

5. Drap, P.; Merad, D.; Mahiddine, A.; Seinturier, J.; Peloso, D.; Boï, J.-M.; Chemisky, B.; Long, L. Underwater Photogrammetry for Archaeology. What Will Be the Next Step? *Int. J. Herit. Digit. Era* **2013**, *2*, 375–394.

6. Buchsenschutz, O. *Images et Relevés Archéologiques, de la Preuve à la Démonstration in 132e Congrès National des Sociétés Historiques et Scientifiques*; Les éditions du Cths (édition électronique): Arles, France, 2007.

7. Harris, E.C. *Principles of Archaeological Stratigraphy*, 1st ed.; Academic Press: London, UK, 1979.

8. Drap, P.; Seinturier, J.; Merad, D.; Chemisky, B.; Seguin, E.; Alcala, F. Système de Prise D'images Permettant L'odométrie en Temps Réel et la Réalisation d'un Modèle Tridimensionnel. France Patent Number 10000249064, 16 July 2014.

9. Lourakis, M.; Argyros, A. SBA: A software package for generic sparse bundle adjustment. *ACM Trans. Math. Softw.* **2009**, *36*, 1–30.

10. Monroy, C.; Furuta, R.; Castro, F. Synthesizing and Storing Maritime Archaeological Data for Assisting in Ship Reconstruction. In *Oxford Handbook of Maritime Archaeology*; Catsambis, B.F.A.A., Ed.; Oxford University Press: Oxford, UK, 2011; pp. 327–346.

11. Doyle, J.; Patil, R.S. Two Theses of Knowledge Representation: Language Restrictions, Taxonomic Classification, and the Utility of Representation Services. *Artif. Intell.* **1991**, *48*, 261–297.

12. Faugeras, O.; Luong, Q.-T.; Papadopoulou, T. *The Geometry of Multiple Images: The Laws That Govern The Formation of Images of A Scene and Some of Their Applications*; MIT Press: Cambridge, MA, USA, 2001.

13. Hartley, R.; Zisserman, A. *Multiple View Geometry in Computer Vision*; Cambridge University Press: New York, NY, USA, 2003.

14. Triggs, B.; McLauchlan, P.F.; Hartley, R.I.; Fitzgibbon, A.W. Bundle Adjustment—A Modern Synthesis. In Proceedings of the International Workshop on Vision Algorithms: Theory and Practice, New York, NY, USA, 21–22 September 1999; pp. 298–373.

15. Xue, J.; Su, X. A new approach for the bundle adjustment problem with fixed constraints in stereo vision. *Optik Int. J. Light Electron Opt.* **2012**, *123*, 1923–1927.

16. Suenderhauf, N.; Konolidge, K.; Lacroix, S.; Protzel, P. Visual Odometry using Sparse Bundle Adjustment on an Autonomous Outdoor Vehicle. In *Tagungsband Autonome Mobile Systeme*; Reihe Informatik aktuell; Springer Verlag: Stuttgart, Germany, 2005; pp. 157–163.

17. Furukawa, Y.; Ponce, J. Accurate, Dense, and Robust Multiview Stereopsis. *IEEE Trans. Pattern Anal. Mach. Intell.* **2010**, *32*, 1362–1376. PubMed]

18. Drap, P.; Seinturier, J.; Long, L. A photogrammetric process driven by an Expert System: A new approach for underwater archaeological surveying applied to the "Grand Ribaud F" Etruscan wreck. In *Applications of Computer Vision in Archaeology ACVA'03*; Monona Terrace Convention Center: Madison, WI, USA, 2003.

19. Seinturier, J.; Drap, P.; Papini, O. Fusion réversible: Application à l'information l'archéologique. In *Journées Nationales sur les Modèles de Raisonnement (JNMR 2003)*. Paris, ed.; Le Chesnay: Institut national de recherche en informatique et en automatique: Paris, France, 2003.

20. Ramon Torres, J. *Las Anforas Fenicio-punIca del MedIterraneo Central y Occidental*; Aix en Provence: Barcelona, Spain, 1995.

21. Grussenmeyer, P.; Drap, P. Teaching Architectural Photogrammetry on the Web with ARPENTEUR. In Proceedings of the XIXth Congress of the International Society for Photogrammetry and Remote Sensing (ISPRS) Geoinformation for all, Amsterdam, The Netherlands, 16–22 July 2000.

22. Curé, O.; Sérayet, M.; Papini, O.; Drap, P. Toward a novel application of CIDOC CRM to underwater archaeological surveys. In Proceedings of the 4th IEEE International Conference on Semantic Computing, Pittsburgh, PA, USA, 22–24 September 2010.

23. Hué, J.; Sérayet, M.; Drap, P.; Papini, O.; Wurbel, E. Underwater archaeological 3D surveys validation within the removed sets framework. In Proceedings of the 11th European Conference on Symbolic and Quantitative Approaches to Reasoning with Uncertainty, Belfast, UK, 29 June–1 July 2011.

24. Sérayet, M.; Drap, P.; Papini, O. Encoding the Revision of Partially Preordered Information in Answer Set Programming. In *Symbolic and Quantitative Approaches to Reasoning with Uncertainty*; Springer: Berlin, Germany, 2009.

25. Seinturier, J. Fusion de Connaissances: Applications aux Relevés photogrammétriques de fouilles archéologiques sous-marines. In *Spécialité Informatique*; Université du Sud Toulon Var: Toulon, France, 2007.

26. Sérayet, M.; Drap, P.; Papini, O. Extending Removed Sets Revision to partially preordered belief bases. *Int. J. Approx. Reason.* **2011**, *52*, 110–126.

27. Amico, N.; Ronzino, P.; Felicetti, A.; Niccolucci, F. Quality management of 3D cultural heritage replicas with CIDOC-CRM. In Proceedings of the CEUR 17th International Conference on Theory and Practice of Digital Libraries (TPDL 2013), Valetta, Malta, 26 September 2013; CEUR-WS.org/Vol-1117, pp. 61–69.

28. Hiebel, G.; Hanke, K.; Hayek, I. Methodology for CIDOC CRM Based Data Integration with Spatial Data. In Proceedings of the 38th Conference on Computer Applications and Quantitative Methods in Archaeology, Granada, Spain, 6–9 April 2010.

29. Xueming, P.; Beckman, P.; Havemann, S.; Tzompanaki, K.; Doerr, M.; Fellner, D.W. A Distributed Object Repository for Cultural Heritage. In Proceedings of the 11th International Symposium on Virtual Reality, Archaeology and Cultural Heritage VAST, Paris, France, 21–24 September 2010.

30. Zhang, C.; Chen, T. Efficient feature extraction for 2D/3D objects in mesh representation. In Proceedings of the 2001 International Conference on Image Processing, Thessaloniki, Greece, 7–10 October 2001; Volume 3, pp. 935–938.

31. Paquet, E.; Rioux, M.; Murching, A.M.; Naveen, T.; Tabatabai, A.J. Description of shape information for 2-D and 3-D objects. *Sig. Proc. Image Commun.* **2000**, *16*, 103–122.

32. Ohbuchi, R.; Otagiri, T.; Ibato, M.; Takei, T. Shape-similarity search of three-dimensional models using parameterized statistics. In Proceedings of the 10th Pacific Conference on Computer Graphics and Applications, Tsinghua University, Beijing, China, 9–11 October 2002; pp. 265–274.

33. Chua, C.; Jarvis, R. Point Signatures: A New Representation for 3D Object Recognition. *Int. J. Comput. Vis.* **1997**, *25*, 63–85.

34. Johnson, A.E.; Hebert, M. Using Spin Images for Efficient Object Recognition in Cluttered 3D Scenes. *IEEE Trans. Pattern Anal. Mach. Intell.* **1999**, *21*, 433–449.

35. Mian, A.S.; Bennamoun, M.; Owens, R.A. 3D Recognition and Segmentation of Objects in Cluttered Scenes. In Proceedings of the Seventh IEEE Workshops on.Application of Computer Vision, Breckenridge, CO, USA, 5–7 January 2005; pp. 8–13.

36. Yi, L.; Hongbin, Z.; Hong, Q. Shape Topics: A Compact Representation and New Algorithms for 3D Partial Shape Retrieval. In Proceedings of the 2006 IEEE Computer Society Conference on Computer Vision and Pattern Recognition, New York, NY, USA, 17–22 June 2006.

37. Sebastian, T.B.; Klein, P.N.; Kimia, B.B. On aligning curves. *IEEE Trans Pattern Anal. Mach. Intell.* **2003**, *25*, 116–125.

38. Johnson, A.E.; Hebert, M. Using spin images for efficient object recognition in cluttered 3D scenes. *IEEE Trans Pattern Anal. Mach. Intell.* **1999**, *21*, 433–449.

39. Rusu, R.B.; Blodow, N.; Beetz, M. Fast Point Feature Histograms (FPFH) for 3D Registration. In Proceedings of the IEEE International Conference on Robotics and Automation (ICRA), Kobe, Japan, 12–17 May 2009.

40. Malassiotis, S.; Strintzis, M.G. Snapshots: A Novel Local Surface Descriptor and Matching Algorithm for Robust 3D Surface Alignment. *IEEE Trans Pattern Anal. Mach. Intell.* **2007**, *29*, 1285–1290. PubMed]

41. Cornea, N.D.; Demirci, M.F.; Silver, D.; Shokoufandeh, A.; Dickinson, S.J.; Kantor, P.B. 3D object retrieval using many-to-many matching of curve skeletons. In Proceedings of the 2005 International Conference Shape Modeling and Applications, Illinois Univ., Urbana, IL, USA, 13–17 June 2005; pp. 366–371.

42. Keselman, Y.; Shokoufandeh, A.; Demirci, M.F.; Dickinson, S. Many-to-many graph matching via metric embedding. In Proceedings of the IEEE Computer Society Conference on Computer Vision and Pattern Recognition, Madison, WI, USA, 18–20 June 2003; Volume 1, pp. I-850–I-857.

43. Sun, Y.; Abidi, M.A. Surface matching by 3D point's fingerprint. In Proceedings of the Eighth IEEE International Conference on Computer Vision, Vancouver, Canada, 7–14 July 2001; Volume 2, pp. 263–269.

44. The Stanford 3D Scanning Repository. Available online: http://graphics.stanford.edu/data/3Dscanrep/ (accessed on 30 November 2015).

45. Chen, D.-Y.; Tian, X.-P.; Shen, Y.-T.; Ouhyoung, M. On Visual Similarity Based 3D Model Retrieval. *Comput. Graph. Forum* **2003**, *22*, 223–232.

46. Mian, A.S.; Bennamoun, M.; Owens, R. Three-dimensional model-based object recognition and segmentation in cluttered scenes. *IEEE Trans. Pattern Anal. Mach. Intell.* **2006**, *28*, 1584–1601. PubMed]

47. Cui, M.; Femiani, J.; Hu, J.; Wonka, P.; Razdan, A. Curve matching for open 2D curves. *Pattern Recognit. Lett.* **2009**, *30*, 1–10.

48. Mahiddine, A.; Merad, D.; Drap, P.; Boi, J.M. Partial 3D-object retrieval using level curves. In Proceedings of the 2014 6th International Conference of Soft Computing and Pattern Recognition (SoCPaR), Tunis, 11–14 August 2014.

49. Aiger, D.; Mitra, N.J.; Cohen-Or, D. 4-Points congruent sets for robust pairwise surface registration. In *ACM SIGGRAPH 2008 Papers*; ACM: Los Angeles, CA, USA, 2008; pp. 1–10.

50. Reuter, P.; Nicolas, M.; Xavier, G.; Isabelle, H.; Robert, V.; Nadine, C. Semi-Automatic 3D Acquisition and Reassembly of Cultural Heritage: The Search Project. *ERCIM News ICT Cult. Herit.* **2011**, *86*, 12–13.

51. Agarwal, A.; Triggs, B. Hyperfeatures–multilevel local coding for visual recognition. In *Computer Vision–ECCV 2006*; Springer: Graz, Austria, 2006; pp. 30–43.

52. Fei-Fei, L. One-Shot Learning of Object Categories. *IEEE Trans. Pattern Anal. Mach. Intell.* **2006**, *28*, 594–611. PubMed]

53. Frome, A.; Huber, D.; Kolluri, R. *Recognizing Objects in Range Data Using Regional Point Descriptors*; Computer Vision-ECCV: Prague, Czech, 2004; pp. 224–237.

54. Fei, B.; Liu, J. *Binary Tree of SVM: A New Fast Multiclass Training and claSsification Algorithm*; Institute of Electrical and Electronics Engineers: New York, NY, USA, 2006; Volume 17, pp. 696–704.

55. Weston, J.; Watkins, C. *Multi-Class Support Vector Machines*; Department of Computer Science, Royal Holloway, University of London: London, UK, 1998.

56. Baker, P.; Green, J. Recording Techniques used during the excavation of the Batavia. *Int. J. Marit. Archaeol.* **1976**, *5*, 143–158.

57. Cederlund, C.O. Preliminary report on recording methods used for the investigation of merchant shipwrecks at Jutholmen and Alvsnabben. *Int. J. Marit. Archaeol.* **1977**, *6*, 87–99.

58. Leonard, P.; Scheifele, S. An underwater measuring device. *J. Marit. Archaeol.* **1972**, *1*, 165–186.

59. Crumlin-Pedersen, O. *Archaeology and the Sea in Scandinavia and Britain. A Personal Account*; Viking Ship Museum Vindeboder 12 DK-4000 Roskilde: Denmark, 2010; Volume 3.

60. Monroy, C. A digital library approach to the reconstruction of ancient sunken ships. In *Computer Science*; Texas A & M University: College Station, TX, USA, 2010.

61. Monroy, C.; Furuta, R.; Castro, F. *Ask Not What Your Text Can do For You. Ask What You Can do For Your Text (a Dictionary's Perspective)*; *Digital Humanities 2009*; Conference Abstracts; University of Maryland: College Park, MD, USA, 2009; pp. 344–347.

62. Monroy, C.; Furuta, R.; Castro, F. Texts, illustrations, and physical objects: The case of ancient shipbuilding treatises. In *Research and Advanced Technology for Digital Libraries*; Kovács, L., Fuhr, N., Meghini, C., Eds.; Springer: Berlin, Germany, 2007; pp. 198–209.

63. Monroy, C.; Furuta, R.; Castro, F. A multilingual approach to technical manuscripts: 16th and 17th-century Portuguese shipbuilding treatises. In Proceedings of the 7th ACM/IEEE-CS Joint Conference on Digital Libraries, Vancouver, BC, Canada, 18–23 June 2007; pp. 413–414.

64. Monroy, C.; Furuta, R.; Castro, F. Design of a Computer-Based Frame to Store, Manage, and Divulge Information from Underwater Archaeological Excavations: The Pepper Wreck Case. In Proceedings of the Society for Historical Archaelogy Annual Meeting, Sacramento, CA, USA, 8–13 January 2008.

65. UNESCO. Convention on the Protection of the Underwater Cultural Heritage. 2001. Avaiable online: http://www.unesco.org/new/en/culture/themes/underwater-cultural-heritage/ (accessed on 30 November 2015).

An Alignment Method for the Integration of Underwater 3D Data Captured by a Stereovision System and an Acoustic Camera

Antonio Lagudi, Gianfranco Bianco, Maurizio Muzzupappa and Fabio Bruno

Abstract: The integration of underwater 3D data captured by acoustic and optical systems is a promising technique in various applications such as mapping or vehicle navigation. It allows for compensating the drawbacks of the low resolution of acoustic sensors and the limitations of optical sensors in bad visibility conditions. Aligning these data is a challenging problem, as it is hard to make a point-to-point correspondence. This paper presents a multi-sensor registration for the automatic integration of 3D data acquired from a stereovision system and a 3D acoustic camera in close-range acquisition. An appropriate rig has been used in the laboratory tests to determine the relative position between the two sensor frames. The experimental results show that our alignment approach, based on the acquisition of a rig in several poses, can be adopted to estimate the rigid transformation between the two heterogeneous sensors. A first estimation of the unknown geometric transformation is obtained by a registration of the two 3D point clouds, but it ends up to be strongly affected by noise and data dispersion. A robust and optimal estimation is obtained by a statistical processing of the transformations computed for each pose. The effectiveness of the method has been demonstrated in this first experimentation of the proposed 3D opto-acoustic camera.

Reprinted from *Sensors*. Cite as: Lagudi, A.; Bianco, G.; Muzzupappa, M.; Bruno, F. An Alignment Method for the Integration of Underwater 3D Data Captured by a Stereovision System and an Acoustic Camera. *Sensors* **2016**, *16*, 536.

1. Introduction

Acoustic and optical 3D systems are widely used to collect 3D information in underwater applications, such as 3D reconstruction of submerged archaeological sites, seabed mapping, or Remotely Operated underwater Vehicle (ROV) navigation [1–3]. Acoustic systems typically give good results in long-range acquisitions and do not suffer from turbidity, but the resulting 3D data are affected by low resolution and accuracy. Optical systems, in contrast, are more suited for close-range acquisitions and allow for gathering high-resolution, accurate 3D data and textures, but the results are strongly influenced by the visibility conditions. Therefore, the integration of 3D data captured by these two types of systems is a promising technique in underwater applications, as it allows for compensating their respective limitations.

Despite the difficulty of combining two modalities that operate at different resolutions, the integration of optical and acoustic systems in an underwater environment has received increasing attention over the past few years, mainly for seabed mapping and egomotion estimation of underwater vehicles [4–9]. Further examples of opto-acoustic integration concern with local area imaging rather than the creation of large area maps. In [10–12], the integration of video and 3D data, acquired through a single optical camera and a 3D acoustic camera (Echoscope 1600) [13], is obtained by geometrically registering such data with respect to a well-known model of the observed scene, while in [14–18] the authors propose a new paradigm of opto-acoustic stereo reconstruction that aims to apply the epipolar geometry to a stereo system composed by an optical camera and a 2D sonar (DIDSON).

Up to now, the works presented in the literature about the integration between several types of sonar (single beam sounder, multibeam, 3D acoustic camera) and optical cameras, adopt a sensor fusion approach, which is mapping-oriented, according to the classification proposed in [19]. This means that the data acquired from the two sensors are described through geometric relationships (position and orientation), and the data integration is performed by means of geometrical correspondences and registration. Data alignment, that is, their transformation from each sensor's local frame into a common reference frame, is a crucial problem of these methods that is usually solved by performing the exterior orientation of the integrated system, *i.e.*, by searching for the rigid transformation between the coordinate systems related to each sensor.

Few works have explicitly treated the alignment problem of opto-acoustic underwater systems. They showed that the methods are highly dependent on the layout and sensors that compose the system, in particular for the type of data structure they provide. In [16], the exterior orientation of the optical camera and the 2D sonar is performed by using a planar grid characterized by considerable optical and acoustic features that are manually associated. Therefore, the relative positions of the cameras are estimated through an optimization algorithm that minimizes the distances between 3D reconstructions of optical and acoustic matching projections.

In [20], the authors propose a method for aligning a single camera with a multibeam sonar on an Autonomous Underwater Vehicle (AUV) using a target placed in a test tank. Assuming a simplified model for the multibeam sonar, the exterior orientation of the proposed opto-acoustic system was compared to the alignment of a laser-camera system, where the method presented in [21] was adopted for its solution.

Finally, a different methodology is presented in [12]. After some pre-processing steps, the acoustic data are registered with respect to a CAD model of a target (an

317

oil rig in this case) using the Iterative Closest Point (ICP) algorithm [22], while the optical alignment is performed by means of the algorithm proposed in [23]. Once the poses of both sensors are calculated with respect to the observed object, the relative pose between the optical camera and the acoustic camera can be estimated without the need to use positioning and motion system equipment.

The Echoscope 3D acoustic camera is an interesting sonar which, unlike other sonars, ensonifies a whole viewing volume with a single ping and outputs 3D data in real-time as 3D point clouds. Therefore, it is suitable to be coupled with optical devices as it provides whole field 3D data from a single acquisition, unlike other devices such as multibeam sonars that acquire multiple slices and stitches them together according to navigation data. In previous works, 3D acoustic cameras have been used for on-line 3D reconstruction of underwater environments from multiple range views [24], or coupled with a single camera to improve the understanding of the underwater scenes and assist the ROV pilots during the navigation [10–12].

In the present work, for the first time, the same 3D acoustic camera used in [10–12] has been coupled with a stereovision system to gather synchronous 3D data and perform 3D opto-acoustic imaging of the acquired underwater scene. The system was conceived to improve the understanding of a human operator guiding an underwater ROV during the navigation in variable turbid water conditions and in operations that require the use of one or two manipulators. The stereo optical camera allows for obtaining a better perception of the scene depth if compared to the use of a single camera as in the actual ROV configuration, while the acoustic camera makes its best contribution in poor visibility conditions. Compared to the similar setup previously described in literature [10–12], the novelty of the proposed approach lies in the adoption of a stereo optical camera that give us the possibility to have a better resolution of the 3D image when the visibility conditions are good enough for optical sensors; moreover, it allows us to overcome the problem of processing heterogeneous data gathered from different sensors by simplifying the correspondence determination in a registration problem of the two 3D point clouds.

The aim of this work is to solve the problem of the automatic alignment of the optical and acoustic 3D images through the definition of a multi-sensor registration approach [25]. The core idea of the method is:

1. obtaining a raw estimation of the unknown geometric transformation through a registration of the optical and acoustic 3D data for each pose of a custom orientation fixture;
2. obtaining a robust and optimal estimation through a statistical processing of the transformations computed for each pose.

Experimental tests have been conducted in laboratory to validate the feasibility and the effectiveness of the proposed method and quantify the accuracy of the integration. These also gave us the opportunity to perform a first experimentation of the proposed 3D opto-acoustic camera, allowing for a better understanding of limitations and drawbacks of the system, and of the problems related to the alignment itself. The experimental results show that our alignment approach, based on several pose acquisitions of an appropriate rig, can be adopted to simultaneously calibrate the stereo optical system and estimate the rigid transformation between the optical and acoustic sensors. The effectiveness of the method has been demonstrated in this first experimentation of the proposed 3D opto-acoustic camera.

2. Relative Orientation of the Opto-Acoustic 3D Camera

To effectively integrate and fuse spatial data from different 3D sensors, the relative position and orientation of their spatial coordinate systems have to be known. The estimate of such spatial relationships can be broken down into two tasks: interior orientation, where internal sensor parameters are determined, and relative orientation, where the position and the orientation of a sensor relative to a given coordinate system are determined.

Assuming that a point $p_o = [x_o, y_o, z_o]^T$ of the stereo-optical reference frame corresponds to a point $p_a = [x_a, y_a, z_a]^T$ of the acoustic reference frame, the main goal of our multi-sensor alignment is to determine the rigid transformation ${}^oT_a$ that relates the two coordinate systems. It may be expressed, in homogeneous coordinates, as:

$$\tilde{p}_o = {}^o\left[\begin{array}{cc} R & t \\ 0 & 1 \end{array}\right]_a \tilde{p}_a = {}^oT_a\, \tilde{p}_a \tag{1}$$

where R is the orthonormal 3×3 matrix that represents the orientation of the acoustic camera to the stereo-optical one, while t is a three-dimensional vector corresponding to their relative positions.

Our multi-sensor alignment method operates in the following way:

1. it executes a synchronous optical and acoustic acquisition of a fixture in several poses;
2. it calculates both the interior and exterior orientation of the stereo optical system;
3. for each pose of the fixture, it calculates the rigid transformation that relates the sensor reference frames among each other through a registration algorithm;
4. it processes the transformation matrices through statistical methods;
5. it calculates the best estimation for the unknown transformation matrix.

2.1. Fixture Design

In the case of a stereo-optical system, the alignment problem can be solved by the optimization of a series of equations in which a collection of correspondent 3D points in both cameras is known. Typically, these data are generated by imaging a fixture that represents a set of 3D points belonging some feature of the target. For example, the centers of a dot pattern and the corners of a checkerboard pattern were used in [26,27], respectively. Such a method is difficult to adopt in the opto-acoustic alignment, because a point-to-point correspondence between 3D points in both representations is not a simple task, *i.e.*, the low resolution and the strong noise component of acoustic data do not allow for precisely localizing a point position as determined for the optical camera, as demonstrated in [2,20]. Therefore, it is necessary to find other features fit for establishing the correspondence between optical and acoustic frames.

Photogrammetric methods that employ a stereo setup acquire the scene by means of two optical cameras with known internal geometric characteristics (principal point, principal distance, and distortion function) and known relative orientation to each other. These are obtained through calibration, according to the selected camera model. Although in underwater environment the well-known pinhole camera limits the reliability and accuracy of the obtained results, as the effects of refraction must be corrected (or modeled) to obtain an accurate calibration [28], we have chosen to use the method proposed in [27] to calibrate the stereovision system. The main advantages of this approach are the simplicity of the calibration fixture and the rapid measurement and processing of the captured images, made possible by the automatic recognition of the checkerboard pattern. However, as reported in [28], this calibration procedure is suitable for applications with modest accuracy requirements, like in our case.

Taking into account that the reflective properties of the optical and acoustic signals vary according to the materials to be used, we have designed a fixture that has to satisfy the following requirements: (a) it allows for detecting geometric features in both systems; (b) it is able to discriminate or highlight areas on the rig; (c) it can be used for both opto-acoustic alignment and optical stereo calibration, simultaneously. Moreover, our underwater fixture has to satisfy several application-specific requirements, including superior visibility of the calibration markers representing object space points, water resistance of the frame and provisions for convenient deployment and retrieval.

The fixture is composed by a checkerboard panel in the center, built from a thin sheet of aluminum Dibond® (an acoustically transparent material) and fixed on an aluminum frame to calibrate the optical cameras, then another aluminum frame is placed around the inner frame to concentrate the acoustic detection along the bars. To highlight the rig areas to be detected form the acoustic system, we have thought to exploit the high reflectivity of the air in water, so the aluminum bars were covered by bubble wrap. The designed rig allows for referring both 3D data on a known-size frame and to determine simple features as centroid, normal to the plane, orientation, perimeter, edges (Figure 1).

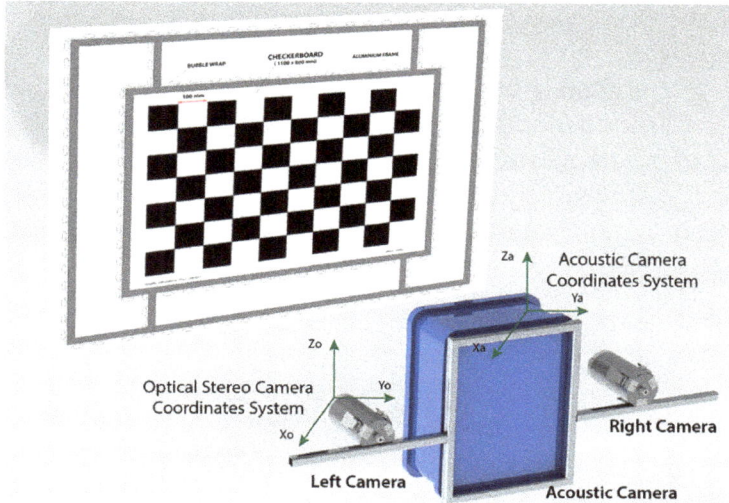

Figure 1. Geometry of the alignment method.

As will be described in Section 3.2, the size of the rig (2 × 2.5 m) was determined through the analysis of the Fields-Of-Views (FOVs) of the sensors in the expected operative range. Although a 3D fixture could be used to obtain a more accurate results, we have chosen to use a 2D rig for our approach, and to move it in a controlled volume. This choice is motivated by the difficulty to handle and move a three-dimensional structure of such size and with the requirements described above, especially in real conditions.

2.2. Optical and Acoustic Data Registration

Since methods that rely on explicit opto-acoustic correspondences have to be avoided [2,20], in our approach the acoustic 3D point clouds representing the orientation rig are matched to the optical counterpart by using the Iterative Closest Point (ICP) algorithm, an iterative least-square technique used for the registration of rigid 3D shapes. This approach eliminates the need to perform any feature extraction or to specify any explicit feature correspondence.

The basic idea behind the ICP algorithm is that, under certain conditions, the point correspondences provided by sets of closest points are reasonable approximations for the true point correspondences. If the process of finding the closest-point sets and then solving the least-square function is iterated, the solution will converge to a local minimum, but there is no guarantee that this will correspond to the actual global minimum.

In our solution, the global convergence is achieved through a data pre-processing to clean up the 3D point clouds from noise or potential outliers, and a coarse registration stage that gives a fairly good initial alignment of the two 3D point clouds.

Therefore, considering the pair $P_o P_a$ of optical and acoustic 3D point clouds, respectively, the associated ^0T_a is determined as a composition of transformations obtained through a coarse and fine registration technique.

Taking as a reference system the local reference frame of the stereo optical camera, the coarse registration stage was carried out through two operations:

1. calculation of ^1T_a by the orientation of the acoustic camera local reference frame, in such a way that the Z axis represents the depth of the scene, in line with the optical system; this step is necessary because the data acquired by the Echoscope are represented in a local reference system in which the depth of the scene is expressed along the Y axis;
2. rough alignment of the pair of 3D point clouds $P_o P_a$ that, through an estimate of the centroid of the two 3D point clouds, determines the translation vector t that relates them (assuming that the rotation matrix R is unitary). As a result of this operation, we obtain the transformation matrix $^2T_1$.

Concerning the step of fine registration, a Matlab® implementation of the ICP algorithm has been applied to the pair $P_o P_a$ aligned in the previous step, to obtain the transformation matrix $^0T_2$. Downstream of the previous operations, the unknown rigid transformation matrix ^0T_a is obtained as (Figure 2):

$$^0T_a = {}^0T_2 \times {}^2T_1 \times {}^1T_a \tag{2}$$

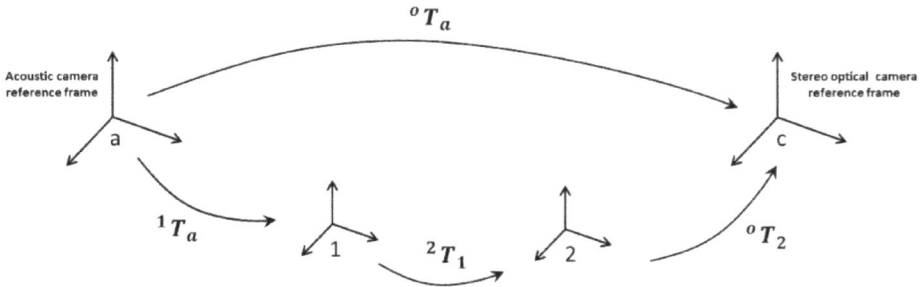

Figure 2. Sequence of transformations for computing the ^0T_a matrix.

2.3. Statistical Estimation of the Geometric Transformation

From the data registration step described in Section 2.2, we have obtained a first estimation of the unknown geometric transformation that is strongly affected by noise and data dispersion. To perform a robust estimation as the working distance changes, we have decided to acquire several data of the transformation for outputting the optimal estimation through a statistical approach.

Three different methods of statistical processing have been implemented and subsequently compared, to estimate the rigid transformation matrix $^0T_a^*$ from the n matrices $^0T_{a,n}$ obtained downstream of the operation of coarse and fine registration applied to n pairs $P_{0,n}P_{a,n}$ of 3D point clouds.

As a first hypothesis, we tried to obtain the elements of the final transformation matrix $^0T_a^*$ through an average calculation on the $^0T_{a,n}$ matrices included in the dataset. However, this operation, if applied directly on the homogeneous matrices $^0T_{a,n}$, compromises the orthonormality of the result, so we decided to apply this solution representing the $^0T_{a,n}$ in the corresponding R_n rotation matrices and t_n translation vectors.

Subsequently, as the mean rotation matrix $\overline{R}$ and the mean translation vector $\overline{t}$ were obtained, we put the final transformation matrix $^0T_a^* = {}^0\left[\begin{array}{cc} \overline{R} & \overline{t} \\ 0 & 1 \end{array} \right]_a^*$. The mean translation vector $\overline{t}$ was calculated as a simple arithmetic mean on the elements of the vectors t_n. Concerning the calculation of the mean rotation matrix $\overline{R}$, there could be multiple solutions. In fact, as reported in [29], there are several formulations in the literature to obtain the mean rotation matrix, either based on Euclidean or Riemannian distance metrics. The same paper also shows how, where data does not present a high variability (as it can be assumed for the present case, since the $^0T_{a,n}$ are estimates of the same transformation matrix), the arithmetic mean applied to the rotation vectors φ_n, obtained in turn from the corresponding rotation matrices R_n, represents an approximate solution to both the Riemannian $\overline{R}_{Riem}$ and Euclidean $\overline{R}_{Eucl}$ averages, and how the calculation of the mean rotation vector $\overline{\varphi}$ leads to different results for values beyond the third decimal place. Based on these considerations, the mean rotation matrix was derived from the mean rotation vector $\overline{\varphi}$: the latter is calculated as the arithmetic mean of the rotation vectors φ_n.

As a second hypothesis, we tried to obtain the rigid transformation matrix $^0T_a^*$ as above, but including in the calculation only one subset of the $^0T_{a,n}$ from the initial dataset. This selection was made to eliminate outliers from the calculation of mean vectors $\overline{\varphi}$ and $\overline{t}$, as these values lead to a polarization of the results obtained from the application of the arithmetic mean to the vectors φ_n and t_n. In particular, an algorithm was implemented in Matlab$^®$ for the automatic determination of the set of matrices $^0T_{a,s}$ with $s \leqslant n$, to be included in the calculation of $^0T_a^*$. It operates in the following way:

1. for each $^oT_{a,i}$ with $i \in [1, n]$, if the RMSE_i (Root Mean Square Error calculated by applying the ICP algorithm in the fine registration stage) is less than RMSE^*, than $^oT_{a,i} \in {}^oT_{a,r}$, with $r \leqslant n$, otherwise it is discarded;

2. for each $^oT_{a,r}$, the translation vectors t_r and the rotation vectors φ_r are determined;

3. for each component x_j of the vectors t_r and φ_r with $j \in [1,3]$, it determines the interquartile range $\text{IQR}(x_j)$ (i.e., the difference between the third $q_{0,75}$ and first quartile $q_{0,25}$ in the ordered series of data);

4. if $(x_j - q_{0,75})$ or $(q_{0,25} - x_j) > 3 \text{ IQR}(x_j)$, then $t_r(x_j)$ (or $\varphi_r(x_j)$) is discarded, otherwise $t_r(x_j)$ (or $\varphi_r(x_j)) \in t_s$ (or φ_s);

5. calculates $^oT_a^*$ from t_s and φ_s vectors as in the first algorithm.

The third proposed solution is based on an algorithm implemented in Matlab®, which automatically determines the final transformation matrix $^oT_a^*$, by selecting it from the $^oT_{a,n}$ matrices included in the initial dataset. It operates in the following way:

1. for each $^oT_{a,n}$, it applies this transformation to the $P_{a,n}$ 3D acoustic point clouds, to align them with the corresponding 3D optical point clouds $P_{o,n}$;

2. for each pair $P_{o,n}P_{a,n}$, the mean distance $d_{oa,n}$ between the points of the 3D optical cloud and the corresponding of the acoustic 3D point cloud is calculated;

3. calculates $\overline{d_n}$ as the mean of $d_{oa,n}$;

4. selects the $\overline{d_{n,min}}$ as the minimum value of $\overline{d_n}$;

5. assumes as $^oT_a^*$ the transformation $^oT_{a,n}$ corresponding to $\overline{d_{n,min}}$.

3. Experimental Setup

3.1. System Configuration

The proposed system is composed of a stereo optical camera and an acoustic camera that will be attached to the ROV rigid frame. We used a Coda Echoscope camera for acoustic sensing. Through the ensonification of the whole viewing volume with a single ping, it uses the phased array technology to process approximately 16,000 beams simultaneously and generate a real-time 3D acoustic image of the entire observed scene. Concerning the optical component of the system, an optical stereo camera, consisting of two ultra-compact digital cameras housed in custom-made waterproof cases, has been developed. The system layout was defined through the use of a CAD model to ensure the maximum overlap of both FOVs at the minimum working distance (about 1 m, minimum working range of the Echoscope sonar) (Figure 3).

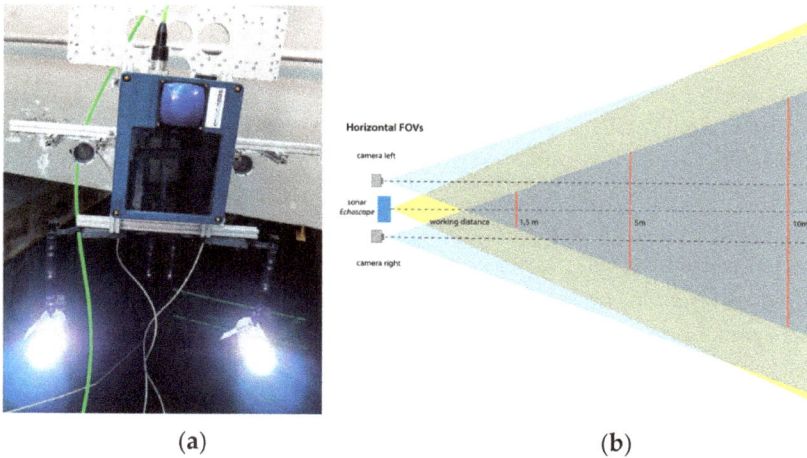

Horizontal FOVs

camera left

sonar
Echoscope

working distance   1,5 m   5m   10m

camera right

(a)   (b)

Figure 3. Layout of the opto-acoustic 3D camera (**a**). A scheme showing the overlap of the FOVs of Echoscope sonar and optical cameras (**b**).

3.1.1. Stereo Optical Camera

The stereo optical camera is the result of a research activity conducted at the Department of Mechanical, Energetic and Management Engineering (DIMEG) of the University of Calabria in the field of the underwater stereo photogrammetry, both for passive and active applications [3,30].

The stereo rig is composed of two ultra-compact digital cameras Point Grey Flea 2, with a Charge Coupled Device (CCD) sensor format of 1/3", a resolution of 0.8 MP (pixel size of 4.65 μm) and a frame rate of 30 fps. The devices are also equipped with a pair of 8.5 mm Pentax C30811TH optical packages.

Since the two cameras were not specifically made for the underwater environment, we have designed and constructed two waterproof housings. The body is made of aluminum to ensure efficient heat dissipation, while the flat port of the camera housing is made of polycarbonate. This solution leads to a reduction of the FOV caused by the refraction of the air-water interface, but its construction is easier. The camera is fixed within the case through an appropriate support that also works as a heat sink.

The system is able to generate a 3D point cloud (about 200,000 points) at a frame rate of 7 fps. The Libelas library is used for the implementation of the stereo matching algorithm and the generation of the disparity map. We have verified that, in real-life conditions (*i.e.*, at sea), the Efficient LArge-scale Stereo (ELAS) algorithm [31] allows for obtaining a more robust and accurate 3D point cloud if compared to that obtained with the library OpenCV [32], which is used, instead, for the rectification of the stereo pair.

325

The main goal of the system is to improve the perception of the underwater scene by providing an output that enabling direct identification of individual objects (rocks, pipelines, walls, archeological artifacts). Their size ranges from centimeters to meters. To identify them with sufficient detail, almost 15 pixels per object are needed [33]. Therefore, a ground sample distance (GSD) less than 10 mm should be assured in the working volume and, consequently, an accuracy better than 10 mm has to be guaranteed. Taking into account the refractive effect of the medium on the camera parameters, the maximum operating distance should range from 8 m to 10 m (depending upon the quality of the acquired images). In this operating range, following previous works [34], we expect an operational accuracy from 0.2% to 0.7% for length measurements in real conditions.

3.1.2. Acoustic Camera

Echoscope is a 3D acoustic camera that provides real-time, high angular resolution images of the acoustic environment. It consists of two distinct subsystems: one containing the acoustic units for transmitting and receiving signals (TX/RX unit), and another for processing the signals to be used in the beamforming process. The head of the acoustic camera has two physically distinct sections, TX array and RX array, both made with conventional piezoelectric sensors.

The acoustic camera ensonifies the volume of observation through a single acoustic pulse and receives the energy reflected from an object that intercepts the propagation through a receiving array of hydrophonic sensors. The TX array for the generation of transmission signals is a wide-beam projector aimed to the ensonification of the environment. The transmission pulses are emitted with a frequency of 610 kHz (or 375 KHz) and a duration of 20 ms: these are generated with a repetition period between subsequent pulses equal to 100 ms, to ensure an adequate number of frames per second.

The receiving section, placed at the bottom of the acoustic camera, is a square planar array made of 48 × 48 analog sensors/channels. It is characterized by an acquisition range between 5 and 10 ms, with a sampling frequency of 10 MHz. Each one of the receiving channels will acquire 1024 samples, for a total of 2304 × 1024 samples. A beamforming process will be conducted on these samples for the generation of high angular resolution acoustic images.

The working range of the Echoscope is from 1 m to 100 m, with a range resolution of 30 mm and a beam spacing of 0.19°. As stated by the manufacturer, the system meets, in real-time, the IHO S-44 Special Order Quality Surveys standard [35], with no post-processing of the point clouds. Therefore, we expect an accuracy of data less than 261 mm at a working distance of 10 m.

3.2. Laboratory Setup

The alignment methodology has been tested at the electro-acoustic laboratory of the Whitehead Sistemi Subacquei S.p.A. (WASS) in Pozzuoli (Naples, Italy), equipped with a water tank for acoustic measurements ($11 \times 5 \times 7$ m) and the necessary tools for handling both the prototype of opto-acoustic camera (telescopic pole) and the target to be acquired (conveyor belt). To ensure a mechanical support for the system, a frame consisting of aluminum profiles has been designed and built. This frame houses the waterproof cases of the two optical cameras and the acoustic section of the Echoscope sonar, the latter connected through a rear plate. The entire structure was then fixed to a support bracket for a mechanical interfacing with a telescopic pole (Figure 4).

The acoustic section is connected, via a proprietary bus, to a Power Supply Unit (PSU), while the workstation (which hosts the software for the management of the entire acoustic subsystem, *i.e.*, the Data Integration Unit or DIU that manages the flow of data coming from the acoustic section with the related I/O controls, and the software UIS Survey Explorer for the visualization of the acoustic 3D point cloud) is connected via a RS232 port (sending and receiving commands) and an Ethernet 10 Mbps interface (receiving data). In turn, the stereo optical subsystem is connected to a PC that allows for the management and the configuration of the acquisition parameters via a Firewire 800 interface, by means of the Flycapture software. The synchronization of the stereoscopic camera is carried out through the trigger function of this software.

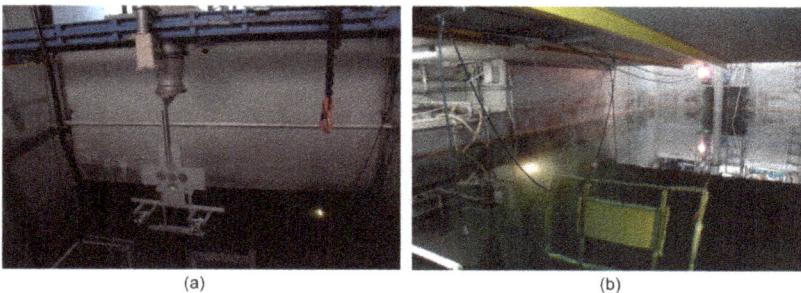

(a)   (b)

Figure 4. The telescopic pole for the handling of the opto-acoustic camera prototype (**a**). Conveyor belt system for the handling of the orientation fixture (**b**).

A schematic layout of the designed laboratory setup is depicted in Figure 5: it shows the connections between the 3D acquisition system (consisting of the acoustic section of the Echoscope sonar and the waterproof cases in which the two optical cameras are housed) and the PC, and the various systems for the handling of the target and the opto-acoustic camera itself.

327

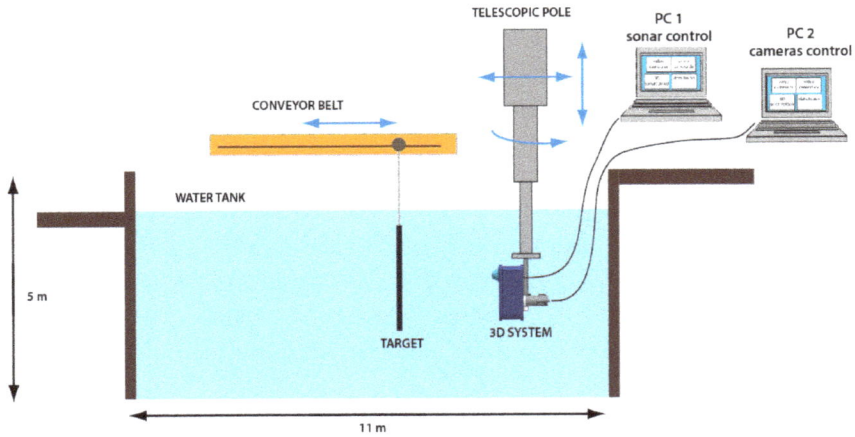

Figure 5. Experimental setup developed at the WASS electro-acoustic lab.

The opto-acoustic system is fixed to a telescopic pole that allows for its immersion in the water tank and offers the possibility of handling it within the four degrees of freedom shown in the diagram, while the target to be acquired is fixed to a conveyor belt that allows for its longitudinal translation.

Figure 6 shows the targets used in various experimental phases. Their sizes were determined through the analysis of the FOV of the two subsystems in the operating range. In addition to the orientation fixture (an aluminum frame of 2 m × 2.5 m with a central checkerboard of 9 × 7 squares, each one with a size of 100 × 100 mm), we built an additional panel with objects of different shapes and materials. In particular, we can see two ceramic vases, marble and tufa tiles, bricks, roofing tiles, and a mask made of terracotta.

(a)   (b)

Figure 6. Orientation fixture (**a**). Target made of objects with different materials and shapes (**b**).

328

4. Experimentation

In this Section we will describe the operations carried out at the different stages of the planned series of tests, to evaluate the performance of the designed opto-acoustic camera prototype. The purpose of the tests is twofold: on the one hand, we want to evaluate which one of the three statistical processing methods ends up to be more effective; on the other hand, we want to get a validation of the alignment method. This has been obtained by estimating the mean distance μ (and the respective standard deviation σ) among the aligned optical and acoustic 3D point clouds of the target panel.

During the selection of the most efficient methodology to be adopted for the calculation of the matrix ${}^{o}T_{a}^{*}$, as discussed in Section 2.3 by comparing the implementation hypotheses, we would assess the complexity of the adopted solution and the quality of the obtained results, considering that there is a limit to the accuracy of the registration of the acoustic and optical 3D point clouds. This limit is due to the error that occurs in the reconstruction of the 3D point cloud by the stereo optical system and the accuracy of the acoustic camera in the reconstruction of the rig. In fact, any result below this threshold would be completely random and linked to the specific dataset used to determine the matrix ${}^{o}T_{a}^{*}$.

4.1. Image Acquisition

Prior to the actual acquisition stage, we carried out some operations to configure the optical and acoustic sensors of the system. In particular, concerning the stereo optical subsystem, we adjusted the focus settings for the cameras in air and performed some acquisition tests underwater, to verify the correct superposition of the FOVs, as provided for in the design of the support frame.

The Echoscope sonar has been adjusted to excite the transmission section with the highest working frequency, corresponding to 610 KHz. In this stage we found that the optical axes had to be kept slightly convergent rather than parallel, to correct the overlap of the FOVs of the two cameras in the range of operation. As for the alignment methodology, we acquired a sequence of 20 different poses of the orientation fixture, by positioning it at a distance varying from about 1.5 m to 10 m and in different orientations. The panel was rolled, tilted and twisted to reduce the correlation of the calibration parameters of the optical stereo camera, while the opto-acoustic camera was clamped to a telescopic pole varying its transverse position and orientation (Figure 7).

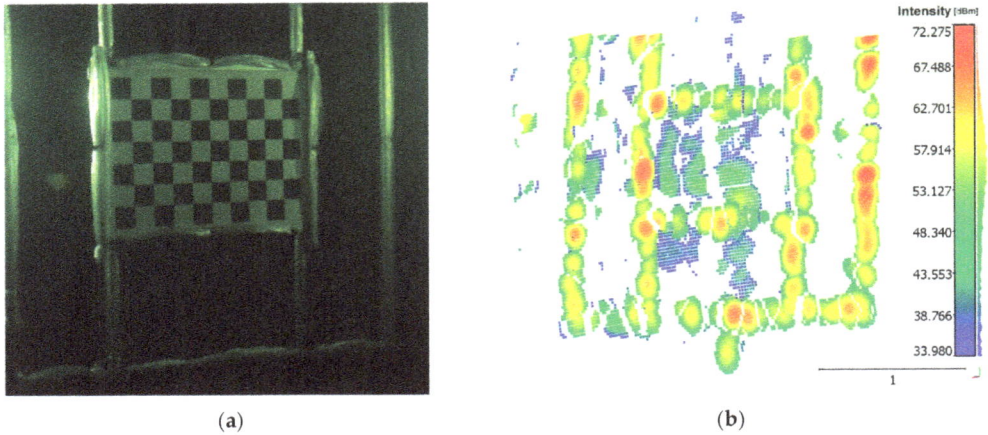

(a)   (b)

Figure 7. Acquisition of a pose of the orientation rig: (**a**) optical; (**b**) acoustic.

Three different poses of the panel containing objects of various shapes and materials were acquired (Figure 8). The panel was positioned on the side edge of the water tank (at a distance of approx. 2.5 m from the opto-acoustic camera) and the camera was handled by means of the telescopic pole.

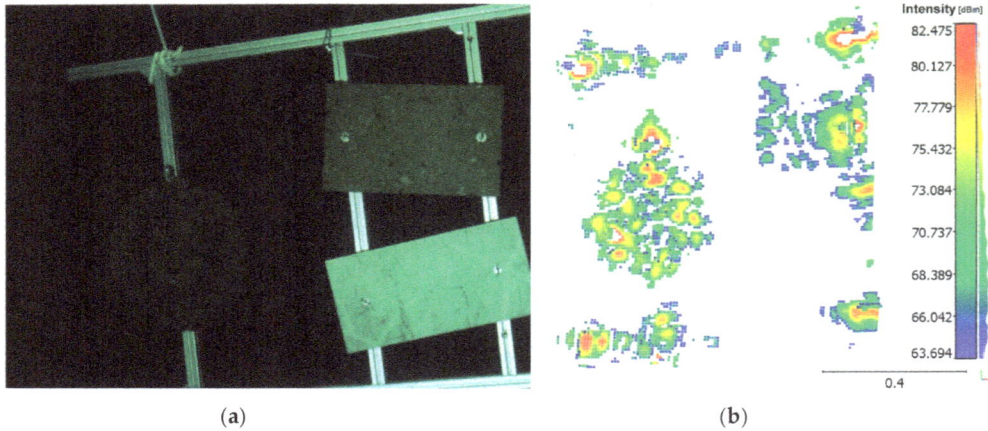

(a)   (b)

Figure 8. Target acquisition with objects: (**a**) optical; (**b**) acoustic.

The output of the acoustic camera, which—in addition to the image of the acquired 3D point cloud—shows the scalar field representing the intensity of the echoes received for each acquired point, has immediately shown the strong limits of Echoscope in close range applications. In fact, contrary to what is stated by the manufacturer, the sonar was not able to process correctly the scattering signal

330

returned by the objects placed at distances less than 2.5 m. This could be due to two causes: either an excessive duration of the transmitted pulse (a pulse with a duration of 2 ms generates a blind range equal to 2 m) or, more likely, the type of beamforming applied, which does not allow for an appropriate phase correction for focusing at close range.

4.2. Data Processing

The processing pipeline of the data acquired through the optical and acoustic sensors is shown in Figure 9. Starting from the synchronous acquisition of the n poses of the orientation rig during the early stage of the process, the system outputs n pairs of 3D point clouds, where the n-th pair is formed by the optical $P_{o,n}$ and the acoustic $P_{a,n}$ 3D point clouds, to calculate, by means of coarse and fine registration technique, n estimates $^{o}T_{a,n}$ of the rigid transformation matrix.

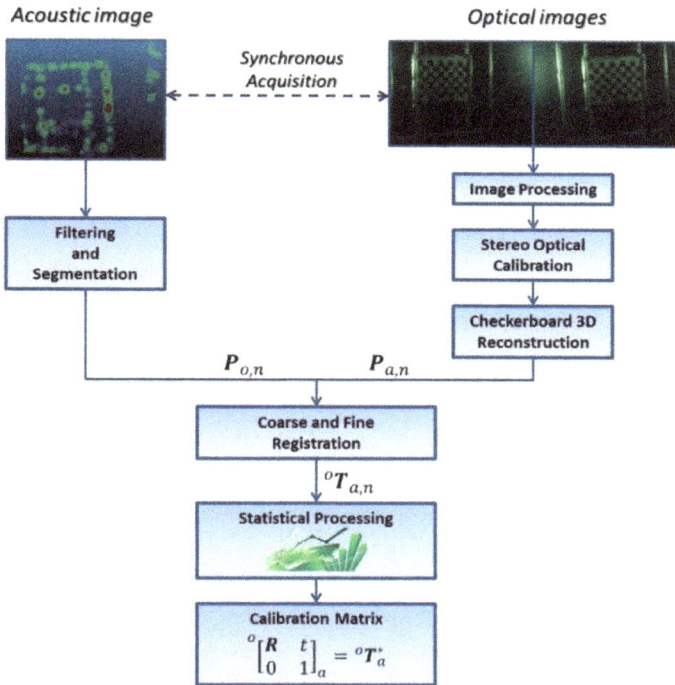

Figure 9. Processing pipeline of acoustic and optical data.

At the end of the process, as described in Section 2.3, the final transformation matrix $^{o}T_{a}^{*}$ is obtained by statistically processing the dataset composed of n transformation matrices $^{o}T_{a,n}$ obtained downstream of the previous registration stage.

4.2.1. Acoustic Image Processing

The 3D data provided by the acoustic camera can be corrupted either by false reflections caused by the secondary lobes of the receiving array or by the noise present in the acquisition phase of the backscattering signals. Although the Echoscope directly performs a low-level preprocessing, it is still necessary to conduct some appropriate operations on the acoustic 3D point cloud, to clean up the images from noise or potential outliers. So it is evident that the operations of filtering (noise reduction and the elimination of possible outliers) and segmentation (differentiation of objects and background in the observed scene) are to be considered as preliminary and mandatory steps for the execution of all integration algorithms to be applied to this specific type of data [36,37].

The solution adopted in this first implementation of the alignment method is based on a thresholding method, performed through the open source software CloudCompare [38].

Given the range and intensity information provided by the acoustic camera, we used two threshold tests to discriminate between actual backscattered echoes and clutters. The method operates in the following way:

1. Test 1: assuming a Gaussian distribution for echo amplitude noise, 3D points with intensity lower than a threshold thr_1 are discarded, while the others are considered as belonging to the target;
2. Test 2: given the histogram of the range values and knowing the geometry of the rig, 3D points outside an interval defined by the thresholds thr_2 and thr_3 are discarded, while the others are considered as belonging to the target.

Threshold values are manually defined, through a vision inspection of the echo and range distribution, respectively (Figure 10), as the implementation of an automated procedure would require further, more focused research. However, different automatic thresholding and filtering methods are described in the literature. A comprehensive review is reported in [36].

Applying the above thresholding method, we have obtained the results shown in Figure 11.

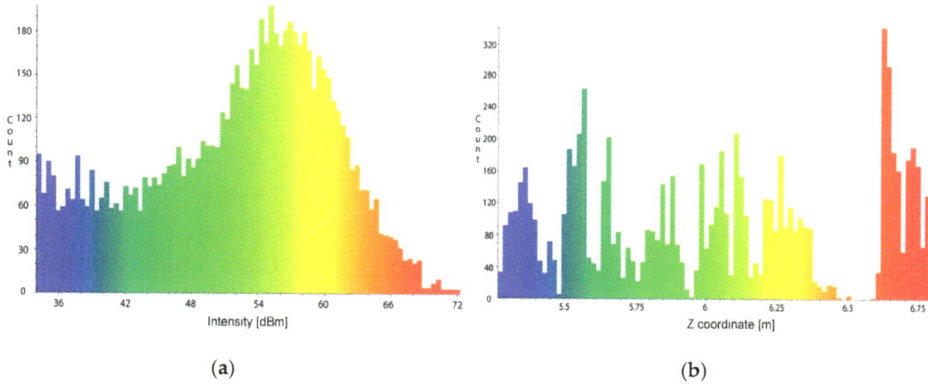

Figure 10. Distribution of the echo intensity (**a**), histogram of the range values (**b**).

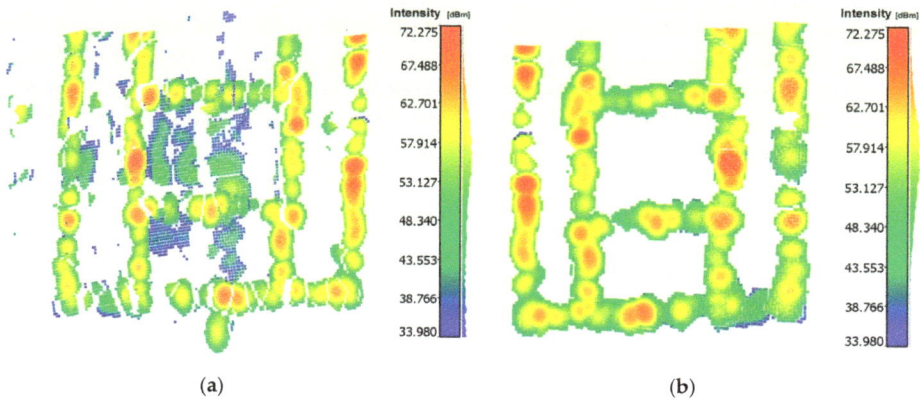

Figure 11. Acoustic 3D point cloud: before (**a**) and after filtering and segmentation stage (**b**).

4.2.2. Stereo Optical Image Processing

To obtain the 3D point clouds of the orientation rig for each pose assumed during the acquisition stage, the stereo optical images are processed through the algorithms reported in Section 3.1.2.

Before 3D reconstruction, the underwater images have been pre-processed to reduce blur due to scattering effects and correct color casts (remove greenish-blue components). This can be done through two different approaches: digital restoration techniques or methods of image enhancement [39,40]. In the proposed alignment method, the adopted solution provides for an image enhancement methodology based on the technique presented in [41].

As a first step, sharp filtering has been performed to remove the fog in the images, due to the scattering effects, which decreased image contrast and increased the blur. Secondly, the images are color corrected through a white balancing in the lαβ color space. It allows for removing the color casts in underwater images that typically suffer from color alterations, by balancing the chromatic components (α and β), while the luminance component (l) is used to improve image contrast by cut-off and histogram stretching (Figure 12). The method is particularly suitable to process our datasets, as it has been proposed for applications like close-range acquisition in nadir direction.

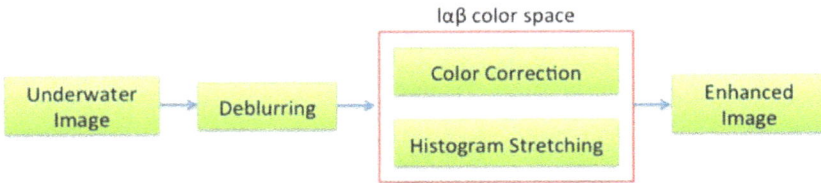

Figure 12. Flow-chart of the image enhancement steps.

Figure 13 shows the results of the algorithm applied to the optical images shown in Figures 7 and 8 respectively.

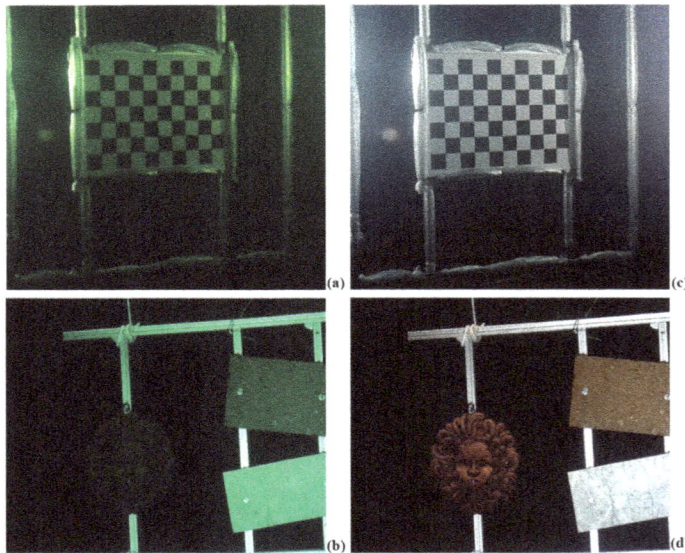

Figure 13. Optical images of the orientation rig (**a**) and target objects; (**b**) after the application of the color enhancement algorithm (**c,d**).

Subsequently, the acquired optical images have been processed to generate the 3D point clouds of the orientation rig and the panel with objects in various poses. The application of the algorithms has allowed for the reconstruction of 11 over 20 poses of the orientation frame, since in 9 poses the stereo matching algorithm was not able to reconstruct the target. A stage of manual cleaning of the raw 3D point clouds performed with CloudCompare, downstream of the reconstruction process, was necessary to eliminate several outlier points. Automatic methods can be used for this stage. For example, the algorithms present in the Point Cloud Library (PCL) [42]. Figure 14 shows two examples of optical 3D point clouds of the orientation frame and the target, respectively.

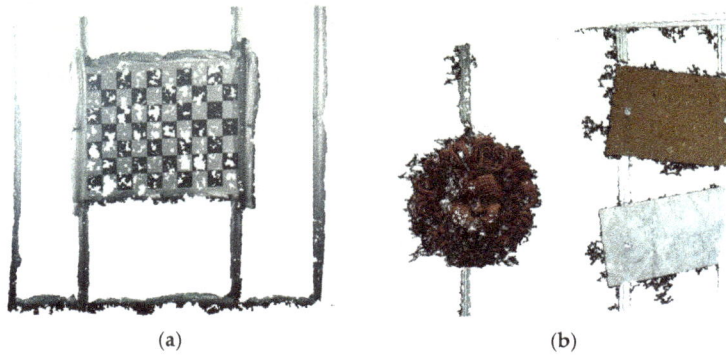

(a)　　　　　　　　　　　　(b)

Figure 14. Optical 3D reconstruction of the orientation rig (**a**) and the target with objects (**b**).

4.2.3. Optical and Acoustic Registration

As described in Section 2.2, the relative orientation between the optical and acoustic system is obtained downstream of a statistical processing of $^oT_{a,n}$ estimates coming from the registration of the n pairs $P_{o,n}P_{a,n}$ of optical and acoustic 3D point clouds. This registration stage was performed for each of the pairs $P_{o,n}P_{a,n}$ with $n \in [1,11]$ derived from the previous step of the implemented alignment method. Figure 15 shows the results of the algorithms on one of the 11 acquired pairs.

4.2.4. Statistical Processing of the Transformation Matrices

The different methods of statistical processing described in Section 2.3 were applied on the dataset composed of $^oT_{a,n}$ with $n \in [1,11]$ estimates of the unknown rigid transformation matrix, obtaining three approximations of the $^oT_a^*$ matrix, *i.e.*, $^oT_{a,1}^*$, $^oT_{a,2}^*$, $^oT_{a,3}^*$ respectively.

Concerning the second methodology, the choice of the RMSE* value was carried out by analyzing the graph in Figure 16, that shows the RMSE values associated

335

with each of the $P_{o,n}P_{a,n}$ pairs of registered 3D point clouds. These were ranked in ascending order along the x-axis, as a function of the acquisition distance by the opto-acoustic imaging system. As can be seen from the graph, the pairs $P_{o,n}P_{a,n}$ with $n = 1, 10, 11$ present a higher RMSE value than the remaining ones, so the RMSE* was set to 65 mm, to eliminate the related matrices ${}^{o}T_{a,n}$ from the average calculation.

(a)   (b)

Figure 15. Optical and acoustic 3D point clouds after coarse (**a**) and fine; (**b**) registration stages.

By applying this solution, five of the 11 estimates ${}^{o}T_{a,n}$ of the rigid transformation matrix have been discarded. It is interesting to note the existing relation among the RMSE values of the ICP algorithm and the different poses of the rig. The sample with index $i = 1$ in the graph (Figure 16) is associated with the only pair of 3D point clouds $P_{o,1}P_{a,1}$ that we were able to obtain from the first nine poses of the orientation frame over a total of 20 poses. These have been acquired at a distance of between 1.5 m and 4.5 m from the opto-acoustic camera. The samples with index $i \in [2,9]$ refer instead to the eight poses acquired at a distance of between 4.5 m and 7 m. Finally, the samples with index $i = 10, 11$ refer to the pairs of 3D point clouds $P_{o,10}P_{a,10}$ and $P_{o,11}P_{a,11}$, obtained from two of the three final poses acquired at a distance of between 7 m and 10 m. The pairs with high RMSE values are associated with the poses acquired at a distance of less than 4.5 m and more than 7 m. Thus, we can deduce that the optimal condition for the opto-acoustic camera, with respect to the specific target, is within distances of between about 4.5 m and 7 m. As one can expect, at a distance of less than 4.5 m, the acoustic component of the system presents the main limitations in reconstructing the acquired target, while for distances greater than 7 m, the stereo optical camera suffers from poor performance, as the stereo algorithm fails to process the acquired images.

336

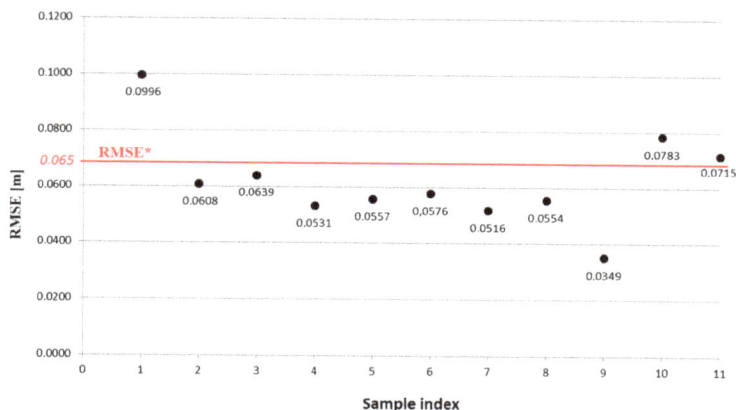

Figure 16. Graph of RMSE values obtained downstream of the registration process (Section 4.2.3) on $P_{0,n}P_{a,n}$ pairs of 3D point clouds.

5. Results

From the three statistical processes we have estimated the rigid transformation matrix that is needed to perform the relative orientation of the opto-acoustic system. Figure 17 shows the relative positions of the coordinate systems related to each device, for one pose of the orientation frame.

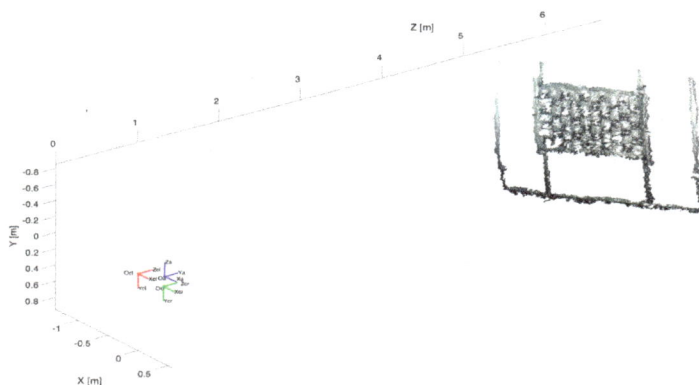

Figure 17. Relative pose of the optical cameras and Echoscope coordinate systems for one pose of the orientation frame.

Table 1 shows the Euler angles computed by the rotation matrix and the norm of the translation vector for each of the three methods. We should notice that there is a relevant difference among the values calculated by the three methods and, in particular, for the magnitude of the translation.

337

Table 1. Euler angles and norm of the translation vector for the three statistical methods.

	I Method	II Method	III Method
α (°)	89.763	87.498	88.394
β (°)	−6.771	−6.154	−5.549
γ (°)	2.869	1.268	0.276
Norm(t) (mm)	0.4920	0.3188	0.2718

We have analyzed the obtained results, in terms of mean registration error, by applying the three transformation matrices on the pair of optical and acoustic 3D point clouds belonging to the panel with objects (Figure 18). Table 2 reports the mean registration error μ and the standard deviation σ.

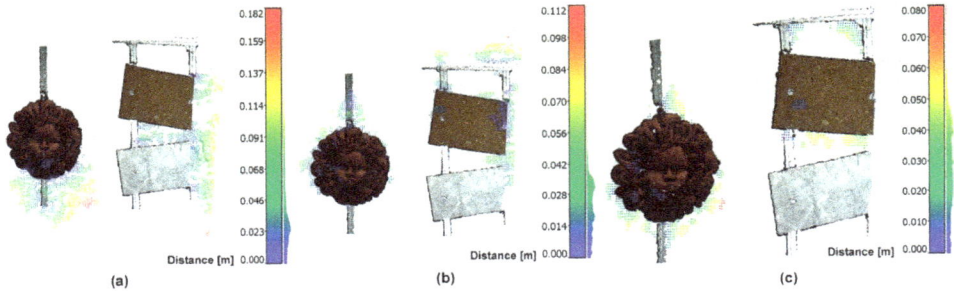

(a)   (b)   (c)

Figure 18. Alignment of opto-acoustic 3D point clouds of the panel with objects, obtained through the application of matrices $^{o}T^{*}_{a,1}$ (**a**), $^{o}T^{*}_{a,2}$ (**b**), $^{o}T^{*}_{a,3}$ (**c**).

Table 2. Mean error of the optical and acoustic registration of the target with objects.

	I Method	II Method	III Method
μ (mm)	46.2	23.4	18.2
σ (mm)	36.5	14.3	13.7

The data confirm that the third statistical method provides the best results. As can be noted in Figure 18c, the values with the greater distance are due to the noise of the acoustic 3D point cloud, especially in proximity of the terracotta mask and the aluminum frame of the panel itself.

To define the accuracy of each system, we consider as ground truth the known-size of the orientation frame. In particular, for the optical system we obtained a value of 6.45 mm as RMSE computed by measuring the diagonals of the checkerboard (of which the dimensions were known) for all poses on the optical point clouds. This is equivalent to a length accuracy of 0.4%. The same procedure was followed for the acoustic system, knowing the external dimensions of the aluminum frame (2 × 2.5 m), and measuring them on the 3D acoustic point clouds for each pose. In the latter case we obtain a RMSE of 15 mm, equivalent to a length accuracy of 0.75%. The obtained results are consistent with the expected ones.

To finalize the validation, we have computed the accuracy of the integrated system by measuring several known lengths on the opto-acoustic 3D point clouds. In this case we have obtained a RMSE of 17.8 mm.

The effectiveness and the potentiality of the system can be further evaluated by analyzing the image in Figure 19. This was obtained by integrating off-line the low resolution acoustic 3D point cloud with the high resolution optical cloud of the acquired scene (Figure 19a). Due to the different materials, the marble tile was present only in the optical acquisition (Figure 19b) while the tufa tile was featured only in the acoustical one (Figure 19c). The integration of the acoustic point cloud (red points) with the optical point cloud (blue points) allows for compensating the errors in the full field-acquisition of the acoustic camera and better discriminate the objects from the background (sky blue points).

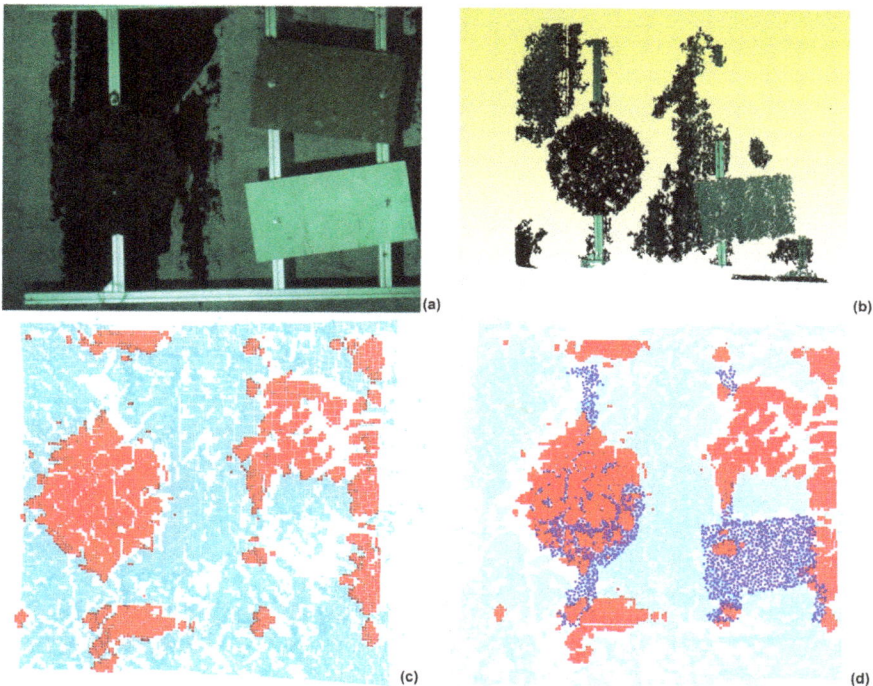

Figure 19. A pose of the target with objects (**a**), optical 3D point cloud; (**b**), acoustic point cloud; (**c**), opto-acoustic reconstruction of the scene (**d**).

6. Summary, Discussion and Outlook

This paper presents a first step towards the realization of an opto-acoustic system for the 3D imaging of the underwater scenes. The system is composed of a stereo optical camera and a 3D acoustic camera and is designed to be installed on a ROV, to improve its control capabilities by the operator during the navigation in turbid water conditions or in close range manipulation operations. The presented opto-acoustic camera is the first underwater systems employing two 3D imaging sensors to obtain a 3D representation of the underwater scene.

In the current work, a multi-sensor registration method has been proposed for the automatic integration of 3D data acquired from the two systems, specifically for this application. We have solved the challenging problem to determine automatically a correspondence between the two different data types, by means of an orientation frame that is able to highlight some geometric features detected from both sensors. Due to the low resolution and the high dispersion of acoustic data, a statistical processing has been performed on several estimations of the geometric transformation that allows for aligning the two datasets. The statistical approach increases the robustness of the estimation by reducing the effect of the noise component. Moreover, we have described in detail the processing stages needed to prepare the data to be integrated. The results demonstrate that the presented methodology is able to return a 3D image made of integrated data. This calls for further investigations in real conditions.

The comparison of technical details and performance capability with other systems based on this technique is not an easy task. To the best of the authors' knowledge, few works in the literature present an evaluation of the registration accuracy of the 3D opto-acoustic data. In [20], an average error of 24.2 mm is obtained when computing the distance of all the reprojected multibeam points to their respective calibration planes while, in [12], the authors report an average registration error between the 3D acoustic data and the CAD model of a pipe of approximately 70 mm. Finally, in [14] the estimated 3D points are within 3.5% of their distances to the optical cameras, utilizing the reconstruction of a plane as ground truth. A maximum error of 63 mm is obtained. These results show how the integration accuracy of our alignment method, as reported in Section 5, has the same magnitude.

The main challenges for the future developments of such a system will be: (a) the search for a 3D acoustic camera that can provide a better performance in the close range to obtain better data on the operating volume of the ROV; (b) the implementation of on-line data integration and visualization techniques, to make the ROV operator able to interpret in the best way the data generated by the integration of optical and acoustic sensors; (c) conducting experimental tests in real conditions.

Acknowledgments: This work has been partially supported by the Project "CoMAS" (PON01 02140), financed by the MIUR under the PON "R&C" 2007/2013 (D.D. Prot.n.01/Ric.18.1.2010).

Author Contributions: Gianfranco Bianco and Fabio Bruno conceived and designed the experiments; Antonio Lagudi, Gianfranco Bianco and Fabio Bruno performed the experiments; Antonio Lagudi and Gianfranco Bianco analyzed the data; Maurizio Muzzupappa contributed materials/analysis tools; Antonio Lagudi and Gianfranco Bianco wrote the paper.

Conflicts of Interest: The authors declare no conflict of interest.

References

1   Johnson-Roberson, M.; Pizarro, O.; Williams, S.B.; Mahon, I. Generation and visualization of large-scale three-dimensional reconstructions from underwater robotic surveys. *J. Field Robot.* **2010**, *27*, 21–51.

2.   Drap, P.; Merad, D.; Boï, J.M.; Mahiddine, A.; Peloso, D.; Chemisky, B.; Seguin, E.; Alcala, F.; Bianchimani, O. Underwater multimodal survey: Merging optical and acoustic data. In *Underwater Seascapes*; Springer International Publishing: Berlin, Germany, 2014; pp. 221–238.

3.   Bruno, F.; Bianco, G.; Muzzupappa, M.; Barone, S.; Razionale, A.V. Experimentation of structured light and stereo vision for underwater 3D reconstruction. *ISPRS J. Photogramm. Remote Sens.* **2011**, *66*, 508–518.

4.   Singh, H.; Salgian, G.; Eustice, R.; Mandelbaum, R. Sensor fusion of structure-from-motion, bathymetric 3D, and beacon-based navigation modalities. In Proceedings of the ICRA'02 International Conference on Robotics and Automation, Washington, DC, USA, 11–15 May 2002; Volume 4, pp. 4024–4031.

5.   Snavely, N.; Seitz, S.M.; Szeliski, R. Modeling the world from internet photo collections. *Int. J. Comput. Vis.* **2008**, *80*, 189–210.

6.   Dissanayake, M.G.; Newman, P.; Clark, S.; Durrant-Whyte, H.F.; Csorba, M. A solution to the simultaneous localization and map building (SLAM) problem. *IEEE Trans. Robot. Autom.* **2001**, *17*, 229–241.

7.   Williams, S.; Mahon, I. Simultaneous localisation and mapping on the great barrier reef. In Proceedings of the ICRA'04 International Conference on Robotics and Automation, New Orleans, LA, USA, 26 April–1 May 2004; Volume 2, pp. 1771–1776.

8.   Kunz, C.; Singh, H. Map building fusing acoustic and visual information using autonomous underwater vehicles. *J. Field Robot.* **2013**, *30*, 763–783.

9.   Grisetti, G.; Kummerle, R.; Stachniss, C.; Burgard, W. A tutorial on graph-based SLAM. *IEEE Intell. Transp. Syst. Mag.* **2010**, *2*, 31–43.

10. Fusiello, A.; Giannitrapani, R.; Isaia, V.; Murino, V. Virtual environment modeling by integrated optical and acoustic sensing. In Proceedings of the Second International Conference on 3-D Digital Imaging and Modeling, Ottawa, ON, Canada, 4–8 October 1999; pp. 437–446.

11. Fusiello, A.; Murino, V. Calibration of an optical-acoustic sensor. *Mach. Graph. Vis.* **2000**, *9*, 207–214.

12. Fusiello, A.; Murino, V. Augmented scene modeling and visualization by optical and acoustic sensor integration. *IEEE Trans. Vis. Comput. Graph.* **2004**, *10*, 625–636.

13. Hansen, R.K.; Andersen, P.A. A 3D underwater acoustic camera—properties and applications. In *Acoustical Imaging*; Springer US: New York, NY, USA, 1996; pp. 607–611.

14. Negahdaripour, S.; Sekkati, H.; Pirsiavash, H. Opti-acoustic stereo imaging: On system calibration and 3-D target reconstruction. *IEEE Trans. Image Process.* **2009**, *18*, 1203–1214.

15. Negahdaripour, S.; Pirsiavash, H.; Sekkati, H. Integration of Motion Cues in Optical and Sonar Videos for 3-D Positioning. In Proceedings of the CVPR'07 Conference on Computer Vision and Pattern Recognition, Minneapolis, MN, USA, 18–23 June 2007; pp. 1–8.

16. Sekkati, H.; Negahdaripour, S. Direct and indirect 3-D reconstruction from opti-acoustic stereo imaging. In Proceedings of the Third International Symposium 3D Data Processing, Visualization, and Transmission, 14–16 June 2006; pp. 615–622.

17. Negahdaripour, S. Calibration of DIDSON forward-scan acoustic video camera. In Proceedings of the MTS/IEEE OCEANS 2005, Washington, DC, USA, 17–23 September 2005; pp. 1287–1294.

18. Negahdaripour, S. Epipolar geometry of opti-acoustic stereo imaging. *IEEE Trans. Pattern Anal. Mach. Intell.* **2007**, *29*, 1776–1788.

19. Nicosevici, T.; Garcia, R. Online robust 3D mapping using structure from motion cues. In Proceedings of the MTS/IEEE OCEANS 2008, Kobe, Japan, 8–11 April 2008; pp. 1–7.

20. Hurtós, N.; Cufí, X.; Salvi, J. Calibration of optical camera coupled to acoustic multibeam for underwater 3D scene reconstruction. In Proceedings of the OCEANS 2010 IEEE-Sydney, Sydney, Australia, 24–27 May 2010; pp. 1–7.

21. Zhang, Q.; Pless, R. Extrinsic calibration of a camera and laser range finder (improves camera calibration). In Proceedings of the IEEE/RSJ International Conference on Intelligent Robots and Systems (IROS 2004), St. Louis, FM, USA, 28 September–2 October 2004; pp. 2301–2306.

22. Besl, P.J.; McKay, N.D. Method for registration of 3-D shapes. *IEEE Trans. Pattern Anal. Mach. Intell.* **1992**, *14*, 239–259.

23. Lowe, D.G. Fitting parameterized three-dimensional models to images. *IEEE Trans. Pattern Anal. Mach. Intell.* **1991**, *13*, 441–450.

24. Castellani, U.; Fusiello, A.; Murino, V.; Papaleo, L.; Puppo, E.; Pittore, M. A complete system for on-line 3D modelling from acoustic images. *Signal Proc. Image Commun.* **2005**, *20*, 832–852.

25. Huang, Y.; Qian, X.; Chen, S. Multi-sensor calibration through iterative registration and fusion. *Comput. Aided Des.* **2009**, *41*, 240–255.

26. Heikkila, J. Geometric Camera Calibration Using Circular Control Points. *IEEE Trans. Pattern Anal. Mach. Intell.* **2000**, *22*, 1066–1077.

27. Zhang, Z. A flexible new technique for camera calibration. *IEEE Trans. PAMI* **2000**, *22*, 1330–1334.

28. Shortis, M. Calibration Techniques for Accurate Measurements by Underwater Camera Systems. *Sensors* **2015**, *15*, 30810–30826.

29. Sharf, I.; Wolf, A.; Rubin, M.B. Arithmetic and geometric solutions for average rigid-body rotation. *Mech. Mach. Theory* **2010**, *45*, 1239–1251.

30. Bianco, G.; Gallo, A.; Bruno, F.; Muzzupappa, M. A Comparative Analysis between Active and Passive Techniques for Underwater 3D Reconstruction of Close-Range Objects. *Sensors* **2013**, *13*, 11007–11031.

31. Geiger, A.; Roser, M.; Urtasun, R. *Efficient Large-Scale Stereo Matching*; Springer: Berlin, Germany, 2011; pp. 25–38.

32. OpenCV. Available online: http://opencv.org/ (accessed on 15 January 2016).

33. Comer, R.; Kinn, G.; Light, D.; Mondello, C. Talking Digital. *Photogramm. Eng. Remote Sens.* **1998**, *64*, 1139–1142.

34. Barnes, H. *Oceanography and Marine Biology*; CRC Press: Boca Raton, FL, USA, 2003.

35. Mills, G.B. International hydrographic survey standards. *Int. Hydrogr. Rev.* **2015**, *75*, 79–85.

36. Murino, V.; Trucco, A. Three-dimensional image generation and processing in underwater acoustic vision. *IEEE Proc.* **2000**, *88*, 1903–1948.

37. Murino, V. Reconstruction and segmentation of underwater acoustic images combining confidence information in MRF models. *Pattern Recognit.* **2001**, *34*, 981–997.

38. CloudCompare. Available online: http://www.danielgm.net/cc/ (accessed on 15 January 2016).

39. Schettini, R.; Corchs, S. Underwater image processing: State of the art of restoration and image enhancement methods. *EURASIP J. Adv. Signal Process.* **2010**.

40. Mahiddine, A.; Seinturier, J.; Boi, D.P.J.; Drap, P.; Merad, D.; Long, L. Underwater image preprocessing for automated photogrammetry in high turbidity water: An application on the Arles-Rhone XIII roman wreck in the Rhodano river, France. In Proceedings of the 18th International Conference Virtual Systems and Multimedia (VSMM), Milan, Italy, 2–5 September 2012; pp. 189–194.

41. Bianco, G.; Muzzupappa, M.; Bruno, F.; Garcia, R.; Neumann, L. A New Color Correction Method for Underwater Imaging. *ISPRS Int. Arch. Photogramm. Remote Sens. Spat. Inf. Sci.* **2015**, *1*, 25–32.

42. PointCloudLibrary (PCL). Available online: http://www.pointclouds.org/ (accessed on 15 January 2016).

DeepSurveyCam—A Deep Ocean Optical Mapping System

Tom Kwasnitschka, Kevin Köser, Jan Sticklus, Marcel Rothenbeck, Tim Weiß, Emanuel Wenzlaff, Timm Schoening, Lars Triebe, Anja Steinführer, Colin Devey and Jens Greinert

Abstract: Underwater photogrammetry and in particular systematic visual surveys of the deep sea are by far less developed than similar techniques on land or in space. The main challenges are the rough conditions with extremely high pressure, the accessibility of target areas (container and ship deployment of robust sensors, then diving for hours to the ocean floor), and the limitations of localization technologies (no GPS). The absence of natural light complicates energy budget considerations for deep diving flash-equipped drones. Refraction effects influence geometric image formation considerations with respect to field of view and focus, while attenuation and scattering degrade the radiometric image quality and limit the effective visibility. As an improvement on the stated issues, we present an AUV-based optical system intended for autonomous visual mapping of large areas of the seafloor (square kilometers) in up to 6000 m water depth. We compare it to existing systems and discuss tradeoffs such as resolution *vs.* mapped area and show results from a recent deployment with 90,000 mapped square meters of deep ocean floor.

Reprinted from *Sensors.* Cite as: Kwasnitschka, T.; Köser, K.; Sticklus, J.; Rothenbeck, M.; Weiß, T.; Wenzlaff, E.; Schoening, T.; Triebe, L.; Steinführer, A.; Devey, C.; Greinert, J. DeepSurveyCam—A Deep Ocean Optical Mapping System. *Sensors* **2016**, *16*, 164.

1. Introduction

When compared to land or airborne topographical survey methods such as radar, lidar or photogrammetry, hydrographic surveys of the structure and texture of the seafloor are considerably impaired by the body of seawater covering it. Water strongly absorbs a wide range of the electromagnetic spectrum. Penetration distances in clear water are around 100 m for sunlight between 350 nm and 550 nm (blue to green) [1]. Suspended particulates and dissolved organic and inorganic compounds reduce the practically visible range much further [2]. Therefore, seafloor exploration is often executed with single beam (1D), multi beam swath echo sounders (MBES, 2D) and imaging sonars (3D) [3]. In addition to the effects of varying sound velocity due to water body stratification, the spatial resolution of acoustic sounding methods in the deep sea is practically limited to one percent of the slant range. The same rule applies to the positioning accuracy of acoustic underwater navigation devices in

344

the absence of GPS signal reception, posing considerable challenges, particularly for incremental or repetitive surveys in very deep waters [4]. A deep seafloor map may have a local precision of centimeters, but can only be referenced with an accuracy of tens of meters in a global context. Landmarks of known absolute position do not exist and cannot easily be generated by geodetic methods.

These restrictions have led to the development of a suite of long range, imprecise and short range, precise survey methods which play out their combined strengths in a nested survey approach across the entire range of scales required in the deep ocean [5,6]: Regional and reconnaissance surveys are conducted using ship-based acoustic methods (side scan sonar or MBES) which provide an absolute geographic reference frame through the incorporation of GPS positioning and—depth-dependent—resolutions on the scale of tens of meters. Based on this information, autonomous underwater vehicles (AUVs) or deep towed sleds employ the same acoustic methods but closer to the bottom and on a smaller area, increasing the resolution to decimeters. Cabled remotely operated vehicles (ROVs) provide enough maneuverability to carry very close-range acoustics and cameras that resolve down to the millimeter range in highly localized areas. Necessarily, there is a trade-off between resolution and achievable coverage in a given amount of time which is why the nested survey approach requires sequential deployments with each finer resolution survey step being informed by a former one, often spanning several seagoing expeditions.

Beyond the information on seafloor geometry (bathymetry), detailed characterization of lithology, structure and habitat requires information on the texture of substrate and organisms. While acoustic methods can deliver a sort of pseudo-texture (by the strength of the backscattered signal) [7], only optical methods can deliver a full color image equivalent to on-land surveys. In the case of seafloor mapping, color information is predominantly important at small to intermediate scales, while most large-scale features can be assessed using acoustic backscatter that has been verified by local observations (e.g., [8]). Depending on their implementation, optical mapping techniques not only deliver a planar (2D) image mosaic (e.g., [9]) to be draped over an acoustic bathymetrical map but also can directly produce the geometry. This technology is well established in photogrammetry and remote sensing on land, e.g., in aerial mapping of the continents, and comprehensive discussions on technology, survey strategies and mathematical concepts for estimation can be found in standard text books such as the *Manual of Photogrammetry* [10]. In the last two decades, these survey technologies have been extended to largely automated systems relying on machine vision and other sensors, being able to cope with huge amounts of data, GPS-denied environments and other less well defined situations such as unordered photos, camera miscalibration or missing ground control points (*cf.*

to [11,12]). A number of approaches have demonstrated the feasibility of AUV-based 3D reconstruction from visual data since then (see for example [13–19]).

Another way in which current optical and acoustical methods differ is whether they produce geometry extruded vertically from a horizontal plane or a frustum (termed 2.5D, such as by MBES or structured light) or whether they yield a complete three dimensional terrain model, allowing for multiple elevation values for any planar coordinate (*i.e.*, overhangs, cavities) and texture coordinates along vertical walls, cliffs or spires. The latter is a particular advantage of photogrammetric reconstruction from area-scan cameras (either through stereo or structure from motion), which provides the richest data set of all above methods, a three-dimensional model with an intuitively understandable full color texture. Unfortunately, the latter method is also most demanding on water quality and has so far only been applied to close-range surveys in the deep sea (e.g., [20]).

Despite rapid advances in the development of instrumentation and robotic platforms in recent years [21], a gap has remained in the range of surveying capabilities, that is, an optical system that can cover a large area on the order of 1 km^2 on a one-day deployment, yielding 3D terrain models at cm-resolution and in full color. We suggest that based on such a model, highly localized observations could be extrapolated with confidence (*cf.* [8]). The larger the coverage of a high-resolution model, the better it can be fit to the absolute geographic reference frame of a low-resolution ship based map, with considerable benefits for time series of studies. These aspirations have led to the development of an AUV based high-altitude color camera system (Figure 1) that allows both for large-scale 2D mosaicking and 3D photogrammetric reconstruction of complex seafloor environments. The details of the system are described below.

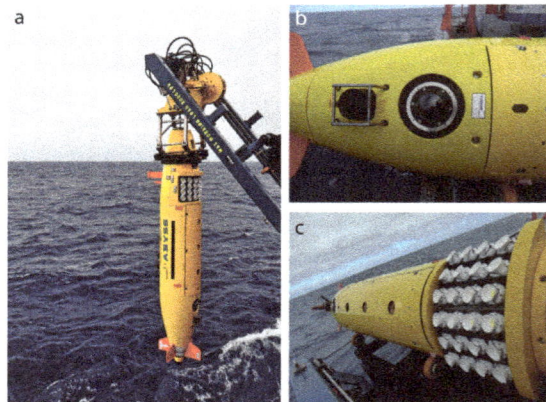

Figure 1. (**a**) The GEOMAR AUV ABYSS prior to launch, equipped with (**b**) a high resolution camera behind a dome port and (**c**) a novel LED flash system.

2. Design Considerations and Previous Work

Situated at a marine research institute, it is important to note that this development has been application driven. The two primary applications behind the design of the system were to map complex morphologies such as volcanic terrain and to efficiently cover large areas e.g., of the abyssal plains in order to conduct large-scale habitat mapping. Necessarily, there is a trade-off between achieved coverage and degree of overlap and (required) redundancy in the data, which may lead to additional dives, depending on the survey strategy. While the surface of flat areas is represented well by a 2D mosaic, for 3D structures with relief more complex representations are required. Although 2D mosaicking is efficient, the distortions induced by high relief (of unknown geometry) bias the scientific evaluations based on such 2D maps, since oblique and sub-vertical areas, cavities or overhangs are poorly or not at all represented. Thus the system would have to be able to supply data suitable for 3D photogrammetric reconstruction. This is based on finding corresponding points in sequences of images in order to compute the (relative) motion of the camera as well as the 3D structure of the environment [11]. While simple geometric statements can sometimes already be made when a seafloor point is observed from two angles, mapping quality is improved and models become denser with a higher number of different observations per point. The simplest hardware setup to achieve this is a single moving camera that takes overlapping images along its way, such that every object is imaged multiple times. Visual reconstruction using a single moving camera, as opposed to using a stereo camera system, requires additional information in order to infer the absolute scale of the scene [11]. In the case of the given system, the AUV's onboard high precision inertial sensor system that has been reported to drift less than 10 m per hour [22] is utilized.

The required number of shots depends on the distance, complexity and also the appearance of the observed scene, but in any case no less than 75% overlap should be aimed for in flat terrain. This number has to be further increased e.g., when moving fauna, particles or smoke can partially occlude the view onto the seafloor. Above flat ground, on a moving platform such as an AUV, the overlap fraction (o_{along}) can be computed from the image capture interval (t), the vehicle speed (v), the field of view (α) and the altitude above ground (h) according to the following relation:

$$o_{along} = \frac{f-d}{f} = 1 - \frac{vt}{2h \tan\left(\frac{\alpha}{2}\right)}$$

which is illustrated in Figure 2.

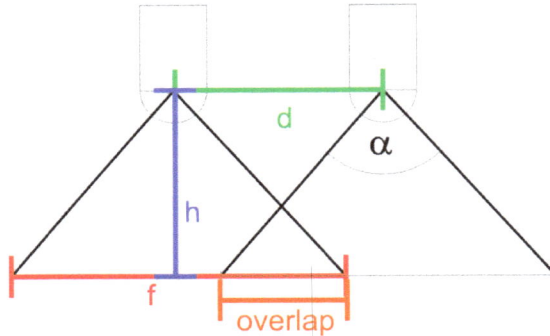

Figure 2. Sketch of image overlap. When a camera with field of view α observes the seafloor from an altitude of h, this creates a footprint of f. After a horizontal movement of d the overlap fraction computes as $(f - d)/f$, where f can be expressed as $2h \times \tan(\alpha/2)$ and d as the product of velocity v and interval t.

Similar considerations can be made for the exposure time. In order to avoid motion blur in the images, the AUV should move less than half the footprint of a single pixel during exposure. This footprint (f_{pixel}) can be computed from the size of a pixel on the sensor (p), the focal length of the lens (l) and the altitude (h) by the intercept theorem as:

$$f_{pixel} = \frac{p}{l}h$$

Common DSLR cameras currently have typical pixel sizes of $p = 6$ μm although novel sensors with larger pixels are available. This is also a limiting factor for the obtainable accuracy of 2D and 3D surveys. A wide-angle lens of $\alpha = 100°$ field of view has a focal length of $l = 15$ mm when using a full-frame sensor. This would lead to a pixel footprint of 4 mm at $h = 10$ m altitude and respective footprints when varying the parameters. The footprints of rectilinear lenses do not change for pixels in the image center or towards the boundaries, those of equiangular fisheye lenses become larger towards the image boundary. Consequently, in downward looking fisheye images above flat terrain motion blur will first be visible in the image center whereas the motion blur is independent of the image position when using a rectilinear lens.

This illustrates that it is attractive to maximize the field of view in order to both cover a large footprint with every image and secure across-track overlap but also to cast oblique rays onto strongly undulating relief which would obstruct the view of a narrowly downward looking camera.

Assuming a carrier AUV that moves at a certain speed ($v = 1.5$ m/s), and targeting at an overlap of more than 75% along track, we arrive at the following constraint:

$$h \tan\left(\frac{\alpha}{2}\right) > 3.75\frac{m}{s}t$$

348

When operating the system using a flash, the recharge time for the flash provides a lower bound on the photo interval t. For instance, when flashing with 1 Hz, the minimum altitude is 3.75 m at a field of view of 90°, or 6.5 m at a field of view of 60° or 14 m for 30°.

For simplicity of presentation in the following we assume the same field of view and footprint width along track and across track. When the AUV goes on a straight line, the covered area A_{line} can be approximated (for readability of equations we will ignore the time taken for the extra "no-overlap" area of the first and the last image, as well as of the first/last line in a pattern and rather consider the asymptotic case of very long and many lines, where these constants can be safely ignored) from the product of the (across-track) footprint and the distance traveled:

$$A_{line} = 2h \tan \left(\frac{\alpha}{2} \right) vt$$

When the survey area is scanned using a pattern of overlapping parallel linear tracks (commonly called a "lawnmower pattern"), the area is reduced by the across-track overlap o_{across} between two subsequent lines. The goal of having this overlap is to generate seamless maps without holes even in presence of small navigation drift, but also to find back corresponding points from the previous line in order to perform re-navigation (bundle adjustment). We suggest having at least 33% overlap.

The maximum coverage achievable in a given amount of time is given by the following relation:

$$A_{area} = 2h \tan \left(\frac{\alpha}{2} \right) vt(1 - o_{across})$$

From this it can be seen that a large field of view is the enabling property for large-scale optical surveying. Cameras for the deep sea have to be mounted in pressure housings that can resist the approximately 600 bar at 6000 m water depth. The camera views the outside world through a port and these ports are typically flat or spherical. Flat ports for these depths have to be several centimeters thick, depending on the diameter of the opening. Refraction at the port will however change the imaging geometry substantially and in particular invalidate the pinhole camera model [19]. Additionally, refraction limits the maximum field of view theoretically possible, e.g., of a fisheye lens behind a flat port, to roughly 96°, and in practice a field of view above 70° is very difficult to realize using a flat port. Dome ports do not limit the field of view and do not cause refraction for the principal rays if the entrance pupil is aligned with the dome port center. However, they act as a lens themselves, shifting focus very close such that lenses behind dome ports have to use an adapted focusing strategy.

A further challenge in underwater imaging is scattering and absorption [1], both of which are complex physical phenomena. The loss of "signal flux" can however be approximately described by an exponential relation:

$$E(x, \lambda) = E(0, \lambda) e^{-\eta(\lambda)x}$$

where E(0) is the irradiance of (parallel) light at some position and E(x) is the attenuated irradiance after the light has travelled a distance of x through the water. η is the wavelength-dependent attenuation coefficient that represents the loss of flux due to absorption and scattering. When viewing an object with a camera, this remaining exponentially attenuated signal is now additionally degraded by light scattered into the viewing path that reduces the signal to noise ratio. Since this scattering happens largely in direction of 180° [1] it is advisable to move the light source away from the camera, such that direct light scattering into the image is avoided. Still, image quality degrades exponentially with altitude, in particular when both the light source and the camera are moved further away from the target.

This implies that: (a) the light source and the camera have to be developed in interdependence of each other and should be separated as far as possible (to limit scattering in the immediate front of the camera) and (b) a wide field of view is problematic, since the light has to traverse a considerably larger distance in the case of peripheral rays, subjected to scattering and absorption. Therefore, an optimum has to be found between the lighting (geometry and wavelength), field of view and survey geometry dictated by the capabilities of the vehicle.

Development of existing deep ocean camera systems and illumination devices has been primarily oriented on the platform they are deployed on. In order to illustrate the spectrum of technical realizations of the optical mapping task, Table 1 gives a non-exhaustive overview of systems developed throughout the past two decades.

The AUV ABYSS at GEOMAR is a type REMUS 6000 flight-class AUV manufactured by Hydroid, Inc. (Pocasset, MA, USA) (Figure 1). It is 4 m long and has a torpedo like shape at a maximum diameter of 0.66 m.

Table 1. A non-exhaustive selection of ocean floor imaging systems of the last two decades reveals the spectrum of applications and solutions, including the original and new camera configurations of the GEOMAR AUV ABYSS. Abbreviations for methods are (M) mosaicking, (SL) structured light, (P) photogrammetry, either using stereo or structure-from-motion. Measures of efficiency are hard to normalize due to non-uniform information on actual image overlap e.g., during nonlinear vehicle tracks. Therefore, the (FoV) field of view, (V) velocity over ground, (t) survey duration or (A) total area covered are reported.

Year Deployed	Research Group, Vehicle	Vehicle Category	Method	Camera Model	Imager	Mode	Depth (m)	Altitude (m)	Cadence (s)	Ground Resolution (mm)	Efficiency	Reference
1996	WHOI, Argo II	Towsled	M	Marquest ESC9100	mono	b/w	1600	10	13	15	3152 m²/h	Escartin et al., 2008 [23]
2007	GEOMAR, Abyss	AUV flight	M	AVT Pike	mono	b/w	6000	4-12	4	6	FoV = 42 × 42°, V = 1.5 m/s	this study
2008	NTNU	ROV	M	Uniqvision	mono	color	500	1.9	4	2.2	800 m²/h	Ludvigsen et al., 2007 [24]
2008	Ifremer, Victor	ROV	M	OTUS	mono	b/w	6000	8-10	10-17	10	6491 m²/h	Prados Gutierrez et al., 2012 [25]
2010	URI, Hercules	ROV	SL/P	AVT Prosilica	stereo	color	2000	3	6.7	2.5	FoV = 35 × 52°/ V = 0.15 m/s	Roman et al., 2010 [26]
2010	WHOI, Seabed	AUV hover	P	Pixelfly	mono	color	2000	3	3		A = 30 × 45 m/ V = 0.25 m/s	Bingham et al., 2010 [27]
2011	Kongsberg, HUGIN	AUV flight	M	Tilecam	mono	b/w	1000	2-5	<1	2.3	FoV = 55 × 35°, V = 2 m/s	Hagen, 2014 [28]
2013	Univ. Tokyo, Hyper Dolphin	ROV	SL	SeaXerocks 2	mono	color	2000	13	5	6.4	8144 m²/h	Bodenmann et al., 2013 [29]
2013	Univ. Girona, Girona 500	AUV hover	P	Canon 5d Mk II	stereo	color	500	3.68	2	0.66	t = 103 min, A = 65 × 20m	Gracias et al., 2013 [30]
2014	NOC, Autosub	AUV flight	M	Point Grey Grasshopper AVT	mono	color	6000	3.2	0.87	0.59	23,000 m²/h	Morris et al., 2014 [31]
2015	ACFR, Sirius	AUV hover	P	Prosilica GC1380	stereo	color	800	3	0.67	2.2	FoV = 45 × 38°, V = 0.5m/s	Williams et al., 2015 [32]
2015	GEOMAR, Abyss	AUV flight	P	Canon 6D	mono	color	6000	4.7-12	1	1.9-5	25,700-166,000 m²/h	this study

The AUV has a practical endurance of 16 h, or 100 km and provides a maximum constant hotel power of 28 V, 6 A available to scientific payload systems. It is dynamically steered by two perpendicular pairs of stern fins, which require a nominal speed of 2.5–3.5 kn. Navigation close to the seafloor is carried out by dead reckoning using a Kearfott inertial navigation system aided by a Doppler Velocity Log, and is subject to a drift of less than 10 m [22] per hour. Additionally, long baseline (LBL) transponder fixes can be incorporated if necessary. An altimeter allows following the terrain contours at a fixed altitude, and a forward-looking pencil beam detects obstacles in the way. Due to the high speed and limited obstacle avoidance capabilities, its main strength lies in acoustic mapping several tens of meters above ground, but navigation closer than 8 m to the seafloor is only possible in well charted, rather level terrain. Payloads can be integrated as part of customized floatation foam segments in the lower bow section (sensor bay) and in a downward oriented cylindrical space located towards the stern of the vehicle (processing electronics). This provides a hard design constraint towards a single light source and a single camera. The original imaging system that was part of the standard sensor package uses a 200 joules xenon flash capable of a 0.5 Hz repetition rate. An Allied Vision Pike b/w, 4 MP camera delivers images with a 41° field of view (FoV). Furthermore, we found the dynamic range of the camera insufficient for the production of complex photographic maps. Given the circumstances we decided for a complete redesign of the camera and lighting unit (Figure 3).

Evaluating the strengths and weaknesses of other systems and with the given constraints of our carrier platform, the design goals were the following:

1. An operating altitude of 5–15 m due to limited maneuverability, requiring strong lighting and a sensitive imager;
2. Provisions to maximize the covered area (e.g., high capacity data storage), in order to create large-area seafloor maps in a single deployment;
3. An image acquisition rate of 1 Hz or better to guarantee multiple (>4) perspectives of complex objects;
4. Color still frame sequences at 12 bit or better, resolving ground features smaller than 1 cm, including automatic adaptation to changing seafloor brightness;
5. A wide FoV on individual images to maximize along- and across track overlap, to compensate for small errors in vehicle navigation running parallel track lines and to cast oblique peripheral rays imaging sub-vertical seafloor features in spite of a vertically downward mounted camera orientation;
6. Adhering to the vehicle's overall depth rating, a pressure resistance of 600 bar plus safety margin.

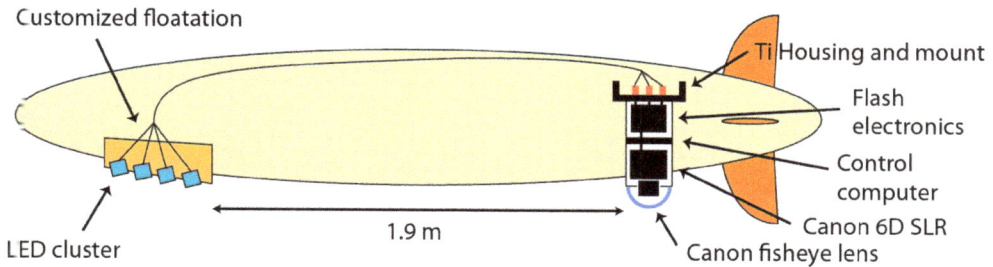

Figure 3. Schematic overview of the components of the newly designed high-altitude camera system for the GEOMAR Remus 6000 AUV.

3. Description of Hardware Components

The overall design can be seen in Figure 3 and is outlined in the following paragraphs.

3.1. Lighting

The strong required illumination ruled out the use of constant lighting in favor of pulsed lighting, so it was decided to use a cluster of 24 LED arrays (Figure 1c). A color temperature of 5600 K was chosen to minimize absorption. By optimizing the power management in pulsed operation mode we were able to increase the light efficiency to 1.6 times the nominal value stated by the manufacturer. Each two arrays are switched in serial configuration, forming 12 parallel pairs driven by capacitor based flash electronics in the rear half of the camera housing.

A 4 Hz maximum operating frequency was found to be a good compromise of image cadence *versus* available power. In order to accommodate the comparatively fast velocity of the AUV, flash durations were kept short and can be varied between 1 ms and 4 ms. A variable trigger delay of 200 μs–12 ms can be manually preset accommodating a variety of cameras. The cluster of 24 individual light sources was mounted flexibly in six rows of four lights, to allow for customized beam forming of the illumination cone. Each light can be tilted individually along the longitudinal vehicle axis and every set of four lights that each can be tilted laterally. This setup allows reacting flexibly to different camera optics or peculiarities of the terrain and visibility conditions. Each LED array is equipped with an external reflector casting a circular light cone with a 74° full width half maximum opening. Using a patented procedure [33], each LED waver is encased in a transparent, pressure neutral resin cast that effectively retains the in-air optical properties of the waver-reflector pair. This solution is not only extremely cost effective, but also light (45 g in water per unit), robust, mass producible, and has since found application in various other

GEOMAR instruments. Provided the lights are operated in water, no thermal issues were found.

3.2. Optical System and Housing

In order to maximize the field of view not just for AUV use but possibly also for ROV deployments of the new camera system, the Canon 8–15 mm f4 fisheye zoom lens was implemented (Figure 4), offering a very wide range of FoV settings without having to change the lens. The fisheye lens design by itself produces a curved focal surface that complies with the dome port optics, and generally offers a large depth of field. In order to keep image brightness and depth of field at a compromise, aperture is set at an intermediate value of f8. In order to match the fisheye FoV, a new camera housing was designed which features a dome port with 100 mm inner diameter and 7.1 mm wall thickness, offering a FoV of 160°. The housing dimensions were maximized to 182 mm inner diameter in order to accommodate the flash and camera electronics.

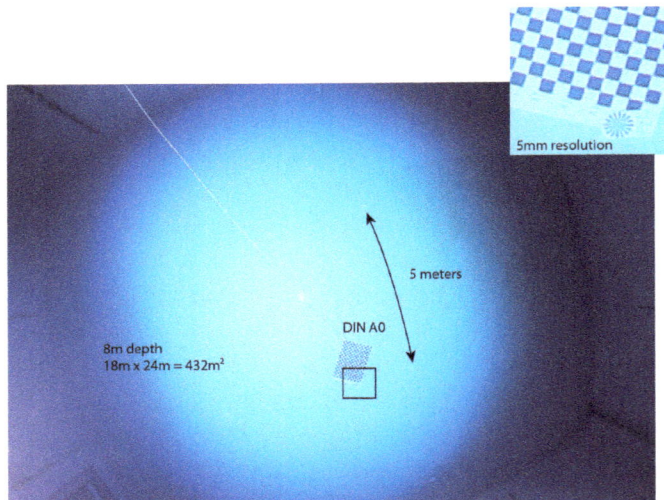

Figure 4. An unprocessed sample image of the system in the test pool of the German Center for Artificial Intelligence (DFKI) in Bremen shows the illumination pattern and achievable resolution in clear water conditions. The inset image shows an enlarged portion of the resolution target in the image center.

3.3. Camera System

Due to the large object distance and limited power supply, the foremost design criterion was the sensitivity of the imager. Following a series of sensitivity tests with machine vision cameras (which are commonly used in deep sea applications), the

Canon EOS 6D full frame SLR camera was selected since it is small enough to fit into the pressure housing with minimal modifications. This choice yielded the best immediately available image quality and avoids image post processing procedures as often required with raw machine vision data. A major drawback of SLR cameras is that they have moving parts (the shutter and mirror mechanics), which are prone to wearing off. In particular, according to the manufacturer the shutters have an endurance of 100,000 exposures, and indeed the shutter of our first test system broke after little more than this figure. Consequently, three spare cameras are kept and proactively exchanged for each cruise, with an average dive producing up to around 40,000 images. All cameras are serviced after a cruise.

The camera is linked to a miniaturized PC workstation through a USB2 connection. Contrary to the original camera system, this new controller is completely autonomous from the AUV and does not receive or send any data in order to avoid interference with the vehicle operations. Accessible over gigabit Ethernet through an external port on the housing, all camera settings can be pre-programmed and scripted through custom developed software employing the Canon Software Developers Kit (SDK). A freely configurable intervalometer can be triggered either by a timer or by depth readings received through a pressure sensor integrated in the camera housing, avoiding unnecessary exposures during descent and ascent. Several different missions can be scheduled in a queue. The aperture is constrained by depth of field considerations on the one hand and the limited available light on the other hand. As there is absolutely no light when the flash is off, the effective exposure time is bound to the flash duration. Consequently, only the ISO sensitivity (amplification) of the sensor can be adjusted to react to brighter or darker environments or to varying distances. Since the inbuilt light metering of the camera does not work (as between the photos the flash is recharging), another way of controlling image brightness is required. As we aim at significantly overlapping photographs (75% and more) it is reasonable to assume that only the topmost small stripe of the image will depict a novel portion of the seafloor, which even often consists of terrain with similar color as in the current image.

Consequently, the brightness histogram of the current image can be exploited for adjusting the ISO setting for the next image, in a way that the dynamic range of the next image will be covered well, but that the image should not be overexposed: In the (most relevant) center region of the image we sort all pixel values and require that the 90th percentile be below 90% of the saturated value (*i.e.*, for 8 bit with maximum pixel value 255, we test whether 90% of the pixels are below 230) and adjust the ISO speed for the next image accordingly. See Algorithm 1 for details.

Algorithm 1. ISO speed determination for next image based on histogram of current image. For presentation reasons the algorithm is oversimplified. ISO steps are not only available as factors of 2, as assumed in the algorithm above, but sometimes also at intermediate steps. We check whether the next ISO speed is likely to still be within the acceptable range, and so the factor of 0.5 resp. 2 in the above has to be exchanged with "factor to next ISO speed".

1. Compute histogram of pixel values
2. Determine 90% quantile: q_{90}
3. if (q_{90}>0.9*MAX_PIXEL_VALUE)
   // image is oversaturated / too bright, lower ISO !
   ISO-speed$_{t+1}$:= 0.5 * ISO-speed$_t$
   else if (2*q_{90}<0.9*MAX_PIXEL_VALUE)
   // image is dark, even with the next higher ISO image would not be oversaturated
   ISO-speed$_{t+1}$:= 2 * ISO-speed$_t$
   else
   // image seems ok, do nothing
   ISO-speed$_{t+1}$:= ISO-speed$_t$

In practice, the maximum cadence is limited by the USB2 connection between camera and PC and by restrictions imposed by the Canon SDK, resulting in 1 Hz and 0.5 Hz frequencies for full resolution jpeg and raw images, respectively. The images are written to a 1 TB solid-state hard drive, which is exchanged or copied over network after the deployment.

3.4. Calibration

So far the camera system is integrated into the AUV such that the camera moves up when the AUV moves forward, and is aligned to the vehicle coordinate system up to a mechanical precision of approximately 1°. Although cameras with fisheye lenses often exhibit a caustic, we approximate the lens as having a single center of projection. It is then mounted such that its nodal point coincides with the dome port center, such that the principal rays are not refracted and the pinhole camera model can be used. For calibration, a one square meter calibration pattern (DIN A0) [34] is presented in 20 different orientations and we obtain the coefficients of the fisheye lens using the OCamCalib [35] software (Figure 5).

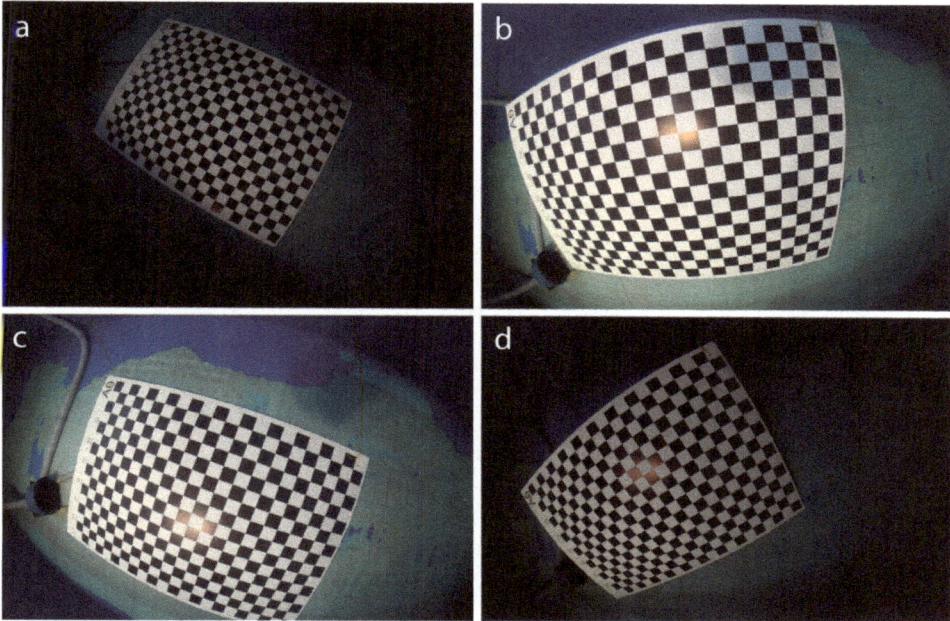

Figure 5. Sample images used in calibration of the camera. The checkerboard is presented in different positions and orientations (**a–d**) relative to the camera in order to obtain the 3D ray associated to each pixel in the image.

4. Field Trials and First Results

The camera system was deployed on two cruises of the German research vessel SONNE to the Pacific Ocean in 2015 (21 photo dives during SO-239 and SO-242 with approximately half a million photos in total). These scientific cruises took place in the context of the JPI Oceans project initiative "Ecological aspects of deep sea mining", which investigates potential impacts of (manganese nodule) mining from the abyssal planes of the Pacific Ocean. The sample imagery of Figure 6 has been captured during cruise SO-242-1 to the DISCOL experimental area offshore Peru. Here, a long-term disturbance experiment has been started in 1989, where an 8 m wide plow had been towed through the seafloor for several weeks, in order to simulate mining activity. Now, after 26 years, the same location has been revisited and the entire area has been mapped using acoustic and visual sensors, in particular also using the AUV camera system discussed in this article. Among other things the desired analyses target the change of the habitat and its potential recovery, the time taken for recolonization, and the risk for extinguishing certain species. Such analyses should be supported by a broad data basis as they might later be used for developing guidelines for future mining activities in the deep sea.

357

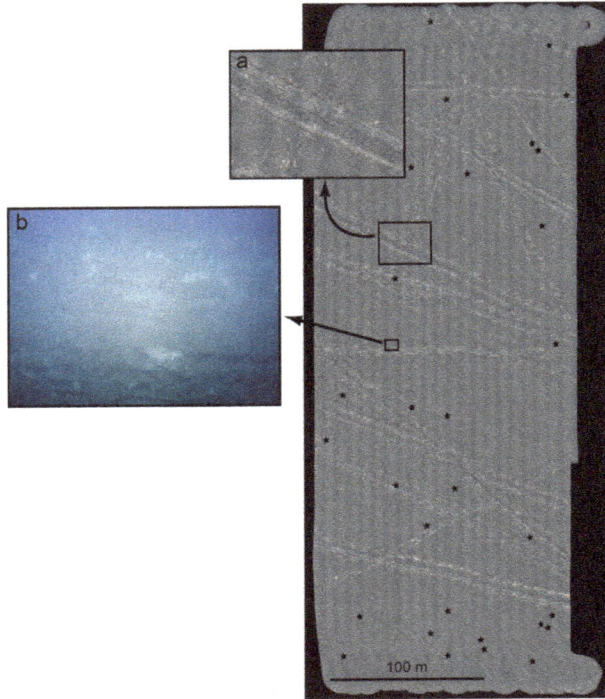

Figure 6. An area of approximately 200 m × 450 m in the DISCOL experimental area of the south-east Pacific ocean offshore Peru. The photo-mosaic consists of 13,000 photos taken from an altitude of 4.7 m on average, captured during 3.5 h by the novel camera system of GEOMAR's AUV Abyss in 4100 m water depth. The tracks are 8 m wide plowmarks (**a**) from a 1989 experiment to simulate deep sea mining and they are well visible in 2015. The resolution of the images captured (**b**, undistorted but not color corrected) allows to systematically evaluate megafauna that recolonized the area. Asterisks mark the positions of manual offset measurements.

A number of post processing approaches have been considered and are still subject to refinement. Figure 6 shows an area of approximately 200 m × 450 m that was mapped in 3.5 h on one of the AUV dives in 4135 m water depth flying at a target altitude of 4.7 m. As a proof of concept, the mosaic has been created by stitching together the individual photographs using navigation data only: Each image was undistorted and empirically normalized using a robust average image that captures vignetting, illumination and attenuation effects. The resulting image was projected onto a plane at 4135 m water depth using the intrinsic camera parameters obtained by calibration and the AUV's navigation (altitude, pitch, longitude and latitude) as exterior orientation. While the absolute positioning accuracy is on the order of

5C m (tied to the water depth), the local fit between tracks (depending on the dead reckoning algorithm of the AUV) is typically better than 2.5 m as determined by 30 manual random spot measurements (see Figure 6). This is not accurate enough for quantitative studies yet sufficient for timely decisions on further actions at sea. The technique allows for rapid results on the ship and generates the mosaic in a few seconds per input image using a multi-band blending strategy (similar to [36]). The result of rapidly merging more than 10,000 images is a virtual seafloor mosaic as one would observe without water from several hundred meters altitude: The 26 year old tracks are still clearly visible and the visual map provides an excellent overview of the situation in the target area. When zooming into the mosaic or looking at individual images on the other hand, the mega fauna that has recolonized the area is clearly resolved.

It is also possible to carry out 3D reconstruction from the images (*cf.* to [37] for details on 3D reconstruction from underwater images). As a proof of concept we have processed two test subsets of roughly 10 images from the DISCOL dataset in the commercial software AgiSoft PhotoScan Pro and obtained the 3D models as displayed in Figure 7.

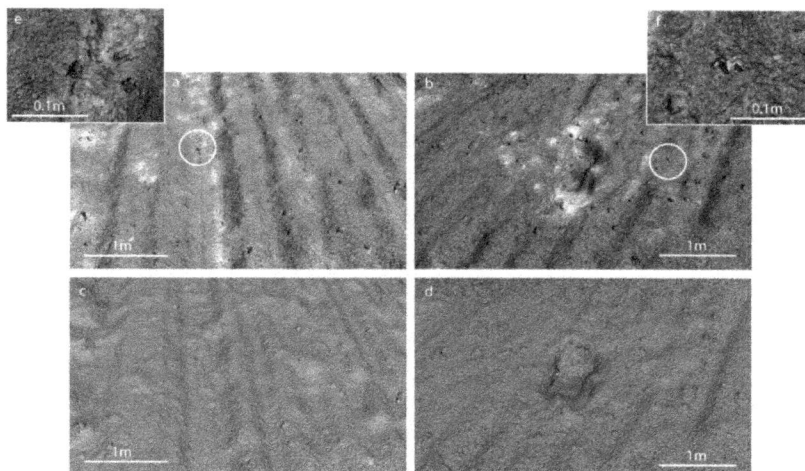

Figure 7. Photogrammetric dense point cloud reconstruction delivers geometry with approx. 1 mm resolution from an altitude of 4.7 m. Renderings show (**a**) artificially perturbated sediment with manganese nodules and holothurian and (**b**) perturbated sediment and the excavation footprint of a box corer. (**c,d**) Show the shaded relief of (**a,b**), respectively, while (**e,f**) show details of epibenthic organisms (white circles) depicted with their rough geometrical shape. Point density equals about one per 2 mm. Compare Figure 6b for a source image of areas (**b,d**).

Scaling and referencing was again carried out using the vehicle navigation record, which provides the motion baseline between adjacent images in the structure from motion reconstruction step (e.g., [38]). The resolution from 4.7 m altitude is sufficient to reconstruct individual manganese nodules, the profile of the trawl-marks and even the geometry of some of the animals. As this article focuses on the design of the camera hardware and since the optimization of the post processing workflow is still under development, a detailed comparison and evaluation of different photogrammetric reconstruction software packages is out of scope of this article. The results demonstrate however that the overlap and the image quality are sufficient for recovering geometry at sub-centimeter resolution. In order to evaluate the feasibility of this reconstruction workflow on the scale of an entire survey and to judge the quality of referencing, an earlier data set of a similar manganese nodule field from the Clarion Clipperton Fracture Zone surveyed during the SO239 cruise was processed (Figure 8).

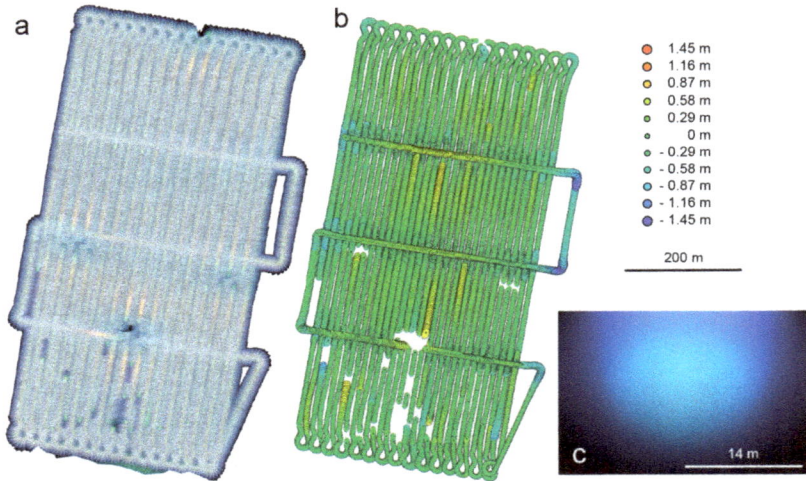

Figure 8. Photogrammetric Reconstruction of a manganese nodule field in the Clarion Clipperton Fracture Zone (CCZ) surveyed during SO239. (**a**) shows the color corrected mosaic without brightness correction to reveal tracks. Colored ellipses in (**b**) mark the extent of lateral positional deviation relative to the acoustic vehicle navigation; the color legend marks the vertical deviation; (**c**) shows a raw fisheye image midway through the survey. Scale bar applies to image center.

It was chosen because the AUV was flown at an altitude of 9 m, which was at the brink of visibility. At 10 m line spacing and an image cadence of 1 Hz an area of 350 m × 700 m was covered. Using Photoscan Pro, an area of 0.24 km^2 was reconstructed from 9635 images with an average nine-fold overlap. The root mean square positional error (*i.e.*, the offset between the original vehicle navigation and the internally consistent camera pose estimation) as illustrated in Figure 8b is 2.96 m (see Table 2 for further statistics, at an average reprojection error of less than 0.6 pixels). This is an error of less than 1 percent with respect to the extent of the survey area and suggests that the scale is correct. At this point accuracy cannot be further improved without known scale and position of seafloor features, which are generally unavailable in our study areas.

Table 2. Statistical offset in meters between acoustically determined navigation and photogrammetrically reconstructed camera poses.

Statistical Measure	Total	X	Y	Z
RMS Error	2.96	1.30	2.65	0.28
Median	2.83	0.09	−0.15	0.01
Minimum	0.02	−6.29	−5.33	−1.45
Maximum	7.77	4.78	7.44	0.76

5. Discussion

The field results show that the system is capable of capturing image data at altitudes from 4 m to 9 m. Compared to the original design goals stated in Section 2, most (such as high resolution, wide field of view and depth rating) were met although our design fell short of some aspects during field tests, foremost the highest possible survey altitude. Figure 9 presents images of the same object at different altitudes. The lowest altitude produces the smallest footprint and the highest resolution image material. Altitudes below 7 m contain color information; above 7 m the red channel is lost.

At least for the Peru basin and CCZ, the initially desired altitude of 15 m thus proved to be too high to yield useable results, although other parts of the ocean may yield better visibility. Small gaps in the southern half of the SO239 survey shown in Figure 8 are due to 3 m deep depressions that increased the camera distance to the ground beyond a maximum acceptable threshold of 9 m, resulting in alignment failure of the respective images. This defines the local maximum working distance determined by water clarity.

We draw the conclusion that realistic working distances of 4–7 m yield color texture information while scans of mere seafloor geometry can still be carried out up to a height of 10 m under optimum conditions. A further challenge is the high amount of light scattered back from the water, an effect that was substantially stronger on the test deployments than in the (actively filtered) test tank. This may require a re-evaluation of light intensity applied and the geometry of the light beam, likely towards a narrower geometry. At the same time, the relatively high ISO settings of typically 6400 to 10,000 required to achieve correct exposure are not optimal in terms of their sensor noise. Thus cameras of even higher sensitivity would be desirable, in order to employ less light. The minimum desired frame rate could be met, although the USB2 connection of the current camera model limits raw capture to 0.5 Hz.

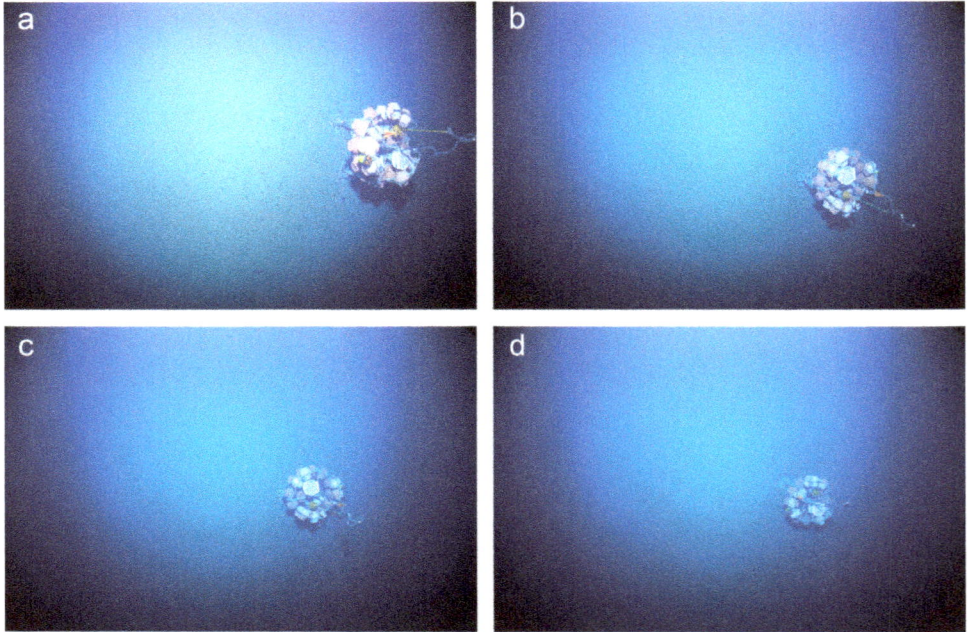

Figure 9. Cropped and undistorted photographs of an autonomous benthic lander (3 m high, 4 m in diameter) photographed from an altitude of (**a**) 6 m; (**b**) and (**c**) 8 m and (**d**) 10 m. Loss in color and light scatter are clearly increasing together with altitude.

The sample data indicate that the system is able to seamlessly survey flat seafloor for mosaicking and that the quality of the INS navigation data is good and in agreement with report of other authors [22]. The results of the photogrammetric reconstruction suggest that even subpixel registration of the images is feasible. Overall, this implies that the accuracy potential of the system is largely bound by the pixel footprint, which is 4 mm at 10 m altitude and 1.6 mm at 4 m altitude.

6. Conclusions

We have presented the technical layout and first results of a camera system for mosaicking and 3D reconstruction that is both capable of operating at advanced depths down to 6000 m and delivers color high-resolution imagery. Together with its carrier platform, the AUV ABYSS, it can deliver seamless seafloor coverage over large areas and currently ranks among the most capable systems available. Nevertheless, current challenges are presented by excess backscatter from the water column, which required the camera to be flown at its nominal lowermost range of 5 m above ground for consistent results. While the LED cluster has proven robust in terms of pressure tolerance, it is too exposed at the keel of the vehicle and negatively affects the hydrodynamic properties of the AUV. A redesign of the flash cluster geometry towards a narrower light cone and better protection is therefore under consideration.

Despite recent success during two expeditions surveying flat seafloors, we have not yet succeeded in demonstrating a case applied in rugged volcanic terrain. In order to do so, the system will be adapted to be flown on ROVs and hovering AUVs to minimize the risk of high velocity collisions. The current long vehicle turnaround time between dives must be shortened by implementation of intelligent data management and automatic image quality control during the mission, for which the system already has sufficient computing power. Another option could be the inclusion of structured light projectors (point and sheet lasers) to directly measure size and to derive high-resolution terrain data even under poor visibility conditions. A long-term goal is the abandonment of SLR technology back towards fully electronic imagers in order to exclude the possibility of mechanical failures.

Acknowledgments: The camera system was developed and funded as part of the Helmholtz Alliance ROBEX. The authors acknowledge the generous support of the Institut für Seenforschung in Langenargen, Lake Constance, during sea trials as well as the support of the crew of R/V SONNE during two expeditions. We would also like to thank our colleagues of the German Center for Artificial Intelligence (DFKI) for providing the test pool for lighting experiments, Thomas Burisch as well as the staff of the GEOMAR Technology and Logistics Centre for their generous technical support. We thank three anonymous reviewers for constructive reviews.

Author Contributions: Tom Kwasnitschka and Kevin Köser specified the overall design requirements and developed the camera and underwater optics. Jan Sticklus designed the LED lighting system and optical experiments. Tim Weiß wrote the camera control software. Timm Schoening managed the two sea trials and data handling. Marcel Rothenbeck, Emanuel Wenzlaff, Lars Triebe and Anja Steinführer oversaw operational feasibility, integrated the camera system into the AUV and operated the AUV. Jens Greinert and Colin Devey advised on scientific application requirements and cruise integration and provided project management support. All authors contributed to the production of this publication.

Conflicts of Interest: The authors declare no conflict of interest.

References

1. Mobley, C.D. *Light and Water: Radiative Transfer in Natural Waters*; Academic Press: San Diego, CA, USA, 1994.
2. Stramski, D.; Boss, E.; Bogucki, D.; Voss, K.J. The role of seawater constituents in light backscattering in the ocean. *Prog. Oceanogr.* **2004**, *61*, 27–56.

3. Kenny, A.J.; Cato, I.; Desprez, M. An overview of seabed-mapping technologies in the context of marine habitat classification. *ICES J. Mar. Sci.* **2003**, *60*, 411–418.

4. Kinsey, J.C.; Eustice, R.M.; Whitcomb, L.L. A survey of underwater vehicle navigation: Recent advances and new challenges. In Proceedings of the IFAC Conference of Manoeuvering and Control of Marine Craft (IFAC 2006), Lisbon, Portugal, 20–22 September 2006.

5. Yoerger, D.; Bradley, A.; Jakuba, M.; German, C. Autonomous and remotely operated vehicle technology for hydrothermal vent discovery, exploration, and sampling. *Oceanography* **2007**, *20*, 152–161.

6. Ballard, R.D.; Yoerger, D.R.; Stewart, W.K.; Bowen, A. ARGO/JASON: A remotely operated survey and sampling system for full-ocean depth. In Proceedings of the IEEE/MTS OCEANS Conference, Honolulu, HI, USA, 1–3 October 1991; pp. 71–75.

7. Le Bas, T.P.; Huvenne, V.A.I. Acquisition and processing of backscatter data for habitat mapping—Comparison of multibeam and sidescan systems. *Appl. Acoust.* **2009**, *70*, 1248–1257.

8. Micallef, A.; le Bas, T.P.; Huvenne, V.A.I.; Blondel, P.; Hühnerbach, V.; Deidun, A. A multi-method approach for benthic habitat mapping of shallow coastal areas with high-resolution multibeam data. *Cont. Shelf Res.* **2012**, *39–40*, 14–26.

9. Barreyre, T.; Escartin, J.; Garcia, R.; Cannat, M.; Mittelstaedt, E.; Prados, R. Structure, temporal evolution, and heat flux estimates from a deep-sea hydrothermal field derived from seafloor image mosaics. *Geochem. Geophys. Geosyst.* **2011**, *13*, 1–52.

10. Mikhail, E.M.; Bethel, J.S.; McGlone, J.C. American Society of Photogrammetry and Remote Sensing and Imaging & Geospatial Information Society. In *Manual of Photogrammetry*, 5th ed.; American Society of Photogrammetry and Remote Sensing: Bethesda, MD, USA, 2004.

11. Hartley, R.; Zisserman, A. *Multiple View Geometry in Computer Vision*; Cambridge University Press: Cambridge, UK, 2003.

12. Szeliski, R. *Computer Vision: Algorithms and Applications*; Springer Science & Business Media: London, UK, 2010.

13. Vincent, A.G.; Pessel, N.; Borgetto, M.; Jouffroy, J.; Opderbecke, J.; Rigaud, V. Real-time geo-referenced video mosaicking with the MATISSE system. In Proceedings of the IEEE OCEANS Conference, San Diego, CA, USA, 22–26 September 2003; pp. 2319–2324.

14. Pizarro, O.; Eustice, R.; Singh, H. Large area 3D reconstructions from underwater surveys. In Proceedings of the IEEE OCEANS Conference, Kobe, Japan, 9–12 November 2004.

15. Singh, H.; Roman, C.; Pizarro, O.; Eustice, R.; Can, A. Towards High-Resolution Imaging from Underwater Vehicles. *Int. J. Robot. Res.* **2007**, *26*, 55–74.

16. Johnson-Roberson, M.; Pizarro, O.; Williams, S.; Mahon, I. Generation and visualization of large-scale three-dimensional reconstructions from underwater robotic surveys. *J. Field Robot.* **2010**, *27*, 21–51.

17. Williams, S.B.; Pizarro, O.R.; Jakuba, M.V.; Johnson, C.R.; Barrett, N.S.; Babcock, R.C.; Kendrick, G.A.; Steinberg, P.D.; Heyward, A.J.; Doherty, P.J.; *et al.* Monitoring of Benthic Reference Sites: Using an Autonomous Underwater Vehicle. *IEEE Robot. Autom. Mag.* **2012**, *19*, 73–84.

18. Smart, C.J.; Roman, C.; Carey, S.N. Detection of diffuse seafloor venting using structured light imaging. *Geochem. Geophys. Geosyst.* **2013**, *14*, 4743–4757.

19. Jordt-Sedlazeck, A.; Koch, R. Refractive Structure from Motion for Underwater Images. In Proceedings of the IEEE international Conference on Computer Vision, Sydney, Australia, 1–8 December 2013; pp. 57–64.

20. Escartin, J.; Garcia, R.; Barreyre, T.; Cannat, M.; Gracias, N.; Shihavuddin, A.; Mittelstaedt, E. Optical methods to monitor temporal changes at the seafloor: The Lucky Strike deep-sea hydrothermal vent field (Mid-Atlantic Ridge). In Proceedings of the IEEE International Underwater Technology Symposium, Tokyo, Japan, 5–8 March 2013; pp. 1–6.

21. Wynn, R.B.; Huvenne, V.A.I.; le Bas, T.P.; Murton, B.J.; Connelly, D.P.; Bett, B.J.; Ruhl, H.A.; Morris, K.J.; Peakall, J.; Parsons, D.R.; *et al.* Autonomous Underwater Vehicles (AUVs): Their past, present and future contributions to the advancement of marine geoscience. *Mar. Geol.* **2014**, *352*, 451–468.

22. Sharp, K.M.; White, R.H. More tools in the toolbox: The naval oceanographic office's Remote Environmental Monitoring UnitS (REMUS) 6000 AUV. In Proceedings of the IEEE OCEANS Conference, Quebec City, QC, Canada, 15–18 September 2008; pp. 1–4.

23. Escartin, J.; Garcia, R.; Delaunoy, O.; Ferrer, J.; Gracias, N.; Elibol, A.; Cufi, X.; Neumann, L.; Fornari, D.J.; Humphris, S.E.; *et al.* Globally aligned photomosaic of the Lucky Strike hydrothermal vent field (Mid-Atlantic Ridge, 37 degrees 18.50' N): Release of georeferenced data, mosaic construction, and viewing software. *Geochem. Geophys. Geosyst.* **2008**, *9*.

24. Ludvigsen, M.; Sortland, B.; Johnsen, G.; Singh, H. Applications of geo-referenced underwater photo mosaics in marine biology and archaeology. *Oceanography* **2007**, *20*, 140–149.

25. Prados Gutiérrez, R.; García Campos, R.; Grácias, N.R. E.; Escartin, J.; Neumann, L. A Novel blending technique for underwater gigamosaicing. *IEEE J. Ocean. Eng.* **2012**, *34*, 626–644.

26. Roman, C.; Inglis, G.; Rutter, J. Application of structured light imaging for high resolution mapping of underwater archaeological sites. In Proceedings of the IEEE Oceans Conference, Sydney, Australia, 24–27 May 2010; pp. 1–9.

27. Bingham, B.; Foley, B.; Singh, H.; Camilli, R.; Delaporta, K.; Eustice, R.; Mallios, A.; Mindell, D.; Roman, C.; Sakellariou, D. Robotic tools for deep water archaeology: Surveying an ancient shipwreck with an autonomous underwater vehicle. *J. Field Robot.* **2010**, *27*, 702–717.

28. Hagen, P.E. Multi-Sensor Pipeline Inspection with AUV. In Proceedings of the Oceanology International Conference, London, UK, 15–17 March 2014; pp. 1–55.

29. Bodenmann, A.; Thornton, B.; Ura, T. Development of long range color imaging for wide area 3D reconstructions of the seafloor. In Proceedings of the IEEE Underwater Technology Symposium, Tokyo, Japan, 5–8 March 2013; pp. 1–5.

30. Gracias, N.; Ridao, P.; Garcia, R.; Escartin, J. Mapping the Moon: Using a lightweight AUV to survey the site of the 17th century ship "La Lune". In Proceedings of the IEEE/MTS OCEANS Conference, Bergen, Norway, 10–14 June 2013; pp. 1–8.

31. Morris, K.J.; Bett, B.J.; Durden, J.M. A new method for ecological surveying of the abyss using autonomous underwater vehicle photography. *Limnol. Oceanogr. Methods* **2014**, *12*, 795–809.

32. Williams, S.B.; Pizarro, O.; Foley, B. Return to Antikythera: Multi-Session SLAM Based AUV Mapping of a First Century BC Wreck Site. In Proceedings of the Field and Service Robotics Conference, Toronto, ON, Canada, 24–26 June 2015; pp. 1–14.

33. Sticklus, J.; Kwasnitschka, T. Verfahren und Vorrichtung zur Herstellung von in Vergussmasse Vergossenen Leuchten. German Patent DE102014118672B3, 1 October 2015.

34. Underwater Camera Calibration. Available online: http://www.geomar.de/go/cameracalibration-e (accessed on 26 January 2016).

35. OcamLib: Omnidirectional Camera Calibration Toolbox for Matlab. Available online: https://sites.google.com/site/scarabotix/ocamcalib-toolbox (accessed on 26 January 2016).

36. Burt, P.J.; Adelson, E.H. A multi resolution spline with application to image mosaics. *ACM Trans. Graph.* **1983**, *2*, 217–236.

37. Kwasnitschka, T.; Hansteen, T.H.; Devey, C.W.; Kutterolf, S. Doing Fieldwork on the Seafloor: Photogrammetric Techniques to yield 3D Visual Models from ROV Video. *Comput. Geosci.* **2012**, *52*, 218–226.

38. Pollefeys, M.; Nistér, D.; Frahm, J.M.; Akbarzadeh, A.; Mordohai, P.; Clipp, B.; Engels, C.; Gallup, D.; Kim, S.J.; Merrell, P.; *et al.* Detailed Real-Time Urban 3D Reconstruction from Video. *Int. J. Comput. Vis.* **2007**, *78*, 143–167.

MDPI AG
Klybeckstrasse 64
4057 Basel, Switzerland
Tel. +41 61 683 77 34
Fax +41 61 302 89 18
http://www.mdpi.com/

Sensors Editorial Office
E-mail: sensors@mdpi.com
http://www.mdpi.com/journal/sensors

www.ingramcontent.com/pod-product-compliance
Lightning Source LLC
Chambersburg PA
CBHW051924190326
41458CB00026B/6405